AF556001

GENOME TO OM

Advance Praise

(Full endorsements available on the website)

"A thought-provoking exploration of the frontiers where science meets spirituality. *Genome to Om* challenges our assumptions and pushes the boundaries of knowledge. A must-read for anyone curious about the future of scientific inquiry. The nine chapters link the reality of the world we live in and the challenges we face, as represented by Genome, which can only be resolved by realizing the true identity, with Om. The authors introduce an interesting concept of the Omcene, the future epoch, with the utopian but feasible goal of a global meta-society focused on universal well-being, finding unity in diversity. A timely endeavour in a world torn apart by wars, disease, and distrust."

Dr K. Kasturirangan, Padma Vibhushan
Former Chairman, Indian Space Research Organisation;
former Director & Emeritus Professor,
National Institute of Advanced Studies, Bengaluru, India; Chairman,
National Education Policy Committee, 2020

"This landmark book bridges 21st-century science with insights from Vedic tradition. A central conundrum concerning the thesis of this book is whether the brain gives rise to consciousness or whether consciousness gives rise to the brain and all matter. The remarkable journey from the genome back to the quantum expression of consciousness, the universal sound or hum, *Om*, is expertly led by the two authors who are deeply immersed in both the science and heritage of their subject. This remarkable book offers a roadmap for the future understanding and exploration of both quantum reality and of consciousness and, indeed, its computability by quantum standards. It is a book for our time."

Dr Gerard Bodeker, MD, PhD
Eminent public health expert, Green Templeton College,
University of Oxford; Adjunct Professor of Epidemiology,
Columbia University; Chair Advisory Group, Western Sydney
University, Australia; Editor-in-Chief, The WHO Global Atlas of
Traditional, Complementary and Alternative Medicine

"*Genome to Om* is a veritable tour de force, an encyclopaedic and thought-provoking view of the interface of science and spirituality. It presents a compelling blueprint for humanity to achieve a bright and vibrant future."

Dr Vikas P. Sukhatme, MD, ScD
Dean and Robert W. Woodruff Professor of Medicine,
Emory School of Medicine; Founding Director,
Morningside Centre for Innovative and Affordable Medicine;
Chief Academic Officer, Emory Healthcare;
former Faculty Dean for Academic Programs,
Beth Israel Deaconess Medical Centre,
Harvard Medical School, Boston, USA

"A brilliant treatise of science, philosophy, psychology. and spirituality. It explores the evolution of S&T towards a meta-scientific awareness of one's Self. The authors take us on an intellectual journey into philosophical and spiritual aspects of existence, consciousness, and the nature of reality. They interweave inspiration from the Upanishads and the *Bhagavad Gita*. The journey in reading this work leads us to appreciate the potential of meta-science as a holistic approach to scientific inquiry, blending ethical, moral, intuitive, and spiritual insights. The book is a poetic and prophetic guide to lead us out of the chaos of today's world toward a future of unity, tolerance, and understanding."

Dr Julia T. Arnold, PhD, MS
Senior cancer researcher; former Program Director and
Special Volunteer, NCI Office of Cancer Complementary and
Alternative Medicine, National Institutes of Health,
Maryland, USA

"My dear friend Bhushan Patwardhan has been doing great work for many years. *Genome to Om* is another feather in his cap. This book, co-authored by him and Indu Ramchandani, is a call to embrace an innovative, inclusive, technologically advanced, yet spiritually enriched and ethically grounded future. It is closely aligned with my views and work. I wish this endeavour great success."

B.M. Hegde, MD, FAMS, MRCP (UK), FRCP
(London, Edinburgh, Glasgow & Dublin), FACC (USA),
PhD (Hon. Causa), Padma Vibhushan
Former Vice Chancellor and Professor of Cardiology, Manipal Academy
of Higher Education, Manipal, India

"Modern science based on Cartesian thinking has produced an enormous amount of knowledge and incredible technological progress. Its focus is on the direct cause-effect principle, mechanistic details and experiments while the consciousness of humans as observers is often disregarded. There is an urgent need to develop new models of thinking based on ethical, moral and spiritual insights and respect the interconnectedness of all aspects of life. This multi-perspective thinking forms the basis for what the authors call meta-science. The wide variety of topics covered shows that this thinking can be applied to all areas of life and can lead to amazing insights. It is a courageous attempt resulting in an impressive outcome."

Dr med. Georg Seifert, MD, PhD
Specialized physician for Integrative Medicine and Professor of Paediatric Oncology and Haematology, Berlin Center for Rare Diseases, Charité University, Berlin, Germany

"An insightful book exploring the intersection of science and spirituality, particularly how yoga, including meditation, can influence our genes and well-being. The book links the Ayurvedic concept of Prakriti with modern science, advocating for the evolution of science into a meta-science termed *Prajnana*. The book draws on the latest scientific research to highlight the significant impact of yoga on physical and mental health. Its engaging writing makes complex topics accessible to both those with a scientific background and those seeking deeper life understanding. The book emphasizes the importance of Yoga as a science of holistic living in daily life and presents practical techniques for integrating these practices. I highly recommend *Genome to Om* to anyone interested in the science of well-being and the power of mindful practices in cultivating peace and wholeness."

Dr H.R. Nagendra, PhD, DSc, Padma Shri, Yoga Shri
Chancellor, Swami Vivekananda Yoga Anusandhana Samsthana University (S-VYASA University), Bengaluru, India;
Acclaimed Yoga guru and pioneer of cyclic meditation

"This compelling book delves into a broad spectrum of modern and contemporary scientific concepts, seamlessly intertwining them with meta-science, spirituality, and the notions of the microcosm and macrocosm. The exploration knows no bounds, like the diverse *darshana*s found in Indian philosophies. The discussion on genetics resonates deeply with ancient Ayurvedic texts. The book portrays a historical narrative of how in the past, science, philosophy, culture, and spirituality were intricately interwoven, mutually enriching each other and evolving. It was great learning to read such a thought-provoking work and believe it would greatly benefit students of Ayurveda. The title, *Genome to Om*, encapsulates the essence of the book intriguingly!"

Vaidya Rajesh Kotecha, MD (Ayu), Padma Shri
Secretary to the Government of India, Ministry of AYUSH;
former Vice Chancellor, Gujarat Ayurveda University,
Jamnagar, India

"Ayurgenomics, which explores and synthesizes Ayurveda with genomics, is a budding discipline in social sciences. Through *Genome to Om*, the authors seek to blend technological marvels with the ancient Indian knowledge system and offer a roadmap for physical and spiritual peace and harmony for the diverse and often conflictual global community. My warm greetings and best wishes to Shri Patwardhan and Indu Ramchandani for this timely endeavour."

Dr Santishree Pandit, MSc, PhD
Vice Chancellor, Jawaharlal Nehru University, New Delhi;
Professor of Political Science,
Savitribai Phule Pune University, India

"*Genome to Om* delves deep into the basis of living, the meaning of life, and its origins. We haven't yet figured out how the whole process from the cell or the cosmos works. 'Om' is the primordial sound of the Cosmos, which pervades the universe. 'Om' allows us to explore the inner reality, which is important for the realization of self. The microcosm and macrocosm are explained very well in the book as one continuum. There is a quote in the *Bhagavad Gita* which says *Anoraneeyan Mahato Maheeyan* which means 'God exists everywhere and manifests as both the infinitesimal and the Infinite.' This book makes a wonderful reading and leads us to realize the purpose of life from science to meta-science."

Dr Carani B. Sanjeevi, MD, MSc, PhD
Professor of Molecular Medicine, Department of Medicine,
Karolinska Institute, Karolinska University Hospital,
Stockholm, Sweden; former Vice-Chancellor,
Sri Sathya Sai Institute of Higher Learning (Deemed University),
Andhra Pradesh, India

"The book, *Genome to Om*, is a journey of the mind of the authors who seek answers to the greater meaning of life. Om is in fact at the core of the genOMe! As scientists we are looking out for evidence-based answers to the questions of health and life, however we reach a dead end when it comes to the role of science in giving us answers to the bigger picture of life. This book raises a number of questions for which some answers are provided on the spiritual and philosophical level while remaining in contact with the reality of science. This book is a must-read for anyone looking for the deeper meaning of life and its connect with the reality of science."

Lt Gen. Dr Madhuri Kanitkar, MD (Paediatrics),
PVSM AVSM VSM (Retd.)
Vice Chancellor, Maharashtra University of Health Sciences, India;
former Dean and Professor of Paediatrics,
Armed Forces Medical College, Pune, India

"If there is a single word that has the deepest meaning in Hindu Philosophy, it is 'Om'. Recordings of the sounds of space, that reverberate through the universe, closely resemble the sound of 'Aum'. *Genome to Om* covers spectacular advances in science and technology as also its perils. In the 21st century, the world will run more on intelligence than industry, and in the next century, the transformation will be complete. It isn't that everyone will be working with AI and ML, but major changes in the world will depend upon the intelligence of machines and not human. To what extent the world will change, is difficult to forecast. Yet one thing shall remain constant, our study of the Genome and our philosophy of Om."

Dr Ravindra Ghooi, MSc (Pharmacology), PhD (Medicine)
Director, Scientia Clinical Services; former Professor of Pharmacology,
Symbiosis International University; Chairman, Ethics Committee,
Sahyadri Clinical Research and
Development Centre, Pune, India

"This book links historical science and knowledge to the present time. *Genome to Om* certainly sheds light on the marvels of science, the perils, and the way forward, and forces all of us to think of a new paradigm that can herald a better future for the benefit of mankind."

Dr Ikhlas Khan, MS, PhD
Director and Distinguished Professor,
National Centre for Natural Products Research;
Research Professor of Pharmacognosy,
University of Mississippi, Oxford, USA

"It is not easy to bridge the worlds of science and spirituality, but what sets this book apart is the care with which there is clearly demarcated reference to each. Like any good exploration, it raises more provocative questions than answers and invites a spiritual and ethical examination – a contemplative inquiry – of our very existence and a path leading to a more enlightened and harmonious future. It is difficult to encapsulate in a few words the content of a book so vast and head-spinning in scope, but I predict it will be an enjoyable and fruitful journey for anyone who embarks on it."

Dr Suresh V. Garimella, MS, PhD
27th President of the University of Vermont; Distinguished Professor Emeritus, Purdue University, Indiana, USA

"*Genome to Om*, is an excellent exploration on evolution and convergence between modern science and meta-science. One is awestruck with the power of observation and deduction of our sages in ancient times that enabled them to understand ourselves and the nature around and their efforts to nurture schools of learning. That seems to have enabled them to conceptualise some of the deepest questions that modern science is still struggling to resolve. Authors deserve our deep appreciation for their efforts in unfolding this journey. My compliments to the authors for putting together this book, it makes a fascinating reading on discovery efforts from ancient to modern times."

Dr Anil Kakodkar, Padma Vibhushan
Former Chairman, Atomic Energy Commission of India; former Secretary, Government of India; former Director, Bhabha Atomic Research Centre; Chancellor, Homi Bhabha National Institute (Deemed to be University), Mumbai, India

"*Genome to Om* by Dr Bhushan Patwardhan and Indu Ramchandani takes us from the cutting edge of modern science to the timeless Vedic sciences of Yoga and Ayurveda, which reflect a higher spiritual science of universal consciousness and *prana* rooted in prime *mantric* knowledge starting with 'Om'. This can heal body and mind and unify society promoting world peace and prosperity. The book projects a new visionary approach of great relevance to all."

Dr David Frawley, DLitt, Padma Bhushan
Acclaimed scholar in Yoga, Ayurveda and oriental philosophy; popular author and the founder of American Institute of Vedic Studies, Santa Fe, New Mexico, USA

"Various levels of knowledge arise from human experience and others are induced through critical observation and experimentation. *Genome to Om* attempts to answer, or at least raises for discussion and conversations to understand the value of Western rationalistic thinking to human existence and the richness that faith born out of human existence can bring to humanity. After all, perhaps the most profound experience of human existence, *Love*, can neither be measured nor even proven. This book is an exploration into the two world views."

Dr Roy Upton, PhD
President, Executive Director and Editor,
American Herbal Pharmacopoeia; adjunct faculty member,
Ric Scalzo Botanical Research Institute; Director of Herbal Research,
Planetary Formulas, Scotts Valley, CA

"The book *Genome to Om: Evolving Journey of Modern Science to Meta-Science* makes an interesting reading. It is contemporary and modern given the background and understanding of DNA, genome sequencing, research in neuroscience, lifestyle changes, and AI and yet aligned to the ancient wisdom of spiritualism, yoga, meditation, minimalism, honesty, ethics, morality, and the goal to live a healthy and happy life. The text is beautifully interwoven with this complex subject and make us think of start to live the Om way, while enjoying the benefits of Science and Technology, save the planet from imminent disaster."

Dr Anil Sahasrabudhe
Chairman, National Education Technology Forum,
National Assessment and Accreditation Council and
National Board of Accreditation; former Chairman,
All India Council for Technical Education, Government of India

"It is a joy to see this meaningful volume being published at a very opportune time when the entire world is in a very paradoxical situation. On the one hand, humanity is enjoying all the luxuries and comforts that emerging technology is providing, while on the other, we are at such a perilous stage that we do not even know whether we will see the end of the 21st century! India is a land of great saints, sages, seers, scientists and philosophers. They gave the world some of the greatest attributes, which include spirituality, Yoga, Aum, and nonduality. Swami Vivekananda had prophesied, "The union of science and spirituality alone will bring harmony and peace to mankind." I endorse and compliment the authors for their diligent effort to reveal the same through *Genome to Om*."

Dr Vishwanath D. Karad
Founder President,
MAEER's MIT World Peace University, Pune, India;
former Professor of Mechanical Engineering,
COEP Technology University Pune

"*Genome to Om* is a rare and much-needed book bringing hope to the future of humanity and science. This book provides insights from science, meta-science, ethics, and spirituality, acknowledges the interconnectedness of all things, and emphasizes a holistic approach to scientific inquiry into understanding life and its pursuits. I feel this is a must-read for everyone."

Dr Amarjeet Singh Bhamra
All-Party Parliamentary Group on Indian Traditional Sciences,
Houses of Parliament, London

"The book resonates deeply with the ground-breaking work of Prof. Brian Josephson. Just as his research has pushed the boundaries of our understanding of mind-matter unification, this exploration of the evolutionary journey from modern science to meta-science promises to challenge existing paradigms and inspire new ways of thinking. Authors have seamlessly weaved together themes of science, consciousness, and metaphysics embracing a holistic approach to knowledge. I am excited to see how these insights contribute to the ongoing dialogue surrounding the interconnectedness of science and spirituality. Congratulations to the authors on creating a work that forges new pathways for future scientific inquiry. It is sure to captivate readers and spark meaningful discussions in the realms of academia and beyond."

Dr Madan Thangavelu
Genome biologist;
General Secretary and Research Director, EUAA
(European Ayurveda Association),
Cambridge, UK

"One of the most perplexing and profound approaches to the understanding of the world around as practiced over ages in India and associated cultures is fundamentally different from Western philosophy. The Indian approach does not consider human beings as separate entities from nature. "Genome to Om" is a well-studied endeavour in which the readers will appreciate the integral approach of science and insightful thought process where the universe in totality is attempted to be understood. I congratulate the authors for putting together such a thoughtful text which makes a very interesting reading."

Dr Shekhar C. Mande, FNA, FASc, FNASc
President, Vijnana Bharati; Distinguished Professor,
Bioinformatics Centre, Savitribai Phule Pune University, Pune, India;
JC Bose Fellow and Honorary Distinguished Scientist,
National Centre for Cell Science, Pune, India;
former Director General, Council for Scientific and Industrial Research;
Secretary S&T, Government of India

"*Genome to Om* is a profound exploration of our current scientific landscape and its potential evolution. This book is a visionary guide, urging us to transcend traditional boundaries of empirical science. It challenges us to consider how we understand the universe and our place within it. Through meticulous scholarship and compelling prose, the authors illuminate the path toward a meta-society that harmonizes technological advancement with timeless values. This book is a manifesto for the future, advocating a balanced integration of cutting-edge innovation and spiritual enlightenment."

Dr Rajiv Kumar
Former Vice Chairman, NITI Aayog;
Chairman, Pahle India Foundation, Delhi, India

"*Genome to Om* is an insightful book that highlights the need for a holistic approach, including ethical, moral, and spiritual insights. It emphasizes the contemporary value of the Upanishads, Yoga, and other ancient knowledge sources in recalibrating our current development path towards a more sustainable environment, peace, and well-being. This motivates readers to think about how technological progress, combined with timeless values, can lead to a future that is not only innovative but also spiritually enriched and ethically grounded. I wholeheartedly endorse this book and believe it will serve as an inspiring guide for those committed to advancing knowledge while honouring the profound interconnectedness of all life."

Dr Suresh Prabhu
Chancellor, Rishihood University;
former Cabinet Minister, Government of India,
with several portfolios including Environment & Forest,
Commerce & Industry, Civil Aviation & Railways

"Vast intimidating canvas reflecting ancient insights and modern discoveries."

Drs Ashok and Rama Vaidya
Eminent physician scientists;
Alumni of Seth GS Medical College and Yale Medical School

"The book provides a path to integrating traditional wisdom with modern scientific discoveries, offering practical applications for personal and societal well-being by delving into the Om Way. This scholarly work is a valuable contribution to the ongoing dialogue between science and spirituality, and I highly recommend it to open-minded scientists and intellectually curious readers interested in exploring profound questions about our purpose and existence."

Dr Vijay Bhatkar, Padma Bhushan
Founder, C-DAC; former Vice Chancellor,
Nalanda University

GENOME TO OM

Evolving Journey of Modern Science to Meta-science

Bhushan Patwardhan and Indu Ramchandani

ISBN: 978-93-6547-823-5

First published in India 2024
This edition published 2025

BluOne Ink Pvt. Ltd
A-76, 2nd Floor, Sector 136, Noida
Uttar Pradesh 201301
www.bluone.ink
publisher@bluone.ink

Printed and bound in India at Nutech Print Services - India

CONTENTS

Foreword

The Sapiens' Crossroads of Redemption

It is with great pleasure that I write a few words about this remarkable volume by Bhushan Patwardhan and Indu Ramchandani. Bhushan and I have been discussing such a book for almost 10 years, and in his preface, Bhushan generously acknowledges my modest contribution in reorienting and strengthening the principal theme of the book, glimpsed in its very title. We have collaborated in the past in publishing *Integrative Approaches to Health* (Elsevier 2015), a well-received and widely read volume. I consider the present work Bhushan's magnum opus, a testament to his deep knowledge.

Genome to Om is an arrow shot in the right direction, one that will further stimulate a growing dialogue around urgent remedial actions required to confront and remedy the existential problems threatening human beings and our planet. In the intricate interphase between science and spirit (metascience as referenced by the authors), *Genome to Om* offers a compelling vision of humanity's potential evolution. The authors' meticulous exploration of the scientific landscape is firmly rooted in their dedication, perseverance, and expertise. Yet, as a student of both the empirical and the esoteric, I find myself wrestling with the nuances of the narrative. While I applaud this clarion call for a transition from Genome to Om—a shift from the dominance of scientific materialism to a world view embracing compassion, ethics, and equanimity—I cannot embrace the notion that this transformation will unfold as a natural progression. The forces of human ego, greed, and self-interest, along with the corruption of vital human institutions, are deeply entrenched in our society. Democracy, our one hope for equality and freedom, is in decline in many of the advanced nations. To transcend these limitations, we require not merely an evolution but a revolution—a conscious, deliberate reorientation of our values and priorities.

Notwithstanding these reflections, the present volume serves as a crucial catalyst for deeper conversations and radical actions that our collective redemption demands. The human story, etched into our very genome, is a tale of astounding paradoxes. We have risen from primitive origins to touching the stars, wielding technologies that reshape reality. We have conquered diseases, built sprawling cities, and enfolded the globe in a digital embrace. Yet, our primal impulses remain, driving us toward conflict, inequality, and environmental devastation. This disconnect between our progress and our peril begs a haunting question: Are we the architects of our own demise?

The Illusion of Progress

The relentless pursuit of material wealth, fame, and fleeting comforts has blinded us to our true purpose. We have prioritized the "I" over the "we", fracturing our communities and neglecting our planet. This blind focus on the individual, while fuelling the ambition necessary for innovation, has also sown the seeds of our discontent. Unbridled consumption and consumerism threaten our very survival, fuelling climate change, ecological collapse, and deepening social divides.

Meanwhile, amidst technological and scientific prowess, we have lost touch with the wisdom of our ancestors. The timeless concept of *Om*, a symbol of universal unity and consciousness, should remind us that we are not separate from the world around us; as our actions ripple outwards, they impact not only ourselves but also the entire web of life.

Genome to Om: A Path to Redemption

The journey from Genome to Om is a call for profound transformation. It is a recognition that our current trajectory is unsustainable both for ourselves and for the planet. It is a call to awaken from our spiritual amnesia and to remember our place in the grand tapestry of existence. This is not a retreat from progress, but a re-directing. We must reimagine our institutions, economies, and educational systems to prioritize collaboration, compassion, and long-term well-being. We must elevate leaders who embody integrity and prioritize the collective good. Above all, we must cultivate a spiritual reorientation that transcends self-interest and embraces our shared humanity.

Each of us holds within us the power to ignite change. While institutions may falter, an individual spark of conscience and courage can inspire a monumental movement. We must amplify these sparks, empowering grassroots initiatives, transforming education, and fostering a new narrative that celebrates cooperation, empathy, and our connection to nature. The challenges we face are not merely technological or economic; they are deeply spiritual. True spirituality is not about dogma or ritual but a lived experience of interconnectedness, compassion, and service to others, recognizing the spark of the divine in every living being and acting accordingly. Each breath, each action, becomes an opportunity to cultivate kindness, wisdom, and reverence for all life. By embracing our spiritual nature, we tap into a wellspring of resilience, creativity, and collective power that can guide us through this perilous crossroads.

The Path Ahead

The path to redemption will not be easy. It demands courage, sacrifice, and a willingness to confront our own shadows. But within us burns the same spark that propelled us from the caves to the cosmos—a spark of ingenuity, resilience, and boundless potential. Let us remember the ancient wisdom that speaks of our shared purpose: to nurture the well-being of all. By integrating the lessons of our genome with the timeless wisdom of Om, we can forge a new path—one that leads to a flourishing planet and a truly awakened humanity. Chanting the *shanti mantra*s (peace aphorisms) is one of the great traditions of India. A *mantra* is a verse, or a set of verses, an uplifting prayer dedicated to the Spirit of Om, the universality of all nature and humanity. Let me wish godspeed to this book's clarion call for the magic and power of the Om paradigm to usher in peace and enlightenment to all sentient beings on our planet.

Om Sarve Bhavantu Sukhinah, Sarve Santu Niraamayaah, |
Sarve Bhadraanni Pashyantu,Maa Kashcid-Duhkha-Bhaag-Bhavet |
Om Shaantih Shaantih Shaantih ||

By the power of *Om* may all sentient beings live in happiness,
may all enjoy good health,
may joyous events fill all our lives, may there be no suffering, may
peace reign over us.

Dr Gururaj Mutalik, MD (Internal Medicine), FAMS
Eminent physician scientist, Yoga and spiritual master active at the age of 95 at Sun City, Florida, USA; formerly professor of medicine and Dean, B.J. Medical College, Pune; former Director, WHO Office at United Nations, New York; former CEO, The International Physicians for the Prevention of Nuclear War (recipient of Nobel Prize 1985)

Foreword

This is a very unusual and interesting book. It covers the transition from modern science to meta-science, embracing holistic wisdom, advocating for a spiritually enriched future, and finally, seeking unity for universal peace and well-being. The authors successfully take the readers on a journey covering a huge landscape of the world we live in with an excellent review of cutting-edge science and technology, its marvels, and perils. The discussion on "what is the mind" brings in advances in brain science, and it also links to consciousness. The difference between religion and philosophy is touched upon as these are questions that arise in every human mind. These are like 'who we are, we are here but do we have a purpose, how do we consciously realize our identity', and so on. These are all discussed deeply and meaningfully. The authors articulate the grand challenge of the search of answers to the perennial questions of life and existence as represented by Om.

The authors succinctly bring out the challenging mission for humanity, namely the connection between modern science and meta-science. The bigger picture is well projected. The discussion on how the centuries-old Indian Knowledge Systems guide these studies is interesting. What I liked especially is the fact that the authors are balanced in their outlook, not veering into rhetoric or religiosity. They are agnostic enough to include some of the key religions and philosophical thoughts from around the world. Finally, the authors make an interesting proposition of a transition from the Anthropocene to the Omcene, as they see the goal of humanity as the quest for universal well-being with unity in diversity. That's a very profound thought indeed.

The overarching message of the book is that interconnectedness, blending ethics with technology, striving for harmony between progress and values, and envisioning a future of unity and well-being will build a better world. And, for this to happen, we must integrate ethical, moral, and spiritual insights into scientific endeavours, fostering a holistic approach while prioritizing harmony between progress and timeless values, and also cultivating unity in diversity for universal peace and well-being.

I congratulate the authors for bringing out this thoughtful scholarly offering, which is the need of the hour in these troubled times for humanity as a whole. While understanding the realities of the external world, where volatility, uncertainty, complexity, and ambiguity have become the new normal, the book helps us to look inward to reinvent ourselves and reconfigure our future.

Dr Raghunath Mashelkar, FRS, Padma Vibhushan
Chairman, Reliance Innovation Council; Chancellor, Jio Institute, India, and Institute of Chemical Technology, Mumbai; former Director General, Council for Scientific and Industrial Research; Secretary, Department of Science and Technology, Government of India; and President, Indian National Science Academy

First Author's Preface

As I reflect upon the inception of *Genome to Om*, my mind goes back to a moment of profound realization on Buddha Poornima, 16 May 2003. The first inspiration came from the Ishavasya Upanishad Shanti Mantra on the concept of Om and completeness:

ॐ पूर्णमदः पूर्णमिदं पूर्णात्पूर्णमुदच्यते । पूर्णस्य पूर्णमादाय
पूर्णमेवावशिष्यते ॥ ॐ शान्तिः शान्तिः शान्तिः ॥

This very significant peace aphorism essentially means, "Om, the outer world is complete with divine consciousness; the inner world is also complete with divine self-consciousness. From the completeness of consciousness, the world is manifested. Taking a part away from the whole, what is taken and what remains is also complete because universal consciousness is infinite. Om Peace, Peace, Peace!" This mantra emphasizes the idea of non-duality, where both the self and universal consciousness are seen as aspects of the same ultimate reality. It also suggests that true peace can be found by understanding this underlying unity.

Reflecting on this great wisdom, when the Human Genome Project was on the horizon, I saw a parallel. The Genome and Om, both are complete as a whole. If we remove a part, both what remains and what is removed can regain completeness. During the same period, my research group was trying to study the genetic basis of the Ayurveda body types known as Prakriti in a search into the concept of AyuGenomics. The intertwining experiences in the laboratory with the intuitive inspiration from *Ishavasya Upanishad* were probably the beginning of a revelation on reinterpreting Om and Genome.

I must admit that this realization has not come out of the blue. I trust that upbringing during childhood plays a major role in shaping epigenetics in the later part of life and career. I grew up in a joint family in the heart of Pune. We did not know what was meant by a cousin brother or sister. At any point in time, about two dozen family members were living together in a modest home, cooking and eating in one kitchen with no independent bedrooms. We used to have a galaxy of diverse visitors, ranging from Sanskrit pundits to illiterate farmers bringing milk from nearby villages. We were meeting, listening to, and occasionally interacting with interesting personalities from the arts, academia, sports, civil servants, and even politicians. Every evening was inspired by discussions with diverse guests on contemporary issues, history, and philosophy, concluding with all of us chanting spiritual prayers together. Every night was a learning experience with my grandfather's astonishing stories from the Ramayana, Mahabharata, Puranas, Upanishads, and Vedas. Every fourth day after the full moon was fasting time to pray to my father's favourite god of knowledge, Ganesha, followed by chanting *Atharvasheersha*. I grew up reading Vivekananda to Osho and listening to Patanjali Yoga Sutras from my mother, who is a yoga teacher. My school, colleges, university teachers, and friends undoubtedly influenced the shaping of my personality. It is difficult to mention all those who directly and indirectly helped me during my journey in the natural and human sciences. I wish to stress the point here that any profound idea or discovery doesn't appear out of the blue. A lot of unfolding remains in the background, which may be unknown to us. I trust the same thing happened to me when the seed of an idea about Genome and Om manifested in my mind.

Although I grew up in a religious family, by nature I had been an atheist since childhood. I used to visit temples and join fasts or prayers because my father was a devotee of Ganesha. My faith was more in my father than in any God. However, passionate mentoring, intense interactions, and insights from Dr Gururaj Mutalik helped me progress to becoming an agnostic. This was probably the beginning of the evolving journey of modern science to meta-science.

Over the last two decades, several interactions with friends and mentors, starting with Alex Hankey, Girish Tillu, Avinash Patwardhan, Ashok and Rama Vaidya, Darshan Shankar, Sharad, and Medha Deshpande, enriched the idea. It became evident that the Genome needed to revert to Om for its evolution to move in the right direction. I realized that the trajectory needed a mid-term correction in the approach and the inclusion of the dazzling advances of science and technology. The book found the right direction after engaging in a series of conversations with my mentor, Gururaj Mutalik. I remain grateful to him for his continuous guidance. My wife, Bhagyada, was not only a sounding board to offer honest reflections on evolving ideas while writing but also took care of all mood swings expected during such a transcending journey.

While engaging with friends and well-wishers across diverse backgrounds, I met my final co-author, Indu Ramchandani, who brought invaluable depth to our exploration, merging her deep grasp of spirituality with linguistic eloquence.

Genome to Om is thus an insightful exploration through a complex terrain, trying to build the theory of knowledge based on the foundational concepts of Vedic philosophy. It is not merely about the confluence of contemporary scientific knowledge with eternal wisdom; it represents a pivotal transformation in our approach to science and technology. This book advocates moving beyond the ever-evolving, temporal knowledge of modern science to meta-science, embracing the timeless and expansive wisdom signified by Om. The narrative traverses the Vedas, the Bhagavad Gita, and the Upanishads, resonating with teachings from various global philosophies and cultures. It critically examines contemporary scientific endeavours, highlighting both the extraordinary potential and the ethical challenges posed by new technologies.

At its core, *Genome to Om* introduces the concept of meta-science inspired by one of the Upanishads' Great Statement *Prajnanam Brahma*, meaning "Consciousness is supreme". If modern science tries to know about Nature, meta-science is about the nature of Nature, where insights, intuition, and wisdom also play pivotal roles. This paradigm shift is vital for aligning scientific progress with universal well-being.

In sum, *Genome to Om* is a narrative of transition, an evolving journey from modern science to meta-science.. It emphasizes the need to address the complexities of the 21st century, ensuring that our scientific advancements contribute positively to sustainable and universal well-being.

Bhushan Patwardhan

Second Author's Preface

My path to co-authoring *Genome to Om* was as unexpected as serendipitous. Unlike Professor Bhushan Patwardhan, I had not envisioned being part of such a project. My journey has been one of continuous learning, far from the traditional academic path. With a background in Economics and English and a career spanning proofreading, editing, book publishing, and writing, I found my passion in the world of books quite early, when it began as a child in my father's bookshop.

Working in a publishing house out of necessity, which helped me build a lifelong career, I navigated through an extensive range of non-fiction, including life sciences and pure sciences. My respect for content deepened during my tenure as South Asia editor-in-chief of Encyclopaedia Britannica. The study of the ancient Vedic scriptures further enriched my understanding. For that, I remain indebted to Pravrajika Vivekaprana ji, who guided me through the study of the primary Upanishads, some works of Adi Shankaracharya, and the *Complete Works of Swami Vivekananda*. My engagement as a freelance editor for the UGC connected me with Dr Patwardhan, and we bonded instantly over our shared interest in the Upanishads.

A chance meeting again with Dr Patwardhan in early 2023 in Pune brought this book into my life. He shared the concept that had been firmly rooted in his mind since 2003, and from that point on, there was no turning back.

I firmly believe life does not operate on coincidences; everything follows a planned course. Thus, my involvement with this book feels like a part of a larger scheme. My connection with the *Ishavasya Upanishad*, especially the first verse, deepened this feeling:

ईशावास्यमिदं सर्वं यत्किञ्च जगत्यां जगत् । तेन त्यक्तेन भुञ्जीथा मा
गृधः कस्य स्विद्धनम् ॥ १ ॥

"All this—whatsoever moveth on the earth—should be covered by the Lord. That renounced, enjoy. Covet not anybody's wealth."

This verse struck a chord in me, revealing the connection between science and spirituality and the underlying unity of all things. It reflected the principle of one-ness, an idea that conveys completeness, linking the Genome with Om. One cannot covet that which is in one-ness. The six enemies of humanity that reside in all of us—lust, anger, greed, attachment or covetousness, ego, and jealousy or envy—exist because we live in duality and multiplicity.

Science and technology have advanced rapidly, focused on fulfilling desires and achieving remarkable feats. However, the relentless pursuit of diversity, innovation, and ownership often lacks a sense of unity. Through the 20th century and in the early decades of the 21st century, we have seen significant upheavals: wars, discrimination, ecological crises, and a relentless push towards industrialization and scientific progress, often disregarding nature. This pursuit has led to a paradoxical world where advancements have not necessarily translated into happiness or satisfaction.

It's time for introspection and reflection, a journey from the external world to the inner reality, from "Genome to Om". This journey isn't merely about finding unity; it's about discovering and realizing it. The one-ness exists, yet it remains veiled by our limited perceptions. *Genome to Om* is an invitation to take this journey of discovery with us through the marvels of science, facing its perils, a flashback to the very beginning, and then moving forward to knowing who I am and why I am here.

Indu Ramchandani

Prologue

Before we begin the journey of *Genome to Om* we must explain what we mean by "Evolving journey of modern science to meta-science". We are exploring the remarkable strides in science and technology, as a tribute to human ingenuity and its relentless pursuit of knowledge, and we are pointing to the potential misuse of these advancements, which forewarn us about a future marred by unchecked ambition and disregard for nature. At this crucial moment in history, when our actions have hastened the onset of the Anthropocene, *Genome to Om* advocates a transformative shift from mere scientific exploration to a meta-scientific approach. In the process, we recall many profound concepts drawn from some key sources of the ancient knowledge systems, mainly the Bhagavad Gita, the Vedas, Patanjali's Yoga Sutras, and the Upanishads. Our initial focus is on the Indian Knowledge Systems because of our familiarity with them. But we recognize and acknowledge the universality of wisdom across global traditions, where, for centuries, thinkers and seers have tried to unveil nature's secrets through inner exploration. We also recognize resonating concepts from Buddhist, Chinese Taoist, Zen, Jain, and other Eastern traditions. The Vedic scriptures were originally written in Sanskrit, although several commentaries and translations are available today in English. To simplify the reading and understanding for those of us who are not comfortable with or aware of Sanskrit, we have avoided using the original terminology in the text, unless unavoidable. But the original terms are alphabetically listed in the glossary for those of us who want to know and possibly learn the original terms.

To know the significance of the term meta-science, we look at the meaning of the prefix meta. In about 40 BCE, the concept of "meta" was created in ancient Greece, and at that time it meant "beyond" or "transcending". Aristotle's works were catalogued almost 300 years after his death, and the terms "physics" and "metaphysics" were used. The initial meaning merely implied that the latter had come after the former. Later, metaphysics came to mean "The science of that which transcends the physical". Before and during the 17th century, meta-science was considered the "higher" science that dealt with more fundamental problems than the original science. The modern usage of "meta" as a prefix indicates self-reference or higher-order abstraction, and it became popular in information and communication technologies. Meta-science, however, also involves philosophical contemplation. Philosophers like Plato and Aristotle used meta-analysis as a starting point for deeper philosophical inquiries. Plato's *Allegory of the Cave* is a philosophical metaphor for the journey from ignorance to enlightenment, emphasizing the nature of perception and reality. Some prisoners are chained in such a way inside a cave that they can only face and see a dark wall. A fire burning behind them casts their shadows on the wall. The prisoners think of the shadows as reality. According to Plato, the cave and the shadows are how we perceive the world through our senses. One prisoner escapes and gradually begins to see the "real" world, and then he realizes that the shadows in the cave are illusions. This is the turning point, from ignorance to knowledge. At times, we tend to deny a reality beyond the experience of the senses, and sometimes this is one of the limitations even of modern science.

Evolution of Modern Science

The evolution of scientific thought has invariably been marked by reductionism, where the focus has been on understanding the fundamental building blocks of nature at the smallest possible scales. This approach, rooted in empiricism and experimentation, has undoubtedly given remarkable insights into the nature of the universe, from the subatomic realm to the cosmos. But reductionism can introduce a narrow-minded view that can miss the context of the whole. Technology as a tool for scientific inquiry further reinforces reductionist tendencies as researchers strive to probe even deeper into the microcosm of atomic particles and molecules. Disciplines such as fluid mechanics, robotics, molecular biology, and nanotechnology have ensured precision and control over matter, but they have also encouraged a myopic focus on individual components, often at the expense of broader systemic considerations.

Evolving Journey from Modern Science to Meta-science

While natural science advanced into modern science, it studied the details of parts. It evolved into the science of atoms, molecules, chromosomes, genes, and genomes. Meta-science is the science that studies science! It is not limited to a form of scientific methodology known as meta-analysis, but an approach to understanding science itself. In our view, *meta-science is the science of science*, especially in trying to know the nature of Nature. We are using meta-science instead of metaphysics because we are making a subtle distinction, which takes the quantitative research of science into the realm of philosophy. This is not just an empirical or scientific pursuit but a deeper exploration of the truth as a direct experience, transcending the boundaries of conventional science and metaphysics. The integration of the microcosm and macrocosm, as symbolized by Genome and Om, respectively, represents a holistic understanding of existence, a transcendent knowledge that bridges the gap between science, philosophy, and spirituality.

To understand the significance of Om, we need to go to ancient Indic knowledge, where we get glimpses into many fundamental and searching questions regarding the universe, nature, and human consciousness. The Upanishads, a collection of ancient philosophical Vedic texts, assert that we are seekers of the truth, and *realizing the truth is a direct experience.* What truth? According to Ramana Maharishi, "The ultimate Truth is so simple. It is nothing more than being in a pristine state. This is all that needs to be said."

A hardcore scientist may question the idea of "direct experience" because it is not empirically measurable. Here we take the great scientific mind Albert Einstein as an exemplar because he talked of intuition. Bob Samples in his book *The Metaphoric Mind: A Celebration of Creative Consciousness* states that Einstein called the intuitive mind "a sacred gift". He also adds that the rational mind is a faithful servant. Nobel Laureate Brian Josephson, from the University of Cambridge, discussing intuition, says, "The importance of intuition as a contributing factor in the process by which knowledge advances needs to be fully acknowledged." The great self-taught German mathematician Carl Friedrich Gauss, who later taught at the University of Göttingen, believed that all his analyses or results *he knew already*, but he did not know how to explain the way he had arrived at these results. He would work in a retrograde action and figure out mathematically what he had already understood or "knew", or as was "revealed" to him. Gauss worked backwards from the transcendent truth, or the state of direct experience, to explain empirically what he already knew experientially. India's Srinivasan Ramanujan, another self-taught math whiz, also revolutionized the field with his intuitive leaps. His ground-breaking formulas, lacking formal proofs initially, spurred mathematicians to bridge the gap between intuition and rigour, ultimately advancing modern knowledge.

Vedic philosophy proposes *anubhava* and *anubhuti* as a unified approach to the quest for knowledge and truth. *Anubhava,* the "experiential knowledge", akin to modern science, emphasizes observation, experimentation, and analysis. *Anubhuti,* the "sensory and emotional response", delves into intuition and realization from that knowledge to get a sense of connection, which we call meta-science.

Two Sides of the Knowledge Coin

Going forward, we want to build the theory of knowledge even beyond meta-science, based on the foundational concepts of Vedic philosophy. This epistemology offers a unique perspective on the theory of knowledge. *Jnana* or *vidya* stands for information and knowledge; *vijnana* or *apara vidya* is specialized worldly knowledge, or empirical science. *Prajnana* or *para vidya* stand for higher-order inner knowledge, wisdom, or Brahman. These terms are interrelated and represent different but complementary types of knowledge essential for a comprehensive understanding of the universe, our lives on Earth, and ourselves.

Worldly knowledge is the study of material, empirical reality. It includes the sciences, mathematics, the arts, and literature—disciplines that engage with the tangible, external world. This knowledge is vital for everyday living. Despite its immense value, it is seen as "lower" because it deals with the transient, ever-changing aspects of reality corresponding to empirical modern science. In contrast, the "inner" knowledge provides higher-order knowledge of the immutable, eternal truths. It gives us the intuitive understanding that transcends the material world and imparts invaluable insights associated with consciousness, self-realization, spiritual awakening, and attaining enlightenment and liberation.

In this evolving journey, we consider "Genome" to symbolically represent modern or New Age science and "Om" to be an expression of the highest form of knowledge—meta-science. There are four "Great Statements" in the Upanishads that give us the essence of the philosophy. One of them, from the *Aitereya Upanishad*, *Prajnanam Brahma* means "Consciousness is Supreme", it is the highest form of knowledge, and it guides our exploration from modern science to meta-science.

While consciousness is accepted *as it is* by many thinkers worldwide, many scientists are connecting it with the brain and mind and seeking to locate it and find where it arises from. Does it arise at all, or is it always present? The connection between quantum physics and consciousness is now very widely discussed. If we talk of quantum entanglement and acausal connectedness, then immediately it takes the mind to two perspectives: First, we venture into science fiction, Isaac Asimov's famous *Foundation Series*, where they discuss how their planet is in a state of super-consciousness. We can say here that science fiction is not "fiction" in the deeper sense when it projects home truths wrapped in fantasy. Many concepts from science fiction have indeed become realities in our lives. The second perspective is based on real, first-hand experiences in grasping the true nature of reality. This means direct experience, stressing how intuition, instinct, sixth sense, or what is known as "gut feeling" are crucial for understanding reality. According to our view, meta-science is the confluence of these perspectives.

Modern science cannot continue to rely *only* on what we can perceive through our senses or measure with the help of machines as their extensions. Good scientific theories are falsifiable; they can potentially be disproved by new evidence. Science strives for objectivity, aiming to minimize bias in the pursuit of knowledge. Scientists also favour simpler explanations and seek to unify different areas of knowledge. Also known as the law of succinctness, Occam's Razor suggests that among competing hypotheses or explanations, the simplest one is usually the correct one. These principles form the foundation for scientific progress and our ever-expanding understanding of the cosmos and life. But these theories are not set in stone and should continue to evolve as new information comes to light.

It is necessary to keep an open mind and consider ideas and knowledge from all sources—whether they are from mythology, philosophy, history, or even fiction. Different civilizations, cultures, and traditions offer diverse views on the creation of the universe or the origin of life. True scientists cannot ignore these by putting them in closed boxes of mythology, religion, belief systems, or spirituality. We must consider how some of these astonishing ideas originated and try to decode them with the help of science.

The Microcosm–Macrocosm Continuum

Vedic philosophy refers to the continuum of microcosm and macrocosm. Genome and Om are the metaphors. This connection between Genome and Om extends to understanding the universe and life at both levels, the cosmic and the atomic. It also represents the principle of non-duality from the Advaita philosophy, which is discussed in various chapters of the book and which presents a universe where separation is illusory or the seeming reality (maya) echoing the holistic and interconnected nature of existence.

What, then, is the story here? The peace aphorism from the *Ishavasya Upanishad*, as mentioned in the first author's preface, resonates remarkably with modern scientific concepts, especially in genomics. The aphorism specifies that the whole is complete, and a part of it is also complete in itself. Applied to the Genome, this verse beautifully parallels the amazing quality that if a part of the genome is extracted, it can still function as a complete unit, capable of producing a whole organism (as seen in cloning or genetic reproduction). Simultaneously, the remaining part of the genome continues to be complete, retaining its potential and integrity. This mirrors the principle that both the part and the whole are interconnected and intrinsically "complete".

Genome to Om is an ambitious journey from specialized external knowledge to supreme inner knowledge that spans the realms of modern science to meta-science, and beyond. This is our endeavour to explore the depths of understanding that combine the tangible world of empirical science with the intangible essence of spirituality. We want to harmonize empirical science and spirituality, seeking an in-depth understanding of life, its meaning, and purpose. We aspire to see a future where scientific progress serves the larger cause, not just the satisfaction of human materialism, comfort, and technological advancement. We are aware of the tremendous challenge of transitioning from modern science to meta-science and progressing with unity in diversity towards universal harmony and well-being. This transition is only possible when modern society evolves into a meta-society for the greater good. *Genome to Om* is our humble effort to catalyse a recalibration in the future direction of science and society by trying to steer towards peace and sustainable universal well-being. This effort of possibly evolving from the "Ascent of man" to the "Ascent of all", from the volatile Anthropocene to a harmonious and all-encompassing Omcene, and creating a global meta-society, is difficult but not impossible. The journey then continues, with humanity united within diversity, progressing in the harmony of the Om Way.

If yesterday's science fiction has become today's reality, we should seriously consider ancient symbolism as a source of ideas and innovations. In doing so, we are not even remotely encouraging pseudoscience. Richard Feynman has used the phrase "cargo cult" while differentiating science from pseudoscience. In our work here, we have garnered and put down all such information that has helped us and, hopefully, will help the discerning reader add value to our beliefs and convictions. We have referred to several works and statements of many thinkers, scientists, and authorities from diverse fields. These tell us what is out there in terms of information and points of view. We are not leading to any judgements, nor are we supporting any biases. We are neither authenticating, reaffirming, nor speculating. We are just saying it as it is to facilitate the evolution of modern science into meta-science.

CHAPTER 1

Genome: The Evolution and Marvels of Science and Technology

The human spirit must prevail over technology.
—Albert Einstein

The journey begins with our earliest ancestors, who innovated as they evolved. Endowed with sufficient cognitive abilities, they learned to adjust, adapt, and communicate. There was continuous change, development, and innovation. What emerged with incredulous results was the advancement in constantly evolving science and technology (S&T), offering new solutions and tools that impact our lives in many ways. Scientific breakthroughs and technological advancements are happening at an incredible pace. Super-intelligent computers powered by AI are learning like never before. Space exploration is no longer limited to science fiction; powerful telescopes like the Hubble capture real images of distant galaxies, while reusable rockets promise more frequent missions and deeper exploration. 3D printing technology allows us to create complex objects layer by layer, revolutionizing manufacturing and prosthetics. In medicine, advancements in AI and genomics are paving the way for treatments tailored to an individual's unique biology, maximizing effectiveness. Nanobots swimming through the bloodstream can perform microsurgeries or deliver medicine directly to diseased cells, and bioprinting is being explored to make human tissues and organs for transplants. Robots are becoming more sophisticated and versatile, with potential applications in precision surgery, manufacturing, and even disaster relief. The fight against climate change is seeing advancements in green energy sources like solar and wind power, alongside technologies like carbon capture to remove harmful emissions from the atmosphere. These are just a few glimpses into the exciting world of S&T, where the impossible seems to become possible with every discovery. It's a future filled with possibilities for solving problems, creating a sustainable world, and pushing the boundaries of human achievement.

With great intelligence, over the centuries, we have experimented, created, and innovated for our comfort. The quest has always been to make life easier, simpler, and far more comfortable, with the hope that it will then, naturally, be happier. The mind's eternal quest, and the instinct or natural urge, is the search for happiness. And happiness is seen in the fulfilment of desires. This gives us momentum and drive—an inherent urge to forge ahead, to improve, and to excel—and this has pushed us towards continuous development and innovation. The first among the "thinking" human species must have been exceedingly curious about the nature around them. Undoubtedly, several other species exhibit thinking abilities, and some animals display impressive cognitive capacities. But the human species has a unique depth and complexity of cognition, abstract thinking, and, most importantly, of self-awareness.

Consciousness and self-awareness, in particular, are distinguishing features of our species since we can reflect on our thoughts, emotions, and identities. The complex sense of self, awareness of the past, present, and future, and the ability to understand and relate to the thoughts and emotions of others are unique aspects of human cognition.

When facing the challenges that nature threw, our ancestors must have developed the impulse to find solutions to deal with unpredictable inclement weather. The effort was undoubtedly to get a little more comfortable in "homes" and to find, grow, and store food. These physiological and biological needs must have prodded the desire to explore, and the natural survival instincts pushed the *Homo sapiens* to search and find. And that has been the journey from the Stone Age to the Anthropocene.

From basic tools to the use of water and fire, to cooking and living together, to housing and sanitation, farming and agriculture, education, and medicine, we have consistently and successfully enhanced our lives. Recognizing emotions and attachments, feelings, and the impulse to create homes and families, realizing the sense of ownership has prompted us to get more out of life, and finally has arisen the ultimate desire: to make this earth our exclusive preserve and permanent home!

One of our species' most natural traits is finding solutions to problems and challenges that are inevitably faced as civilizations grow. From the uncontrolled and unpredictable life in the wilderness to "planning" a way of life undoubtedly posed issues for which solutions were needed, and this gave the impetus to innovation and to scientific and technological advancements. Competition, collaboration, and the pursuit of a better quality of life have driven us to continuously improve, excel, and develop in the field of S&T. The underlying drive of all advancement has been, and shall continue to be, the quest for happiness. And happiness was and is assumed to be dependent on physical and material comfort, exactly the direction in which we have moved.

But with these developments have also come challenges and the unending race for the survival of the fittest. Surely, our key questions today are exactly as they must have been in the minds of the "first humans". In awe of nature, of the surrounding elements, of the mysteries of light and sound, of day and night, they must have wondered as we still do. But we see a subtle change in the nature of the basic questions. Initially, when our ancestors were surrounded by the incredible miracles of creation, they were perhaps more conscious of their link with nature; they maintained a deep connection with the universal being. Every element was ruled by a god, and humans propitiated or worshipped these gods. They knew that "angering" the gods invited their wrath. That explained the unpredictable turbulence of nature, the vagaries of weather, and all the uncertainties. However, once we became more "aware" or "educated" and as the scientific mind began its own fascinating and brilliant journey, in the process of evolution, we gradually found explanations for the mysteries or answers in scientific discoveries. The mysteries of nature are now scientifically explained. Along with the discoveries have come inventions in the form of technological gadgets to change the discoveries into useful "partners" in this fast-paced journey of progress.

The questions that now lurk are: How far have we come in this journey, and finally, where are we heading? Are we losing track of who we are and of the real goal and purpose of life, and do we even know why we are here? As these questions arise in the conscious and subconscious mind, we propose the need for the transcendence of modern science to meta-science. Here lies the map and track sheet of our journey. We thus start this part of our story with the genome, reiterating that it encompasses modern science and technology.

Genome: An Emblem of Modern Science

In this intricate story of life, the genome emerges as a remarkable *emblem of the conscious microcosm*, intertwining the marvels of modern science with the profound wisdom of Vedic knowledge. This might sound like a metaphorical or even poetic way of describing the genome, but this is how we perceive it. The genome plays a definitive role in shaping living organisms, including human beings. Scientifically, the genome is the complete set of genetic material (deoxyribonucleic acid, DNA) that is present in an organism. It contains the instructions necessary for the development, functioning, and maintenance of that organism. The human genome is said to consist of approximately 20,000–25,000 genes that code for proteins and regulate various biological processes. The study of the genome is rooted in empirical science, which relies on observation, experimentation, and evidence-based understanding. It exemplifies the scientific method, where hypotheses are tested through rigorous experiments, data is analysed, and conclusions are drawn based on observable facts. The mapping of the human genome has provided an unprecedented understanding of the genetic factors that govern life, disease, and evolution.

Inside every cell of our body lies a hidden library, a treasure trove of instructions on how to build and maintain. These instructions are stored in forty-six chromosomes, thread-like structures made of DNA. Each chromosome acts like a giant book, containing thousands of genes and the recipes for proteins that run the show. Why are we detailing all this information? Matt Ridley in his book *Genome: Autobiography of a Species* listed the twenty-three pairs of chromosomes, and next to each, he listed themes of human nature! We imagine these chromosomes as twenty-three pairs of partners. They carry a mix of genes that we inherit from our parents. The first twenty-two pairs, called autosomes, aren't concerned with sex. Chromosome 1, the biggest of the bunch, holds around 2,000 genes, including those for bone development and even some cancer risks. Chromosome 4, with roughly 1,100 genes, has the blueprint for a protein involved in Huntington's disease, while chromosome 6, containing over 1,200 genes, carries the instructions for the immune system and a susceptibility gene for Type 1 diabetes. Chromosome 11, with about 1,400 genes, even has the ABO gene that helps determine our blood type. Instructions for the colour of the eye come from chromosome 15, and genes from chromosome 17 govern how our body builds muscles.

The sex chromosomes are where things get interesting. The Y chromosome stands out in the sex chromosome pair for its miniature size. Compared to the X chromosome, which boasts around 1,000 genes, the Y chromosome is a relative dwarf, containing only about 50 genes (see Fig. 1.1). This significant size difference has puzzled scientists for some time, and it has yet to be resolved. This size difference is a scientific mystery. The X chromosome's larger size might allow more diverse gene expression, potentially explaining why the X chromosome influences various functions beyond sex determination. While emotions are complex, the X chromosome's wider range of genes could contribute to emotional intelligence and resilience. So, while men may be seen as physically stronger, women are taken as the emotional powerhouse.

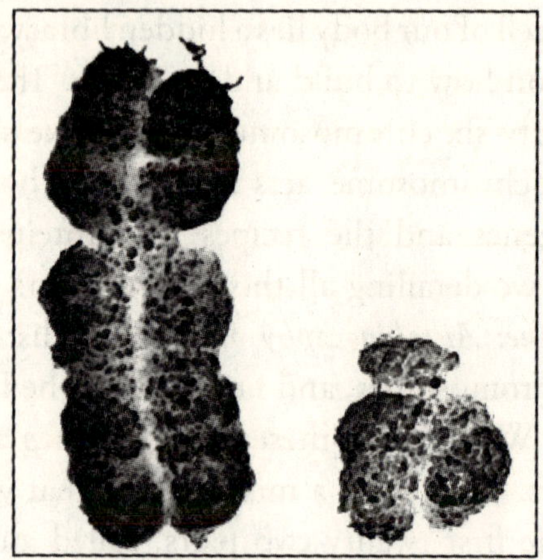

Fig. 1.1. X and Y chromosomes.

Of course, these are just a few examples. Each chromosome is a maze of intricate codes; some sections are well understood, others are an ongoing scientific mystery. But one thing is for sure: these forty-six chromosomes hold a fascinating story, from our physical traits to our hidden predispositions. They are the building blocks that have come together to create us and continue to guide each function of the body, making us what we are.

We may assume that, as humans, as the most evolved species, we would have the largest genome. That is not so. There are species with much larger genomes. Among vertebrate animals, the smallest genomes are found in ray-finned fish and birds, whereas the largest are found in lungfish and salamanders. The mudpuppy salamander, for example, has a genome size nearly thirty times the size of the human genome. We do not know the reasons for such wide variations between species, as researchers continue to explore the relationships between genome size, complexity, and biological traits. However, we do know that the genome serves as a comprehensive blueprint for building and maintaining a living organism. It contains all the information that is needed to develop and regulate the complex structures and functions of cells, tissues, and organs. This is only its physiological identification, as it plays a significant role in determining the traits, characteristics, and predispositions of an organism. The genome carries the evolutionary history of a species. It contains the genetic code that has come down through generations, as it reflects the changes and adaptations that have occurred over time. The unique combination of genes in each individual's genome contributes to the diversity of life. Along with all physical attributes, it affects all aspects of the human personality, behaviour, and susceptibility to various diseases.

Ongoing research and the ambitious projects of genome research in the 21st century have highlighted what defines an individual—genetics, microbial communities, and environmental factors. They have collectively broadened and deepened our perspective on biological identity. Some of the significant outcomes have been:

The Human Genome Project is a comprehensive reference for the human genetic code; it can map and sequence the entire human genome, which is the complete set of DNA in the human body. It is enabling significant advances in genetics, medicine, and biotechnology. It helps us to understand genes associated with diseases, genetic diversity, and evolutionary relationships, offering a more comprehensive view of our genetic identity. The 1000 Genomes Project maps and sequences the genomes of a diverse group of individuals from different populations around the world. This project has been invaluable for research in genetics, helping to identify genetic variations associated with diseases and facilitating the development of personalized medicine, risk prediction, and disease prevention. This project has emphasized that genetic identity is not a uniform concept but varies significantly among individuals and populations.

The genome, the complete set of genetic material within an organism, transcends its physical entity to embody attributes of consciousness, universality, immortality, and completeness. While the specific relationship between the genome and consciousness is complex and not fully understood, genes play a crucial role in the development and function of the brain. We are veering away from hardcore scientific understanding and analysis and are trying to understand in a deeper or, shall we say, different way. We want to explore the genome's microcosmic nature and its alignment with its macrocosmic counterpart, Om.

Conscious Genome: The genetic code entrapped in DNA is made of special building blocks called nucleotides. These have cool names: Adenine (A), Thymine (T), Cytosine (C), and Guanine (G). This code isn't just ours. It's found in every living thing, from the tiniest bacteria to the tallest tree. It's like a universal language—a shared story passed down for generations, hinting at a common ancestor for all life on Earth. This code acts like a self-replicating recipe book, making sure the instructions with A, T, C, and G are copied accurately and transcribed on messenger ribonucleic acid (mRNA). The genetic code is a set of three-letter words, like ATT or GCG. Each word tells the body how to build a tiny building block called an amino acid, like isoleucine or alanine. These translated amino acids then come together like Lego bricks to form proteins, the workhorses that run everything in our bodies. So, the code, basically, tells the body how to build itself using A, T, C, and G! While the code with A, T, C, and G isn't a conscious entity, it is remarkably linked to both stability and change. The core instructions remain interestingly similar across all living things, ensuring the basic building blocks of life (amino acids made from the code's words) are similar. But the code also allows for variation and mutations, leading to the incredible diversity of life on Earth. This process of transcription and translation of the genome looks like a Christmas tree under a high-resolution electron microscope. Astonishingly, the genetic code is common for all living things on the planet, and it continues to be alive after having evolved over millions of years. We call this microcosm a conscious genome.

The exploration of the genome through the lenses of both modern science and Vedic philosophy, the ancient truths that represent the universal perspective, reveals the deep unity between the microcosm and macrocosm, between empirical knowledge and higher-order wisdom. The genome is not just a biological entity; it is a dynamic, self-regulating system reflective of a conscious microcosm. However far we venture or deeply we explore, the core of our search is based on the awareness of "Consciousness Is", which means that we are seeing the microcosm and macrocosm as one. Each nucleotide base pair and triplet in the genetic code, running uniformly through all species from amoebas to apes and viruses to vertebrates, gives us a universal language of life. This universality highlights the genome's comprehensive and all-encompassing nature.

Epigenetics: In the world of genetics, scientists used to think that our DNA, like a set of instructions, determined everything about us. But epigenetics shows us that it's not just our DNA that decides our fate; it is also how our genes interact with the world around us. Epigenetics is derived from the Greek prefix "epi-" meaning "over, outside of, around". It refers to the changes in organisms caused by modifications of the "expression" of genes rather than by altering the genetic code itself. Environmental, psychological, and social factors such as the air we breathe, the food we eat, and the way we think, behave, and interact with others can and do cause changes in the DNA, which then initiate changes in the expression of genes. These modifications or changes sometimes get passed down to subsequent generations, affecting their health and behaviour without necessarily changing the underlying sequence of the DNA. Lifestyle and environmental factors can and do directly influence genetic expression and, potentially, the health and development of offspring. The relationship between the genome and epigenome is complex and not fully understood. But both play important roles in shaping our lives. At the outer or physical level, they ensure good health and a sense of well-being, and at the inner level, they create a calm, peaceful, and generally happier demeanour.

One of the key players in epigenetics is DNA methylation. We can think of DNA methylation as a little tag that can be added to our DNA, like a "post-it" note. These tags can either turn genes on or off, like switches, influencing the traits we inherit from our parents and how our bodies respond to our environment. Epigenetics reveals how environmental factors, such as diet, stress, and lifestyle, can modify gene expression through DNA methylation and modifications of histone, a protein that provides structural support for a chromosome, leaving lasting imprints on our biology. Gene regulation is another important player in epigenetics. It helps decide which genes are active and which ones stay quiet. Together, DNA methylation and gene regulation show us that our biology is much more dynamic and flexible than we ever imagined, influenced not only by our genes but also by the world we live in.

AyuGenomics: A new field pioneered by the first author, as mentioned in his preface, is trying to bridge the gap between tradition and innovation through AyuGenomics (also Ayurgenomics). Ayurveda and Yoga are ancient traditional practices that provide a holistic approach to well-being, treating the body as an interconnected system. In these systems, the term used for good health and well-being is *swasthya*, meaning "balanced within the self". It transcends the absence of disease, emphasizing a balanced mind, body, and spirit. *Swasthya* acknowledges the interconnectedness of these aspects, with practices like yoga promoting mental tranquillity and a focus on individual needs to achieve a dynamic state of thriving, not just surviving.

Ayurveda, "the science of life", emphasizes understanding our body type or unique constitution, known as its prakriti (nature). This can be considered a personalized blueprint for health. Three major prakriti types or *dosha*s (the humors) are known as *kapha*, *pitta*, and *vata*. *Dosha*s represent the five primordial elements, roughly equivalent to earth, water, fire or energy, air, and ether. *Kapha* is characterized by the qualities of earth and water. People with a predominant *kapha* constitution are typically described as calm, compassionate, and nurturing. When *kapha* is balanced, it provides stability, strength, and immunity. Physiologically, *kapha* types often have a sturdy build, smooth skin, and slow digestion. Pitta is associated with the qualities of fire. *Pitta*-dominant individuals are described as ambitious, focused, and passionate. *Pitta* governs digestion, metabolism, and the regulation of body temperature. Physically, *pitta* types tend to have a medium build, a fair or ruddy complexion, and a strong appetite. *Vata* is characterized by the qualities of air and ether (space). Those with a dominant *vata* constitution tend to be lively, creative, and enthusiastic. *Vata* governs movement in the body, including circulation, breathing, and elimination. Physiologically, *vata* types typically have a light build, dry skin, and irregular digestion. Each of us has a unique combination of these *dosha*s, with one or two typically being dominant. Recent studies have shown that the *dosha*s have a genetic basis. The important field of AyuGenomics explores the potential link between prakriti, the three *dosha*s, and genetics.

Recent studies using genome-wide analysis suggest specific genes might influence individual prakriti types, potentially explaining the link between our genes and inherent traits. But AyuGenomics goes beyond genetics. By understanding how genes, prakriti, and epigenetics interact, AyuGenomics can pave the way for a more direct link to health. It is expected to play a very significant role in personalized medicine. In essence, prakriti provides a framework from Ayurveda, genetics offers a lens to see the underlying influence, and epigenetics explains the impact of *samskara* on overall personality.

Samskara: In the Vedic systems, *samskara*s are impressions or imprints left on the mind by experiences. We have discussed how the environment and lifestyle can influence gene expression through epigenetics. This also includes upbringing, nurturing, and the concept of *samskara*, which refers to imprints left on the mind-body by education, upbringing, and experiences. The imprints on the mind can be positive or negative and are believed to shape our predispositions, behaviours, and even future life paths. *Samskara*s begin right from the prenatal stage and are cultivated or acquired through family upbringing, cultural practices, religious rituals, education, and ethical living to generate positive impressions that ensure a virtuous life. While *samskara* encompasses a broader metaphysical, spiritual, and philosophical dimension, epigenetics provides the scientific framework that explains how such impacts can be mediated and how changes in gene expression can be inherited. *Samskara* involves a more metaphysical transmission through *karma* and rebirth. Exploring the connections between *samskara* and epigenetics, we see the intricate interplay between nature and nurture in shaping human existence.

Immortal Genome: Remarkably, the genome shows the properties of immortality and completeness. If a part of the genome is removed, both the original and the separated parts have the inherent capacity to regenerate into their complete forms. This characteristic is not just a biological wonder; it reflects a deeper, intrinsic quality of consciousness and self-regulation within the genome. It points out that within our genetic makeup lies a reflection of the vast, interconnected web of life, resonating with universal consciousness. The smallest is united with the greatest to give us the "whole". This concept of the peace aphorism referred to in the preface by the first author gives us the image of the whole. What is visible is the infinite or the complete, the whole. What is invisible is also the infinite or the whole. This finite is part of that infinite, and this too is complete in itself. In the domain of modern empirical science, the study of the genome stands as a quintessential representative.

Our story begins with knowing who we are physiologically. With the knowledge of our genetic structure and as we explore Genome and the marvels of S&T, we find that it all resonates strikingly with the ancient Vedic system concepts of worldly knowledge, *apara vidya.* And then we need to take that leap to higher knowledge of the Self, *para vidya.* The very sublime continuum, *Genome to Om*, is the thread that binds together all our thoughts. But to understand this concept and to know the complete picture of the macrocosm and microcosm, we turn the pages back to explore the rapid evolution of modern science and its dramatic advances through recorded history.

From Caveman to Anthropocene

The ascent of humanity from the Stone Age and the Bronze Age to the modern era of Industry 5.0 is marked by incredible leaps in S&T, each a stepping stone that propelled us into new realms of possibility. The Stone Age lasted for roughly 3.4 million years and ended between 4000 BCE and 2000 BCE with the advent of metalworking.

This journey, with the stories of brilliant minds and their astounding discoveries, tells us of the indomitable human spirit and its ceaseless quest for knowledge. Science and technology have galloped and continue to evolve, but the roots can be traced to the cumulative contributions and shifts in thinking over many centuries. At the same time, we are aware of the perils, the dangers, and the negative impact of misuse or over-ambitious pursuits.

History of Scientific Development

Necessity is indeed the mother of invention, and while it is believed that the first spark of fire was discovered in the Stone Age, one does not know whether it was the accidental rubbing of two stones, the act of nature with lightening setting trees on fire, or the rubbing of wooden branches that brought the miracle of this amazing phenomenon to human awareness. As the virtue of the use of fire was appreciated, the "god of fire" became most relevant. Almost every ancient civilization, whether African, Egyptian, Chinese, Indian, Japanese, or Greek, among others, worshipped the god of fire. A volcanic eruption meant that God was angry and needed to be propitiated. We are not implying that the fears of gods or such related "superstitions" need to be reinstated! We are just recalling that the natural elements had to be understood and respected, and it was important to "live with nature". The early humans systematically developed their methods of harnessing this elemental force. They learnt to cook, ward off predators, and lighten up the darkness. The mastery of fire can be seen as humanity's first true technological leap.

We move forward in the Stone Age to the emergence of the wheel sometime during 3500 BCE, a pivotal moment that revolutionized transportation and machinery. The wheel's invention not only made movement and trade easier but also laid the foundation for future technological advancements in mechanics and engineering. The earliest development of the scientific method was a gradual and cumulative process that has evolved over centuries. It was shaped by contributions from various thinkers, philosophers, scientists, and practitioners. There could never be a single inventor or creator in S&T or in the scientific method. The methodology of scientific study had to emerge as a set of principles and practices that revolutionized the way knowledge was acquired and tested. From the very beginning, we come to the comparatively more "recent times".

The Ancient Legacy

Writer and researcher Graham Hancock has proposed that an advanced civilization existed before ancient Mesopotamia, Babylonia, and Egypt. He believes that this civilization was wiped out by a comet strike around 12,000 years ago, leaving only the faintest of traces. Hancock's theory of advanced ancient civilizations has ignited curiosity and very heated debates. Scientific evidence, combined with intriguing archaeological finds, has given rise to a compelling narrative that encourages us to reconsider the achievements of ancient cultures. Although many scientists have discarded Hancock's theory, the supporting archaeological and scientific evidence prompts us to at least reconsider the intellectual and technological capabilities of the ancient civilizations. The legacy of ancient science and technology is incredible. From the fertile floodplains of different parts of the world, brilliance did emerge. These early civilizations, from Mesopotamia's ingenious irrigation to the Maya's mathematical brilliance, the Vedic civilization's scientific contributions, and the Roman Empire's engineering marvels, laid the foundation for scientific progress that continues to influence us even today. The early inventions and discoveries testify to human ingenuity and the enduring power of knowledge. The megalithic wonders of the ancient world, including the Great Pyramid of Giza and the Sphinx, endorse the expertise in ancient engineering, astronomy, and mathematical complexity, seeing the precision in construction and the alignment with celestial bodies. The timelines of major ancient civilizations are given in Fig. 1.2.

Influence of the East

The Eastern influence on the development of S&T in the West is a fascinating tale of intellectual exchange, cultural interplay, and the shared human quest for understanding the truths of existence. This story begins in the ancient Vedic era, with the Sindhu–Sarasvati and other civilizations spanning several centuries, and how Eastern wisdom shaped and inspired some of the greatest minds in Western science and philosophy.

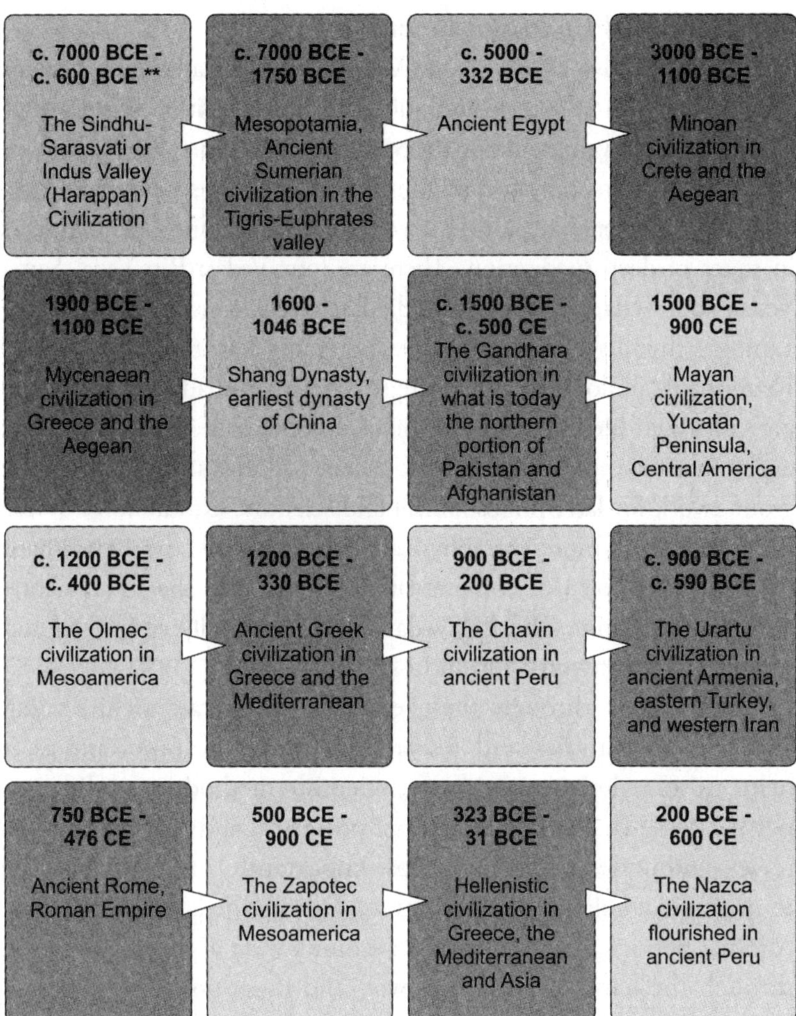

** The civilizations of the Indian Subcontinent, where the Vedic, Sindhu-Sarasvati, Indus Valley-Harappan civilizations thrived are dated to over 8000 or more years ago as asserted by several historians and thinkers. Although some historians continue to place the Indus Valley-Harappan civilization in 3000- 1900 BCE

Fig. 1.2. Timeline of ancient civilizations.

The Vedic and Upanishadic Era

The Vedic era, also known as the Vedic period, is dated by historians roughly between 1500 BCE and 500 BCE. However, it is considered to be much older, even predating the Sarasvati civilization. The Vedic texts were passed down orally and are admittedly the oldest scriptures known. They are considered authorless, revealed to the ancient sages and seers in states of deep meditation. There are four Vedas: Rig Veda, Sama Veda, Yajur Veda, and Atharva Veda. Each Veda consists of four parts: Samhitas (hymns), Brahmanas (rituals), Aranyakas (meditations), and Upanishads (philosophical teachings). The Upanishads are an integral part of Vedic literature with their philosophical depth. These texts explore the concepts of self, consciousness, and the universe, laying the groundwork for later philosophical and scientific explorations. While the approach is more metaphysical than empirical, the Upanishads demonstrate a form of contemplative inquiry into the fundamental aspects of existence. The Rig Veda was systematically composed and written down between 1500 and 1200 BCE.

The journey through these evolving historical, architectural, and scientific studies and research, however, is important as it helps us to review new evidence, scientific methods, and different perspectives. These are all dynamic processes, and it is up to us to review and understand them. Most importantly, we must continue to maintain a balance of perspectives. During the later Mughal and colonial periods, some historical accounts were apparently written through the lens of those in power, and these possibly generated biases and misconceptions. Modern-day efforts are challenging and revising these historical accounts. The objective is to present an accurate depiction of the Indian subcontinent's rich and complex history, acknowledging indigenous perspectives and dismantling any biases that may have coloured narratives in the past over the centuries or might still be colouring our present understanding if due diligence is not adhered to. This is what meta-science must ensure—an unbiased and "as true as is known" narrative so that we can transcend and move forward.

The advanced achievements in urban planning, engineering, and astronomy prompt us to unravel the mysteries of those historic wonders. We must view the narrative of human history with a more expansive and open-minded exploration without getting deviated by any biases or preconceived notions. This balanced approach will hold us in good stead as we transcend from science to meta-science. Possibly, ongoing research will someday be endorsed by empirical evidence. But what most historians, for the moment, agree on is the existence of the Sarasvati–Sindhu or Indus Valley Civilization, popularly known as the Harappan Civilization.

The Indus Valley Civilization, an ancient culture, thrived in the Indian subcontinent, what is now Pakistan and northwest India. It was a Bronze Age civilization and is dated from 3300 BCE to 1300 BCE, although there is a strong contemporary belief that this is one of the earliest, if not the earliest, and can be as ancient as 8,000 years old. This civilization was a beacon of urban planning and architectural ingenuity. Archaeological evidence suggests a level of sophistication in various aspects, including town planning, engineering, and possibly early forms of measurement and mathematics. Excavations at the sites of the cities of the Indus Valley, Mohenjo-Daro and Harappa, display that level of advanced urban planning. The streets were well laid out, and the construction of the cities was carefully organized and, obviously, well-perceived. The builders of the cities had a very practical understanding of urban sanitation. They had set up advanced and complex water supply, drainage, and sanitation systems with a focus on wastewater disposal in the cities, including covered drains and public baths. The Great Bath of Mohenjo-Daro, a large, public bathing area, is an example of the architectural and hydraulic sophistication of the civilization (see Plate 1). Excavations at sites like Rakhigarhi, one of the largest Harappan sites, reveal a similar level of urban planning with well-planned roads and drainage systems.

The seals with inscriptions and depictions of various animals suggest a system of writing or symbolic communication Our understanding of the Indus Valley Civilization is limited by the lack of deciphered written records. The Indus script apparently still remains undeciphered and, therefore, is a tantalizing mystery, hinting at a rich linguistic and cultural legacy yet to be fully understood.

The details of their scientific and technological knowledge are not fully known. Despite this limitation, the archaeological evidence and all the analyses of the excavated sites indicate a certain degree of scientific and engineering advancement in the Indus Valley Civilization.

The Sarasvati Civilization stood parallel to the Indus Valley Civilization, flourishing along the banks of the now non-existent Sarasvati River. One fascinating mystery surrounds the Sarasvati River, whose existence was denied till recent archaeological findings. Satellite images have confirmed its existence and highlighted its significance. Harappa, one of the important centres of the Sarasvati Civilization, was apparently located on the confluence of the rivers Sarasvati and Drishadvati. Scholars suggest that the river supported civilization, agriculture, trading, and thus the development of their cities. When the rivers dried, that was possibly the turning point when the civilizations must have died out or the population gradually migrated.

Recent research on DNA analysis of skeletal samples from the Rakhigarhi archaeological site of the ancient Harappan Civilization located in present-day Haryana, India, dating back to the Bronze Age shows that the estimated age of the Rakhigarhi era could be more than 8,000 years old. Excavations in northwest India in Rajasthan have unearthed a 5,000-year-old industrial production centre featuring furnaces, hearths, and mudbrick structures . This specialized centre suggests a network beyond large cities and compels a re-evaluation of the social and economic organization potentially linked to the Sarasvati Civilization. This period corresponds to the mature phase of the Sarasvati–Sindhu or Indus Civilization, characterized by advanced urban planning, sophisticated architecture, jewellery making, and a complex socio-economic structure. Such studies and emerging archaeological evidence are likely to bring more surprises to our understanding of this rich ancient era.

One controversy that has raged over the Indus Valley Civilization has been the Aryan Invasion Theory, a long-standing explanation for the development of the ancient civilization of the Indian subcontinent. It was proposed in the mid-19th century by German linguist and Sanskrit scholar Max Muller. Recent archaeological and genetic research has challenged this theory, suggesting that the belief in a large-scale Aryan migration does not align with the genetic continuity and linguistic patterns observed in the Indian subcontinent. David Reich is a geneticist and professor at Harvard Medical School. He has done immense work on ancient DNA analysis and has contributed significantly to understanding human migrations and population movements in ancient times. In his book, *Who We Are and How We Got Here: Ancient DNA and the New Science of the Human Past*, Reich discusses the genetic evidence related to ancient migrations, including those in the Indian subcontinent. The ancient DNA and the evolution of languages, especially Prakrit and Sanskrit, indicate a more gradual and indigenous development of the Indo-European languages. A senior Indian academic N.S. Rajaram states that the "Rigvedic people were already established in Bharat (India) no later than 4000 BCE ..." We are not getting into this ongoing debate here because, from what we understand, neither point of view—the one that insists on an invasion and the other explaining another perspective—has complete concurrence and endorsement. But there are enough scholars today who support the idea of indigenous cultural developments in the Indian subcontinent and of a very ancient Vedic civilization and culture.

As we move eastward, the architectural marvels across the Indian subcontinent tell their tales. The periods of the great epics from ancient India, the Ramayana and Mahabharata, steeped in mythology, are associated with an array of monumental structures. But the underwater city of Dwarka off the coast of Gujarat in India provides another intriguing piece of evidence. This submerged city is no longer a mythological story as narrated in the Mahabharata. The National Institute of Oceanography (NIO) and the Archaeological Survey of India (ASI) have discovered very old structures submerged off the coast of Dwarka through their marine archaeological explorations (see Plate 2). It is claimed that carbon dating on debris recovered from the site, including pottery, sections of walls, beads, sculpture, and human bones and teeth, indicate Dwarka as nearly 9,500 years old. Such ongoing studies on various archaeological sites using advanced genetics, carbon dating, and other reliable scientific research also make us wonder if the story of the Great War of the Mahabharata is just an epic or an integral part of ancient history.

The Kasar Devi temple in Almora in Uttarakhand, India, is said to have a unique connection with Machu Pichu in Peru and Stonehenge in England. These spots are said to be positioned on the Earth's Van Allen Belt, a zone of energy-charged particles. The precise alignment of such structures with celestial events suggests an intimate knowledge of astronomy among ancient builders. Further north, in the region of Kashmir, ancient temples like the Martand Sun Temple showcase a blend of architectural styles that reflect the diverse cultural influences of the region. In Nepal, the stately pagodas and stupas, rich in detail and history, again reflect a vibrant cultural and religious heritage. The Nalanda Mahavira was a Buddhist monastery and monastic centre and a very ancient university, situated in Bihar in northeast India. It is said to have been the first residential university in the world. While archaeological findings date it to having been built in the 5th century CE, it is also talked of as having existed much earlier. It was ransacked in the 12th century in invasive raids in Bihar and never recovered. Renowned centres of learning like Nalanda and Takshashila attracted scholars from far and wide.

The regions from Afghanistan through Pakistan, India, Nepal, Cambodia, and Thailand have their stories to tell. The Sun Temple in Konark, the Brihadeeswara Temple in Thanjavur, the temples of Khajuraho, and the caves of Ajanta and Ellora in India, for example, are rich with intricate stone-carving techniques and architectural marvels. All these monuments reflect the advanced knowledge that these civilizations had in mathematics, physics, engineering, and technology.

Angkor Wat in Cambodia is majestic. The monumental structures display advanced architectural and engineering skills. Built sometime between the 9th and the 12th centuries, this temple complex is the largest religious monument in the world. Initially a Hindu temple dedicated to Vishnu, it later became a Buddhist site.

Angkor Wat, with its intricate carvings and majestic towers, is a marvel of Khmer architecture and symbolizes the zenith of ancient Southeast Asian civilization's architectural achievements.

Thailand has the ruins of Ayutthaya and Sukhothai, which tell stories of the splendour of a bygone era. The towering stupas and intricate carvings echo the architectural and spiritual influence of the Indian subcontinent. Afghanistan is where the ancient Buddhas of Bamiyan once stood. Carved into the cliffs, these colossal statues were not only marvels of art but also the remnants of Buddhist monasteries and stupas that bore witness to the region's rich history. These ancient structures told the tale of the exchange of architectural and spiritual ideas along the Silk Road. The tragic destruction of the Buddhas of Bamiyan is a poignant reminder of the fragility of cultural heritage and of the vagaries of human minds.

Spanning millennia and encompassing a vast expanse from Afghanistan to Southeast Asia, we see a picture of a world rich in diversity yet connected by the threads of shared human experiences. This historical reminiscence and the journey across ancient civilizations and their architectural achievements are not just a chronicle of stone and mortar. It is not just about the monuments and ruins; it's a story of cities built and lost, of empires that rose and fell, and of ideas that transcended time and space. We can see the enduring spirit of exploration, creativity, and the quest for meaning that defines human civilization. This story is about the people, the ideas, and the advancements that shaped these civilizations. It mirrors the human soul's aspirations and struggles, a continuous dialogue between our ancestors and us, and a reminder of the legacy we inherit and the future we are to shape. It displays human ingenuity, artistic expression, and the relentless pursuit of excellence. Each monument and ruin tells a story of a past where architecture was not merely about building structures but was a reflection of their knowledge about the cosmos and the marvels of life on Earth. At Angor Wat, there is an interesting carving of a dinosaur. The intriguing fact is that it was only in 1824 that professor of geology William Buckland described and named the first known dinosaur, although the Angor Wat was built in the 12th century. So we wonder, how did they know about this extinct species at that time? (see Plate 3).

From the philosophical depths of the Upanishads to the mathematical and astronomical insights of Aryabhata, from the medical knowledge in Sushruta's compendiums of Ayurveda to the political treatises of Chanakya in the *Arthashastra*, the intellectual and spiritual achievements of these times were phenomenal. Subsequent Indian dynasties, especially the Mauryan (322–185 BCE) and Gupta (320–550 CE) empires, brought in a golden age of scientific and philosophical advancements. The Gupta period, in particular, witnessed remarkable achievements in mathematics, astronomy, literature, and the arts. Scholars like Panini, Patanjali, Charaka, Sushruta, Aryabhata, and Varahamihira made significant contributions, advancing the understanding of linguistics, medicine, surgery, astronomy, and mathematics. Parallel to these, Southeast Asian empires like the Khmer and Srivijaya made significant contributions to architecture, art, and the spread of Buddhism.

In the realm of S&T in the East, it can be argued that there existed a harmonious balance between scientific understanding and spiritual insights. This synthesis of knowledge suggests the presence of a meta-scientific approach, a holistic perspective that embraces both the empirical and the metaphysical aspects of existence. This nuanced approach prevailed during a period that gradually waned with the Age of Enlightenment and the Renaissance in the Western world, eventually laying the foundations of modern science.

The Shift to Western Dominance

The decline of the Eastern empires and the disruption caused by invasions and colonization started the gradual erosion of these centres of learning and knowledge. The Renaissance in Europe, ignited by the rediscovery of classical knowledge, was the beginning of the West's ascendancy in science and technology. The Scientific Revolution, characterized by Copernicus, Galileo, and Newton, among others, established a method of inquiry based on empirical evidence and experimentation that became the foundation of modern science.

Mediaeval Period: Scholasticism and the Seeds of Empiricism

During the mediaeval period (11th to 17th centuries), scholastic philosophy predominated, with knowledge often based on authority and deduction. However, 13th-century scholar Roger Bacon argued that empirical observation and experimentation were essential for scientific understanding. His suggestion for a more empirical approach to science was a crucial step towards modern scientific methodology.

The Renaissance, flourishing from the 14th to the 17th centuries, was a period of significant cultural and intellectual transformation in Europe. Beginning in Italy and spreading across the continent, it initiated the revival of interest in classical learning, art, and literature, along with a renewed emphasis on the study of classical texts, human potential, and the importance of reason. Along with the artistic movement, there was a scientific revolution. Visionaries like Leonardo da Vinci and Galileo Galilei were the spirit of the Renaissance with their insatiable curiosity and the way they succeeded in blending art, science, and technology. Departing from the sole reliance on ancient texts, the focus on empirical investigation was renewed. During the 16th and 17th centuries, scientists were more interested in understanding the mind and human behaviour. They intertwined psychology with philosophy. In the 17th century, thinkers like René Descartes, the French scientist–philosopher, and John Locke, the British philosopher and medical researcher, contributed significantly to early psychology. René Descartes was the first person in that era, in the West, to talk of body–mind dualism—that the mind and body are indeed two different entities, distinct from each other. He voluntarily exiled himself from his country because he advocated free speech and religious tolerance. John Locke emphasized the importance of critical thinking and the scientific method. He proposed the idea of *tabula rasa* (the mind as a blank slate), which influenced later psychological thought. John Locke argued that knowledge comes from sensory experience and reflection. Baruch Spinoza and David Hume were also among the key figures in shaping the debate between rationalism and empiricism. David Hume stressed the importance of sensory experience and empirical observation.

The end of the 16th century or early 17th century up to the 18th–19th century was a period of rapid scientific progress and is thus referred to as the Scientific Revolution. It was perhaps an era of transformation in the history of science. There was a shift in thinking and analysis. Thinkers like Copernicus, Galileo, Kepler, and Newton challenged traditional views. There was a shift towards empirical observation, direct experimentation, and developing systematic methods for understanding the natural world, which had profound implications for philosophy, particularly in the scientific method. The importance of empirical data and mathematical analysis became the norm. The inductive reasoning methods were used to generate the empirical evidence through observation, experiments, systematic data collection, analysis, and rigorous testing using deductive reasoning. Machines for manufacturing, the growth of factories, and modern production methods transformed societies and economies. These revolutions set the stage for the modern world, with a focus on efficiency, mechanization, and large-scale production.

In the 18th century, the French Enlightenment writer and philosopher who took up the name Voltaire, emphasized reason, empiricism, and the scientific method across various fields. Voltaire advocated freedom of speech. His famous statement was, "I may disapprove of what you say, but I will defend to the death your right to say it." This era introduced a broader acceptance and application of scientific principles, solidifying the foundations of modern science. The evolution of scientific thought, from its philosophical origins in ancient Greece through the critical developments in the Islamic Golden Age and the transformative shifts of the Renaissance and Enlightenment, illustrates a dynamic and ever-progressing quest for knowledge. Each era built upon the works of its predecessor, creating a rich legacy of scientific inquiry that laid the foundation for the modern understanding of the world. This historical journey tells us of the collective human endeavour to comprehend and engage with the natural world through observation, experimentation, and reason. The developments in S&T were now on the fast track. The world of Genome was expanding exponentially.

The quest for knowledge—for finding answers to the mysteries of the world—has always been the driving force for seekers, thinkers, researchers, and scientists. Reaching out beyond oneself and one's vested interests has inspirational motivation. The great scientists through the ages also saw the macrocosm behind the microcosm. That was undoubtedly the driving force behind the thinking of those incredible researchers who invented the microscopes and telescopes. The early microscope was built by Zacharias Janssen in 1590, followed by a compound microscope by Galileo Galilei in 1609. Living cells were first observed by Antonie van Leeuwenhoek in 1676. With the invention of the transmission electron microscope and phase contrast microscopy in 1931, our knowledge of bacteria, viruses, and living cells was completely transformed, leading to breakthroughs in the biological sciences and medicine. But the dream was to probe the expanse of the universe and the minutest microbes together. Such inventions have always emerged from an inherent curiosity and thirst for answers. The telescope evolved dramatically with Galileo, but in the 17th century itself and at almost the same time, Johannes Kepler, Thomas Harriot, and Christopher Scheiner developed the telescope with their adaptations or innovations. Later advancements introduced more sophistication in designs.

The scientific revolution galloped through the late 18th and 19th centuries with significant developments in chemistry, medicine, and other diverse fields. The Scientific and Industrial Revolutions had profound implications for various fields of study, influencing not only science but also philosophy, politics, and society as a whole. The phenomenal change in the method of work, shifting from farming, art and craft to industrialization and mechanization, was the transformation of the 18th century. One can say that James Watt and his steam engine ushered in this phase. The engine became the heart of the industry, and factories, locomotives, and later automobiles followed soon after. It is fascinating that in the 18th and 19th centuries, there were some very significant innovations made by some immensely motivated human beings whose scientific method or path was inspired by much more than the expected norms. Such discoverers or innovators perhaps touched new levels of intuitive and aspirational insight to be motivated the way they were. When we see these leaps in science and technology, we are convinced that such inspiration and excellence come to the fore not merely because of education. It comes from the depths of human awareness and realization. The thread of *Genome to Om* strings together the events. It is not a mere coincidence that in the East, before S&T could begin its story, religion, philosophy, understanding of human psychology, and the pertinent questions about life and its meaning were in the hands of thinkers and sages who sought nothing but the ultimate answers in their quest for the truth. An insight into the lives of these brilliant discoverers and inventors gives us the conviction of their unique pursuits.

George Stephenson, with no formal education, was the one to develop the steam locomotive. Michael Faraday, perhaps the greatest scientist of the 19th century, was a chemist whose later, most significant, contribution was in the field of electricity and magnetism. Faraday was born in England, grew up in abject poverty, and got a very basic education at the Sunday school. Samuel Morse, an inspired and committed artist for whom painting had been a way of life, developed the Morse Code, a system of transmitting telegraphic messages with coded alphabets, numerals, and punctuation! Alexander Graham Bell's mother was almost deaf, and his father taught the art of elocution to the deaf. The son was inspired to become a teacher for the deaf later in his life. Alexander did not have proper schooling, as the parents moved homes often. Pursuing a teaching career, Bell got interested in the burgeoning telegraph industry. He invented the working telephone in 1876. Despite suffering mental and physical breakdowns from overwork, Nikola Tesla in 1891 invented the Tesla Coil, a high-frequency transformer, and laid the groundwork for the modern world's electrical infrastructure. Thomas Edison was almost deaf from a very young age due to some congenital issues. He barely attended school because the education system bored him, and he was a misfit in school due to his hearing disability. He became a voracious reader, learnt telegraphy, and invented the phonograph. Edison also invented a safe, mild, and inexpensive electric light that would replace the gaslight. A science graduate, Francis Upton, joined the laboratory later that year and provided the mathematical and scientific methodology needed. Edison later revealed, "At the time that I experimented on the incandescent lamp, I did not understand Ohm's Law." On another occasion, he said, "I do not depend on figures at all. I try an experiment and reason out the result, somehow, by methods which I could not explain." Karl Benz invented the first practical automobile in 1885, while the Wright Brothers achieved the first controlled, sustained aeroplane flight in 1903. Later in 1913, Henry Ford installed the first moving assembly line for the mass production of cars, which also revolutionized industrial manufacturing.

These highlights collectively represent a significant departure from conventional views and methodologies, ushering in a new era of scientific inquiry and simultaneously developing the Age of Enlightenment, also known as the Age of Reason, which spanned roughly from the late 17th century to the late 18th century. With this phase of a scientific revolution, there was a simultaneous and significant evolution, creating a compartment between science and the humanities, especially in philosophy and psychology in the West. This paved the way for transmuting holistic science into reductive modern science. This phase was characterized by a focus on reason, rationality, empiricism, experimentation, individual rights, and scepticism of traditional authority. While brilliant individuals were inventing almost miraculous gadgets to make life easier and science and technology were leaping forward, enlightened thinkers championed reason and empirical evidence as the primary sources of knowledge.

Introspection and Turbulence

Running almost parallel to the story of the Genome with its remarkable developments at the level of the microcosm, the human quest to seek answers to questions arising from greater inner depths was spreading far and wide. Introspection has prompted philosophers and thinkers to pose the same perennial questions: how did it all begin, and why? The questions persisted and that gave impetus to the rise of religions and belief systems. Religion is the belief in a power above and beyond human cognition that explains natural phenomena and their purpose. It provided moral and ethical guidelines as societies formed and human interaction with each other and with nature grew. Proto-religious practices must have existed for several ages prior to the modern age, but they found expression with the evolution of civilizations in different parts of the world. Faith and belief systems, the growth and impact of religions on people, and the deeper understanding of spirituality were growing at the same pace as S&T.

At the peak of the S&T revolution, the transformative impact of these technologies on society and the ethical and existential challenges they posed were noticed. The turbulence that surfaced had very widespread ramifications. The Russian and French Revolutions were two major political and social upheavals that occurred in the late 18th and early 19th centuries. They were distinct events with different causes and effects, but both were responses to oppressive political systems, social inequality, and economic hardships. France was facing a severe financial crisis due to the opulence of the monarchy and expensive wars. The overburdened lower classes faced economic difficulties with high food prices, unemployment, and poverty. There was immense unrest. With the Age of Enlightenment, the ideas of liberty and equality became popular. Intellectuals and philosophers were calling for social reform. The revolution was volatile, with widespread violence and executions. Ultimately, Napoleon Bonaparte seized power to become emperor in 1804, and the revolutionary era ended.

At the beginning of the 20th century, Russia's participation in World War I placed a heavy burden on the economy and led to significant loss of life. Military failures, food shortages, and discontent among soldiers and civilians contributed to the revolutionary atmosphere. Vladimir Lenin led the Bolsheviks through a series of strikes and protests, which were followed by their seizing power in the October Revolution of 1917. The revolution was followed by a bloody civil war between the Red Army (Bolsheviks) and the White Army (anti-Bolshevik forces) and their allies.

The era of the 19th and 20th centuries perhaps wrote the most complex chapters in the story of the evolution of S&T with the emergence of nuclear science, which changed the face of humanity altogether. World War I from 1914 to 1918 had devastating consequences, and it set the scene for future conflicts in the now-divided Europe. The most significant facts to note are the festering anger, frustrations, distrust, and the need to be more powerful than others. As long as the power of the "other" looms large in the human psyche, there can never be any peace and harmony. Regardless of "Genome's" rapid advances, the 20th century certainly seemed to cloud the relevance of "Om". Human persecution, colonization, slavery, and rampant discrimination, almost at a global level, reigned supreme. And in this environment, almost paradoxically, the development of nuclear science had already begun.

Nuclear Revolution

The late 19th and early 20th centuries marked the genesis of another "revolution" with the emergence of nuclear science. Physicists Henri Becquerel, Marie Curie, and Pierre Curie were at the helm. Henri Becquerel's accidental discovery of radioactivity in uranium in 1896 paved the way for further ground-breaking research. "When Henri Becquerel investigated the newly discovered X-rays in 1896, it led to studies of how uranium salts are affected by light. By accident, he discovered that uranium salts spontaneously emit penetrating radiation that can be registered on a photographic plate. Further studies made it clear that this radiation was something new and not X-ray radiation he had discovered a new phenomenon, radioactivity." In 1898, in their research on pitchblende, Marie and Pierre Curie discovered radium and polonium. A year after isolating radium, they shared the 1903 Nobel Prize in physics with Henri Becquerel for their unique investigations of radioactivity.

Ernest Rutherford's research helped to develop the nuclear model of the atom in 1909. The important discovery was that atoms have a dense nucleus in the centre. In 1913, Niels Bohr proposed a model of the atom that laid the groundwork for understanding atomic structure, thus laying the foundation for modern atomic physics. James Chadwick discovered the neutron in 1932. This was a particle with no electric charge. He added another crucial piece to the puzzle of nuclear science, shedding light on the composition of the nucleus and nuclear reactions. Otto Hahn, Fritz Strassmann, and Lise Meitner discovered nuclear fission in 1938. This was a pivotal moment, revealing the ability to split an atomic nucleus and opening doors to the immense potential of nuclear power. The world was all set for the revolution of nuclear energy, and its potential to help mankind was unfathomable at the time. It also set the stage for the development of nuclear weapons.

The Ethical Dilemma: The Manhattan Project and Beyond

World War II started in 1939. The Allied forces feared that Nazi Germany was developing an atomic bomb. While this was not true, in the sense that Germany's progress in this area was not as fast as feared, it triggered the Manhattan Project in the United States in 1939. The project aimed to develop nuclear weapons. Led by physicist J. Robert Oppenheimer, this project culminated in the first successful test of an atomic bomb, codenamed "Trinity", which took place on 16 July 1945, in New Mexico. Robert Oppenheimer was a keen reader of the Bhagavad Gita. When he saw the large mushroom of the first test of the bomb, two of Sri Krishna's statements came to his mind instantly:

दिवि सूर्यसहस्रस्य भवेद्गपदुत्थिता | यदि भा: सदृशी सा
स्याद्भासस्तस्य महात्मन: || 11: 12||

"If the splendour of a thousand suns were to rise up simultaneously in the sky, that would be like the splendour of that Mighty Being."

He referred to the brilliance of the light that emanated from the explosion of the trial bomb. At the same time, a feeling of dread arose, and he questioned their act of having developed this instrument of death.

कालोऽस्मि लोकक्षयकृत्प्रवृद्धो लोकान्समाहर्तुमिह प्रवृत्तः।
ऋतेऽपि त्वां न भविष्यन्ति सर्वे येऽवस्थिताः प्रत्यनीकेषु योधाः || 11: 32||

> "I am the mighty destroying Time, here made manifest for the purpose of unfolding the world. Even without thee, none of the warriors arrayed in the hostile armies shall live."

Oppenheimer was very shaken seeing the potential of the atom bomb. He cautioned against its use. He believed that he had created that mighty, destroying weapon that would "infold" the world. Most people around him in the 1940s in America did not have a clue as to what he was referring to. His reference to some Hindu scripture was clouded in mystique. Despite his pivotal role in the Manhattan Project, Oppenheimer expressed deep reservations about the use of nuclear weapons, particularly against civilian populations in Japan. His warning was not heeded. The atomic bombs were dropped on Hiroshima and Nagasaki. The result was incomprehensible devastation, as hundreds of thousands of civilians were killed, and the terrible impact of the radiation is talked of even today. The formidable nuclear age began as World War II ended. The United Nations was established in 1945 to help prevent future global conflict and promote international cooperation. The bombings of Hiroshima and Nagasaki ushered in the Cold War between the United States and Russia. This "war" triggered the arms race, and more powerful nuclear weapons were developed, each with a suspicious and untrusting eye on the other. Finally, the Nuclear Non-Proliferation Treaty (NPT) was signed in 1968, aimed at preventing the spread of nuclear weapons and promoting peaceful nuclear energy.

Peaceful uses developed as vast nuclear plants were set up to harness nuclear energy. These plants helped generate huge quantities of electricity. However, nuclear disasters like Chernobyl in Russia in 1986 and Fukushima in Japan in 2011 highlighted the ongoing risks associated with nuclear technology.

Marie Curie's pioneering research in radioactivity also had profound implications in medicine, particularly in developing radiation therapy for cancer treatment. Radioactive substances, including radium, were initially used to treat cancer, although the long-term consequences of radiation exposure were not fully understood at the time. Curie herself was extensively exposed to ionizing radiation during her research, which ultimately killed her.

Modern Era of Scientific Discoveries

In the 21st century, the unbroken progress of research, development, and innovation in S&T has yielded unique contributions from the basic sciences, showcasing remarkable discoveries and advancements that weave a narrative rich in innovation and exploration.

Physics, the realm where microscopic and cosmic dimensions converge, has seen the theoretical field of quantum mechanics transition into practical applications. Quantum computing, a major leap in computing and information technology, holds the promise of revolutionizing fields like cryptography, materials science, and complex system simulation. Companies such as Google and IBM are actively engaged in developing quantum computers, ushering in an era of unimaginable computational power with profound implications for medicine, environmental and meteorological sciences, and material engineering.

Advancements span from the infinitesimal scale of quantum particles to the vast expanse of cosmology. The decade began with rovers exploring the Martian landscape, uncovering secrets of an ancient past. The European Space Agency's Gaia mission mapped the stars with unprecedented accuracy, fuelling dreams of future interstellar travel. Images from the James Webb Space Telescope expand cosmic understanding and drive advancements in diverse fields. Astrophysics, delving into the vast expanses of space, celebrated a crowning achievement with the first-ever image of a black hole captured by the Event Horizon Telescope. The Hubble Space Telescope, which can see objects over 13 billion light years away, has given us a "bird's eye view" of when the universe was only 400 million years old, while the transmission electron microscope can magnify a specimen 50 million times. Sample pictures of the Butterfly Nebula by Hubble (see Plate 4 a–c) and the Novel Corona Virus (see Plate 5) testify to the beauty and power of these technologies.

Organic chemistry experienced a renaissance with advancements in catalytic methods, enabling the creation of complex molecules more efficiently. These breakthroughs extend beyond academia, holding the key to new pharmaceuticals, cleaner industrial processes, and a more sustainable future. Chemical engineering, particularly in nanotechnology, challenges the limits of nature, revolutionizing drug delivery systems for more efficient and site-specific delivery of active pharmaceutical ingredients.

The modern era, applauded for its incredible discoveries, unfolds a luminous chapter in human progress. Precise measurements, faster data processing, interdisciplinary collaboration, and the democratization of information through digital platforms characterize the unprecedented pace of scientific advancements. The digital age has intricately woven the world into a web of information and communication, while laboratories globally contribute to fascinating areas like genetic engineering and space exploration.

The rapid pace of discovery brings both opportunities and challenges. Innovations in renewable energy, such as efficient solar panels, advanced battery storage, and electric vehicles, signal a commitment to a greener future amidst serious challenges like climate change. RNA medicine, synthetic chemistry breakthroughs, limb regeneration research, and progress in nuclear fusion present awe-inspiring developments with profound ethical considerations and societal impacts. As we navigate through these advancements, a balanced and conscientious approach is crucial. The dual nature of technological progress demands reflection on its unparalleled benefits and the moral questions it raises. Responsible handling of these advancements, aligned with ethical values, ensures a better future for all. The power and responsibility accompanying these disruptive innovations emphasize the need for a balanced approach, acknowledging the promises and perils of this transformative era.

Biotechnology and Its Transformative Impact

Biotechnology has transformed remarkably, marked by a series of phenomenal advancements that contribute to a captivating narrative of human ingenuity and scientific exploration. These developments have left an indelible mark on health, medicine, and agriculture, shaping the trajectory of biotechnological evolution through high-impact research projects and global initiatives.

There was monumental progress in molecular biology and genomics, with the completion of numerous genome-sequencing projects, including the human genome. This achievement has given us a profound understanding of the blueprint of life, paving the way for personalized medicine tailored to an individual's genetic makeup. Scientists now possess a detailed map of the human genome, offering critical insights into the genetic basis of diseases, human development, and the intricate interplay between genetics and the environment. Early in the decade, next-generation sequencing technologies, exemplified by projects like the Human Cell Atlas, have revealed the diverse complexity of human genetics and cellular biology, providing fresh perspectives on our biological makeup. Simultaneously, ecology has gained urgency in addressing climate change, unravelling complex ecosystem interactions crucial for shaping policies and actions to preserve our planet.

Biotechnology in Agriculture

The impact of biotechnology on agriculture has rewritten the narrative of food production and sustainability. There's a fascinating new development in the world of food—scientists are figuring out how to grow meat in labs! Researchers like Dr Shulamit Levenberg at Tufts are pioneering this technique. Instead of raising livestock, they take a small sample of animal cells and cultivate them in a controlled environment. This could be a game-changer for the environment, potentially reducing greenhouse gas emissions and land use associated with traditional animal husbandry. Beyond meat, scientists are also using biotechnology to improve familiar foods. Researchers in Switzerland, led by Dr Natalie Rovers, have successfully increased the iron content in cow's milk, addressing a common nutritional deficiency. This is just one example of how biotechnology can enhance nutrition in food. In another area, scientists have changed the ways of agriculture through genetic modification. For instance, some researchers are developing rice varieties enriched with essential micronutrients.

Transgenic technology is used to create organisms whose genetic material has been artificially manipulated through genetic engineering. This technology allows scientists to move genes from one species to another, creating organisms with desired traits such as increased resistance to pests or herbicides, enhanced nutritional content, or improved crop yield. The application of transgenic technology spans agriculture, pharmaceuticals, industrial biotechnology, and more.

Genetically modified organisms (GMOs) are known to have shown substantial benefits in agriculture. Increased agricultural productivity has been one of the most significant advantages, with GM crops often yielding more per acre while requiring fewer inputs such as water and fertilizers. This promises better food security, especially in developing countries where these resources are scarce. GMOs can also be engineered for nutritional enhancement. They are designed to resist pests, diseases, and extreme weather conditions and bolster crop yields while reducing reliance on chemical pesticides. Biofortification, exemplified by innovations like Golden Rice, addresses global health issues by enhancing crop nutritional value.

Revolution in Health and Medicine

Biotechnology has emerged as a beacon of hope in health and medicine, driven by "OMICS" technologies—genomics, proteomics, and metabolomics. The ability to decode the genetic blueprint is facilitating early and accurate disease diagnoses, with mRNA vaccines showing the agility of biotechnology in responding to global health crises during the COVID-19 pandemic. Immunotherapy and personalized medicine, tailored to individual genetic profiles, are transforming cancer treatment, offering new avenues in the ongoing battle against this complex disease.

Microbiome: The microbiome is the collection of all microbes, such as bacteria, fungi, viruses, and their genes, that naturally live on and within our bodies. In a way, we are in the minority because there are more microbial cells than human cells in our body. The microbiome is crucial to our health and wellness, affecting many aspects of our physiology, immunity, metabolism, and behaviour. The microbiome can help us digest food, produce vitamins, fight infections, regulate inflammation, and even modulate our mood and cognition. Microbiome transplants and targeted probiotics offer novel therapeutic strategies. However, the microbiome can also be disrupted by factors such as diet, antibiotics, stress, and environmental toxins, leading to dysbiosis, or an imbalance of microbes.

The Human Microbiome Project (HMP) was launched by the National Institutes of Health (NIH) in 2007. It characterizes the microbial communities (microbiota) living in and on the human body (human microbiome) and aims at better understanding the role of these microorganisms. By studying the human microbiome, HMP has provided valuable insights into how these microbial communities influence physiology, immunity, and various health conditions, broadening our perspective of what contributes to our biological identity.

Stem Cells and Regenerative Medicine: Stem cells are undifferentiated cells that can self-renew and differentiate into specialized cell types. Stem cells can be procured from embryos and various tissues in the body. The induced pluripotent stem cell (iPSC) technology holds the promise of personalized regenerative therapies, offering potential cures for a range of diseases, including Parkinson's and Alzheimer's. Stem cells can be used to regenerate damaged tissues and organs. Clinical trials are focused on using stem cells for heart regeneration, repairing spinal cord injuries, and treating various degenerative diseases. This aspect of biotechnology holds great promise for a future where organ transplants are no longer reliant on donors and chronic injuries and diseases are treatable. Despite the promises, however, ethical debates persist, particularly concerning the use of embryonic stem cells and the potential for human cloning. Organoids are tiny, three-dimensional structures grown in a lab from stem cells that mimic the function and structure of real organs. Researchers get to test a novel therapy for different diseases with these structures.

Cloning is creating an organism with a genetic makeup that is identical to another organism. In 1996, a team of scientists led by Dr Ian Wilmut and his colleagues at the Roslin Institute in Scotland cloned a sheep named Dolly by using the somatic cell nuclear transfer technique. This was the first mammal cloned and was a sensational breakthrough in seeking cures for genetic diseases (see Plate 6). Cloning technology and stem cells have several applications in regenerative medicine—growing tissues or organs for transplantation without the risk of rejection.

Tissue Engineering: The progress in tissue engineering, marked by 3D bioprinting and biomaterial developments, holds promise for regenerative engineering. Growing functional human tissues and organs in the lab, including skin for burn victims and lab-grown organs for transplantation, exemplifies tangible clinical applications. While challenges remain, there is potential for a future where organ transplants are independent of donors and chronic injuries and diseases are treatable.

Non-coding RNAs (ncRNAs): These fascinating groups of RNA molecules that don't directly make proteins were considered junk DNA. But scientists underestimated nature. There is nothing junk in nature. Every creation has a role. It may take time to understand it. RNA's primary role is indeed protein synthesis; however, ncRNAs defy this norm. They come in various forms, each with unique functions. The microRNAs (miRNAs) regulate gene expression. They work as molecular traffic cops, deciding which genes get translated into proteins and which don't. The long non-coding RNAs (lncRNAs) orchestrate complex cellular processes by acting as scaffolds, bringing together proteins involved in gene regulation. Small interfering RNAs (siRNAs) play a role in gene silencing. When a virus attacks, siRNAs step in like molecular ninjas, shutting down viral genes. Piwi-interacting RNAs (piRNAs) are like the genome's bouncers to maintain genome stability, ensuring order in the DNA. The small nucleolar RNAs (snoRNAs) modify other RNAs, ensuring their proper function. The ribosomal

RNAs (rRNAs) are the adaptors bridging the gap between the genetic code and the amino acids needed to build proteins. So far, a total of five siRNA-based therapies have been approved by the US FDA for the treatment of a variety of diseases. On the other hand, miRNA therapeutics remain in their infancy, and none have received approval so far.

Advanced research on ncRNA has opened exciting possibilities in medicine and therapeutics. For instance, miRNAs and lncRNAs impact glioblastoma onset, progression, and recurrence. The ncRNAs also help in forensic medicine to identify tissues, analyse the cause of death, estimate time-related factors, and even distinguish monozygotic twins. Thus, ncRNAs, once considered silent players, now reveal their importance in health, disease, and forensic investigations.

CRISPR-Cas9

In genomics and biotechnology, the revolutionary CRISPR-Cas9 technology has empowered precise editing and modification of DNA, providing unprecedented control over the genetic code. The 2020 Nobel Prize in chemistry was awarded to Emmanuelle Charpentier and Jennifer Doudna for their work that unlocked the potential of CRISPR (clustered regularly interspaced short palindromic repeats) as a precise and powerful gene-editing tool. This has transformed genetic research, allowing scientists to edit genomes with unparalleled precision, holding immense promise for treating genetic disorders and reshaping medical treatments and agriculture. In doing so, such technologies can be easily misused if put in the wrong hands. One can imagine what can happen if gene editing becomes like book editing. The natural process of evolution and the existence of life forms can be taken over by artificially created species, chimaeras, zombies, and uncontrollable monsters. The ethical considerations surrounding CRISPR-Cas9, particularly concerning the editing of the human germline, necessitate responsible and transparent practices to ensure the ethical use of this powerful technology.

Chimaeras and hybrids blur the lines between species, creating fascinating biological combinations. Chimaeras, with their mosaic of DNA, offer a unique window into cellular interaction and development. The green-eyed chimaeric monkey, a marvel of genetic engineering, serves as a prime example. Scientists were able to introduce glowing stem cells into embryos, resulting in an organism with a significant portion of its cells derived from the donor. This breakthrough opens doors for studying complex human diseases in organisms with close biological resemblance to humans. Hybrids, on the other hand, showcase the wonders of natural reproduction between different species. Mules, zebroids, and wholphins all exhibit a captivating blend of traits from their parents. In agriculture, hybrids like Beefalo (buffalo and cattle) demonstrate the practical applications of this genetic merging, leading to animals with desired characteristics. These advancements in genetic understanding not only hold promise for medical breakthroughs but also push the boundaries of scientific exploration. With CRISPR technology, several new exciting possibilities are emerging for applications in the medicine and agriculture sectors.

Nanotechnology is the science of the extremely small. Imagine a world where machines are so tiny that they can manipulate individual atoms and molecules. This is the promise of nanotechnology, where scientists and engineers dive into the minuscule to create materials and devices with extraordinary properties. At its core, nanotechnology involves working with structures at the nanoscale, typically measuring in the range of one to 100 nanometers. To put that into perspective, a single strand of human hair is about 80,000 nanometers wide. This level of precision allows scientists to engineer materials with unique properties, opening up unimaginable possibilities. The miniaturization of electronic components has led to more powerful and efficient devices. The chips inside our computers and smartphones are proof of the incredible capabilities of nanotechnology, enabling us to carry the computing power of entire rooms in the palm of our hands. Nanomaterials exhibit remarkable properties, such as increased strength, flexibility, and conductivity. These materials find applications in everything from lightweight and durable sports equipment to more efficient solar panels and advanced coatings. One of the most exciting aspects of nanotechnology is its potential application in medicine. Imagine a tiny machine, smaller than a blood cell, navigating through our veins, delivering drugs precisely to targeted cells or repairing damaged tissues.

Synthetic biology, marrying biology with engineering principles, has witnessed significant strides, challenging conventional notions of life and creation. Synthetic biology is revolutionizing biomolecule and material production and manufacturing a spectrum of products, from therapeutics to sustainable fabrics. Innovations in single-cell metabolomics offer deeper insights into cellular functions, potentially revolutionizing disease diagnosis and treatment.

Evolution has taught us that life is not contained in one territory in nature. It crosses barriers, facing several challenges. Life breaks free and finds a way. This was a message from *Jurassic Park*, the 1993 science fiction film. After three decades in real science, evolutionary biologist Jay Lennon, a professor of biology at Indiana University Bloomington, actually reported that life does indeed find a way. His team found that the synthetic cell can evolve just as fast as a normal cell, demonstrating the capacity for organisms to adapt, even with an unnatural genome (see Plate 7).

The creation of synthetic life forms, such as the first synthetic bacterial cell, opens avenues for biofuel production, bioremediation, and pharmaceutical synthesis, raising ethical debates on the manipulation of life. However, as of now, scientists have not been able to create "life" even in a single cell.

Confluence of Technologies

At the heart of this technological renaissance is the seamless integration of diverse fields. Integration with artificial intelligence (AI) and machine learning (ML) creates sophisticated machines capable of complex tasks, from autonomous driving to medical diagnostics. AI and ML are the brains, equipping machines with the ability to learn from and interpret the world in ways that mirror human cognition. Robotics provides the body, offering the physical form and function necessary to interact with the physical world. Quantum computing promises to be the supercharged heart, pulsating with the power to process information at speeds unfathomable to classical computers, while biotechnology and nanotechnology act as the nervous system and the cellular machinery, enabling intricate, precise manipulations at the smallest scales of life and matter. Nanobots, combining nanotechnology and robotics, offer unprecedented capabilities. The convergence of biotechnology, nanotechnology, information technology, and air and space technology amplifies their potential and reflects human aspirations—to heal, nourish, and understand life's mysteries. The perils that emerge side-by-side also project an important image, which cannot be neglected.

World Wide Web

A British scientist, Tim Berners-Lee, invented the World Wide Web (WWW) in 1989 while working at CERN. The need for automated information-sharing between scientists in universities and institutes around the world was the trigger. The web was originally conceived and developed to meet this demand. From being an invaluable tool for gathering information to the development of portals for communication, the digital age democratized access to information, making knowledge more accessible than ever before. Search engines and social media platforms provide instant access to information on virtually any subject, empowering individuals with the tools to learn, grow, and make informed decisions. Today, we are caught in the web of social media. The impact of social media and technology has touched every aspect of our lives, from communication and connectivity to politics, education, and mental health. Platforms such as Facebook, Twitter (X), Instagram, WhatsApp, and LinkedIn have made it possible for everyone to stay in touch with everyone out there—friends and family, business associates, and social contacts—across the globe in real time. This has not only made personal relationships more accessible but also opened up new avenues for professional networking and collaboration, breaking down geographical and cultural barriers.

Social media is the way to engage and communicate today. Everyone is reachable from anywhere. The huge, diverse planet is now a global village with no barriers to outreach. The digital age has completely changed our outlook. All information, whether at the level of governments, society, or individuals, is accessible—literally, at the fingertips. It is also a powerful tool for political engagement and social activism. It has facilitated the organization and mobilization of protests, allowing movements to gain momentum at an unprecedented pace. The Arab Spring and the #MeToo movements are examples of how social media can amplify voices and bring about societal change.

Technology has transformed the educational landscape; it has enabled remote learning, digital classrooms, and access to a wealth of online resources. Education is now more accessible to people around the world. Massive open online courses (MOOCs) are free online courses made available to anyone with an internet connection and a computer. These online education programmes don't require tuition or an academic programme commitment. These developments have broken down traditional barriers and made the culture of lifelong learning a reality. The rise of social media and technology has spurred economic growth and innovation; new industries have sprouted and traditional ones have been transformed. E-commerce, digital marketing, and the gig economy are just a few examples of how technology has reshaped the economic landscape.

Artificial Intelligence and Machine Learning

Finally, we are at this juncture, as "Industry 5.0" is here—the era when humans and robots work side by side. We are witnessing the integration of advanced technologies like AI, robotics, and the Internet of Things (IoT) into manufacturing. This new industrial era promises a future where smart technology and human ingenuity can create sustainable, efficient, and highly personalized production.

The story of AI and ML over the last ten years has been one of rapid transformation and unprecedented growth. From being a niche area of computer science, AI has become almost an omnipresent force, reshaping industries, businesses, and even human everyday life. Machine learning algorithms have become extremely sophisticated, learning from vast datasets to make astonishingly accurate predictions and decisions. In healthcare, AI systems are diagnosing diseases, sometimes with greater accuracy than human doctors. In finance, they're revolutionizing trading and risk assessment. AI and ML have evolved from theoretical concepts into the cornerstones of modern technological advancement. The development of neural networks and deep learning algorithms has significantly enhanced AI capabilities, leading to the kind of practical applications that were once only seen in science fiction. The integration of AI and ML into various sectors has revolutionized the way data is analysed and decisions are made. This has helped to devise hitherto unimaginable innovations. AI in everyday devices, from smartphones to home assistants, has made itself an invisible yet integral part of daily life. To what extent is this a welcome change? Only time can tell, but telltale signs are visible even today. Most importantly, whether AI is permitted to become as invisibly integral to daily life as "Genome to Om" is, depends entirely on human intelligence, perception, and conscious decision-making. There is no scope for "artificial" intelligence here.

In the digital realm, the IoT is a network of interconnected physical devices that use the Internet to communicate with each other and exchange data. These devices, embedded with sensors and other technologies, collect and share information to enable various applications and services. The primary goal is to create a more efficient and seamless integration between the digital and physical worlds. Cybersecurity has emerged as a critical frontier, with advances in encryption and network security becoming of paramount importance in protecting our increasingly digital lives. Cyber fraud is as much the modern-era S&T industry as any other, especially as the rollout of 5G technology has promised a new era of connectivity, paving the way for smart cities, the IoT, and autonomous vehicles. While IoT and the glimpses of the future, as they are rolling in now, do claim to bring several benefits of increased efficiency and improved conveniences, which will bring far more comfort to the people living in smart cities, for example, or to the farmers in cultivating their fields with maximum efficiency, it raises serious challenges related to security and privacy.

Ensuring the security of IoT devices and the data they collect is a vital consideration as technology continues to evolve. The world of Darknet and scamsters is growing, perhaps at a faster rate than the technology that facilitates them.

Augmented reality (AR) and virtual reality (VR) have rewritten the rules of engagement and experience in the digital world. They represent the cutting-edge of immersive technology. VR creates a completely artificial environment, while AR overlays digital information onto the physical world, enhancing real life with digital details, be it in gaming, navigation, or education. These technologies have not only become tools for entertainment but also powerful instruments for education, training, and therapeutic purposes. VR creates immersive, simulated training environments for surgeons and pilots. In healthcare, VR and AR are being used for surgical simulations and patient rehabilitation. In education, they provide immersive learning experiences that enhance understanding and engagement.

Humanoids and Social Robots

The story of robotics, automation, and the evolving human-machine nexus has unfurled like a science fiction novel. This is a world where machines not only complement human efforts but increasingly integrate with them, blurring the lines between manual and automation.

Robotics has stepped out of industrial settings and entered our everyday lives. In agriculture, robotics is used for tasks like harvesting and monitoring crop health, improving efficiency, and productivity. In labs, humanoids are taking their first steps, learning to navigate complex environments. In homes, robotic vacuum cleaners and lawnmowers have become commonplace, simplifying mundane chores. In healthcare, robotic assistance in surgeries has enhanced precision, reduced recovery times, and improved outcomes. Industrial robotics has also transformed. Collaborative robots, or "cobots", are ready to work alongside humans. Unlike their predecessors, who were confined to safety cages, these cobots are equipped with sensors and AI, allowing them to understand

and react to their human counterparts, making industrial processes more efficient and safer. Social robots are being tested in homes and public spaces, learning to interact with people of all ages. Quantum computing research is accelerating, promising to unlock new potentials in AI and robotics, while advances in biotechnology and nanotechnology are opening up new possibilities for enhancing these robots' functionalities and interactions. This vision is no longer distant. Around the world, researchers are diligently working on projects that edge us closer to this future.

Humanoids and social robots embody the ambition of S&T to not only replicate human appearance and behaviour but also to forge meaningful connections with us. They are being designed to understand and engage with our emotions, to learn from and adapt to our social cues, and to perform tasks that range from the mundane to the complex, all while seamlessly blending into our daily lives. There could come a time when a humanoid robot could, possibly, learn to sense our mood from our voice and facial expression and go to the extent of offering words of comfort or a joke to lighten our mood! Or social robots could serve as companions for the elderly, providing both care and companionship and reducing the sting of loneliness. Humanoids would soon be working alongside humans in factories, hospitals, and homes, not as lifeless machines but as entities capable of learning, adapting, and even empathizing. The development of humanoid robots, although still in its infancy, hints at a future where robots could become our companions and helpers. Do we assume that empathetic human companions and helpers will be hard to find?

In an interesting aside, we look at teleportation and holoportation, which are becoming realities and matters of concern today and which were known in ancient mythology! Just as space travel was the fascinating "dream" of science fiction at some point in time. Ancient cultures have had the most colourful and engaging stories of travelling in space! The people flew with wings or with carriages through the heavens. In Greek mythology, Icarus and Daedalus attempt to escape from prison by flying with wings made of feathers and wax, but Icarus flies too close to the sun and falls to his death. In Hindu mythology, gods and celestial beings can very easily transport themselves from one realm to another. Just by the power of the mind, they travel from the heavens to earth and the netherworld. In the epic Ramayana, Ravana kidnaps Sita and transports her to his kingdom on an airborne chariot.

Jules Verne fired a projectile from a giant cannon in the 1865 book *From the Earth to the Moon*. In 1898, H.G. Wells had Martians attacking Earth in *The War of the Worlds*. They travelled in a cylindrical aircraft. Possibly Georges Méliès (in 1902) got inspiration from Verne for his silent film *A Trip to the Moon* (1902), in which he launched a group of astronomers to the Moon from a giant cannon in a spacecraft.

Drones, also called Unmanned Aerial Vehicles (UAVs) or Unmanned Aircraft Systems (UAS), were once a novelty and are now being used for everything from delivery services to agricultural monitoring, chemical spraying, and video photography. But most importantly, drones are also being used now for secret surveillance, remote warfare, and targeted strikes as they can locate and wipe out high-value targets. Some drones are equipped for electronic warfare, disrupting or disabling enemy communication systems, radar, and other electronic equipment. This capability can be crucial in degrading the effectiveness of an enemy's defences. This is the big picture of the present and futuristic scenario. And once again, we realize that as long as vested interests focus on the survival of the fittest, the smartest, or the foxiest, and if that remains the goal, a futuristic scenario of harmony and camaraderie remains a pipe dream. The need for the path to veer towards meta-science gets reasserted at every turn.

Automation and AI have created systems capable of learning, adapting, and making decisions. Self-driven cars, that is, cars not driven by human beings, were once a dream and are now navigating real roads, albeit in limited capacities. The implications are far-reaching, promising not just convenience but a fundamental shift in urban planning, logistics, and even lifestyle. The pertinent question is, *what about conscious decisions?* Chatbots and virtual assistants, powered by natural language processing, are providing customer service that could become increasingly indistinguishable from human interaction. But the unique human ability of conscious questioning, debating, and even vacillating in the quest for reasoning, is that also becoming automated?

At this juncture, all the questions that have risen in our minds as we have come along this incredible journey of human evolution riding on the path-breaking developments in S&T are being discussed with growing debates, particularly around the potential for conscious machines. Advancements in AI have led some to speculate about a future where machines not only mimic cognitive functions but also develop consciousness. While this remains a subject of many philosophical and scientific ongoing debates, the very discussion points to the immense impact of these technologies on our understanding of consciousness and intelligence. The marvels of science, the advances of technology, and the impressive successes that have resulted from the immense efforts of the dedicated men and women of science do not take away from the misuse. Overuse or self-centred use have devastating consequences. We have to recognize the perils and face the challenges that we are now faced with. This is the focus of the next chapter.

CHAPTER 2

Science and Technology: The Perils of Misuse

With great power comes great responsibility.

—Voltaire

In the Anthropocene, the epoch of the immense impact of fast-paced human activity on the Earth's geology and ecosystems, we face multiple challenges that span ethical, economic, ecological, and societal realms. These challenges have become even more severe with the rapid advancements in S&T and the significant geopolitical and socio-economic developments. Technological advancement in the past century in particular, in the realms of nuclear science, biotechnology, nanotechnology, CRISPR gene editing, artificial intelligence (AI), machine learning (ML), and virtual reality (VR), stands as proof of the incredible human ingenuity and the relentless pursuit of progress. These innovations have undeniably ushered in an era of unprecedented change, offering solutions to age-old challenges such as disease, hunger, and environmental degradation. We now have a revived and intricate web of consequences that these technologies entail, both beneficial and hazardous. When we review these issues, we become aware of the immense complexity of navigating the modern world's opportunities on the one hand and perils on the other. The perennial debate that has raged over centuries and certainly over recent decades remains: advances in S&T are needed, research is imperative, but the balance has to be maintained. Achievements and their consequences, advances and challenges need to be faced and addressed.

The Challenges

Advances in S&T provide tremendous power, as discussed in Chapter 1, but they also entail a huge responsibility to avoid their perils. There are diverse challenges to be addressed here: ethical, ecological, societal, and economic, to name a few key areas. There is a distinct overlap between these because some of the great innovations have raised red flags in more than one of these four areas. What is ethically unviable, for example, is very likely to have an adverse impact, both economically and socially. To address these challenges, we need to carefully consider the implications of technological advancements and proactive measures. We, as human beings and as the caretakers of the planet, need to remedy the negative impacts while maximizing the benefits for society and all living beings. Collaboration among policymakers, technology developers, civil society organizations, and the public is essential to ensuring that the advancements in S&T contribute to overall well-being and prosperity. The challenges highlight the importance of responsible use of technology, equity, and the moral obligation to future generations.

Ethical Challenges

These are the moral dilemmas and considerations that arise in the context of advancements in S&T. These challenges stem from the actual or potential adverse impacts on individuals, societies, and the environment, and they often involve complex questions about right and wrong, fairness, justice, and responsibility.

Pharma and agro chemicals

In the 20th century, especially with the World Wars and innumerable skirmishes and other wars, we saw several shades of human creation and destruction. Many historians refer to the period as the chemists' war because of the scientific discoveries and engineering advances. Ironically, true to the adage "necessity is the mother of invention", it was necessary for industry to repurpose many toxic chemicals they had produced during wartime.

The origin of many toxic pharmaceutical drugs and poisonous agrochemicals used even today can be found in the post-World War crisis. Chemical weapons and gases such as chlorine, phosgene, and mustard were used to attack civilian or military populations. Several synthetic chemicals were used primarily for military needs. Advances in scientific research, especially during the late 19th century, led to several discoveries and inventions. In this "chemical revolution era", Bayer AG was founded in 1863 primarily as a dyestuff factory. It also produced wonder drugs like aspirin, lifesaving drugs like sulphonamide, and addictive drugs like barbiturates. Bayer went through several transitions, including an alleged association with the Nazis and the Holocaust. If we study the history of synthetic dyes, drugs, disinfectants, fertilizers, pesticides, and possibly other toxic chemicals, many known big "agro" and "pharma" industries appear to be connected. The power of big companies gets more consolidated through mergers and acquisitions. Over the years, Bayer alone acquired a total of thirty-six companies across various sectors, including life sciences, agri-tech, and crop technology, including Monsanto in 2018, the largest acquisition in its history. And there are many others like them in this business.

Our world is filled with chemicals and drugs that offer benefits but can come with a hidden cost. Since 1953, the World Health Organization (WHO) has warned against the toxic hazards of pesticides and other chemicals. A case in point is Thalidomide, a drug that was widely used in the late 1950s and early 1960s for treating nausea in pregnant women. The Thalidomide disaster is considered one of the most tragic drug-related medical events in history. It is estimated that more than 10,000 children were born with severe deformities, and almost 20,000 embryos were affected. Unfortunately, about 40 per cent of these children died at, or shortly after, birth. The disaster also resulted in thousands of miscarriages. This drug was completely withdrawn by its manufacturer, Chemie Grünenthal, from Germany on 26 November 1961. Today, survivors continue to live with the consequences: in Germany, over 2,200; in the UK, 450; in Spain, 250; and 146 as currently registered with the Australian Thalidomide Survivors Support Program. On 29 November 2023, Prime Minister

Anthony Albanese rendered a national apology to all Australians impacted by the Thalidomide Tragedy (see Fig. 2.1).

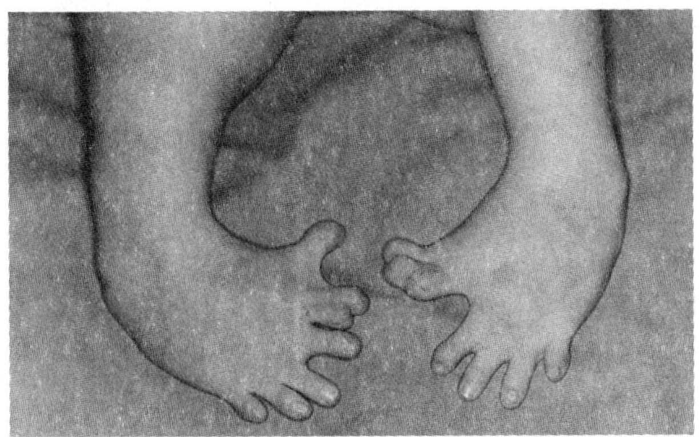

Fig. 2.1. Thalidomide disaster.

https://commons.wikimedia.org/wiki/File:NCP14053.jpg

Cancer as a case in point

In the past five decades, a very concerning picture has emerged: global cancer rates have climbed alarmingly, with estimates suggesting a nearly fourfold increase since the 1970s. Yes, diagnostics have advanced exponentially and have undoubtedly played a crucial part in this statistical rise by catching previously missed cases. A finger is increasingly being pointed at the dramatic surge in chemical use. Agricultural chemicals like fertilizers have seen a staggering sixfold increase since the 1960s, and some of these chemicals are suspected to have carcinogenic properties. Studies suggest they can disrupt hormones and damage DNA, possibly causing cancers like leukaemia and lymphoma in highly exposed populations, like farmers. The WHO has reported that the global cancer burden is increasing. The GLOBOCAN 2020 data show that 19.3 million new cancer cases were reported in 2020 as compared to about 4.7 million in 1970. Even after adjusting to the population increase, the estimated new global cancer cases seem to have risen by a staggering 310.64 per cent between 1970 and 2020. The data suggests a worrying trend: the largest increases are in low- and middle-income countries from Asia, Africa, and Latin America (see Fig. 2.2). This can be attributed to factors like limited access to preventive care and treatment, higher exposure to risk factors, including chemical and environmental pollution, and unhealthy food and lifestyles due to poverty.

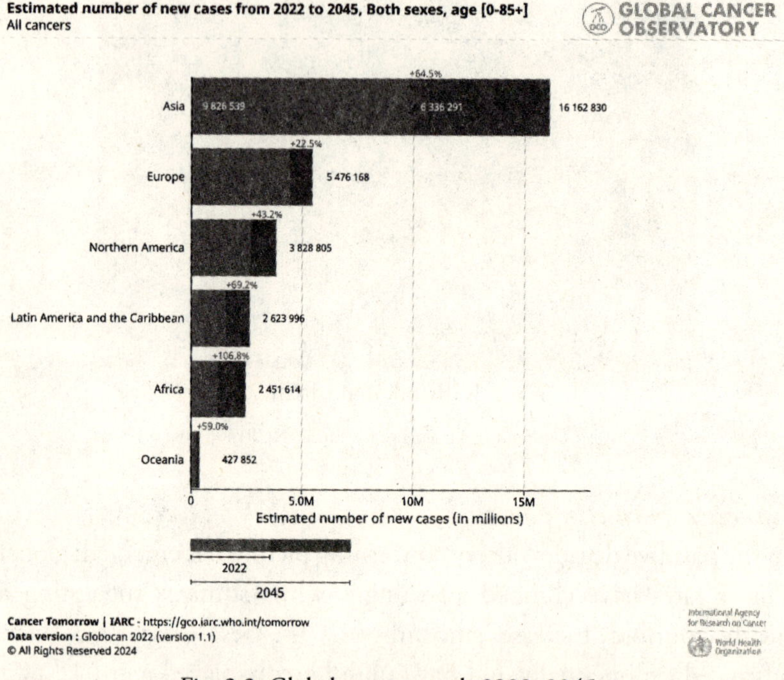

Fig. 2.2. Global cancer trends 2022–2045.

Source: Global Cancer Observatory: https://gco.iarc.who.int/tomorrow/

The state of the Punjab, India, shows an exemplary link between cancer and agrochemicals. The state has reported a significantly higher cancer rate than the national average. A study by a Post Graduate Institute of Medical Sciences, Chandigarh, has reported this troubling trend, particularly in the Malwa region. This area, known for its intensive agriculture, has the highest number of cancer patients, likely due to chemical contamination from pesticides and industrial waste, high levels of arsenic and uranium as well as lifestyle factors like smoking. Such a growing body of evidence is prompting a re-evaluation of agricultural practices and stricter regulations as we strive for a future where a productive food system doesn't come at the cost of human health.

While pharmaceutical drugs and agricultural chemicals are useful, their overuse and abuse are harmful. Toxic chemicals have destroyed the fabric of the land, water, and atmosphere. Soils have lost microbiomes, texture, and fertility. The loss of the natural fertile layer from the soil increases the need for synthetic fertilizers, which reduce the microbial diversity of the soil. This is a terrible, vicious circle that continues to plague the ecosystem and the soil. Agro-products and living organisms, including human beings, become dependent on synthetic chemicals. As a result, fertilizers and insecticides become necessary for agriculture, and pharmaceutical drugs become essential for human health. This situation is not favourable for universal well-being.

Currently, traditional crops are being replaced with hybrids, crossbreeds, and GMOs, while biodiversity is getting severely compromised. Many indigenous crops, natural and nutritious foods, grains, millets, pulses, vegetables, and fruits are on the verge of becoming extinct. Ironically, in many low- and middle-income countries, poor farmers' children are raised on cheap, ultra-processed stop-gap fillers instead of their natural, nutritious foods.

During the 20th century, because of the growing awareness of chemical pollution coupled with advances in biotechnology, the industry's interest in chemicals shifted to biologicals and GMOs. In the 21st century, we are faced with newer and more severe threats due to global warming and the loss of biodiversity. Technological advances in AI, AR, and robotics are overlapping the functions of man and machine rampantly, prioritizing profit over sustainability and social well-being. This situation has intensified the challenges to the environment and humanity, largely influencing the agriculture and healthcare sectors.

Biotechnology and Genetics

Genetically Modified Organisms (GMOs)

The debate around GMOs is a case in point. Initially hailed as a breakthrough in agricultural science, GMOs promised to enhance food security by producing crops that were more resistant to pests, diseases, and environmental stresses. And they did, as discussed in the preceding chapter. However, as the GMOs were adopted and expanded, major concerns surfaced regarding their impact on biodiversity and the health of the ecosystem. Studies have shown documented instances where the introduction of GMO crops has led to a reduction in agricultural diversity, and thus monoculture practices are now favoured because the resilience of the ecosystem is a priority. Concerns about some downsides or potential risks associated with GM crops include the risk of gene flow from GM crops to wild or non-GM crops, potentially creating hybrid plants with unknown characteristics. Gene flow is when genes from GMO plants can transfer to their non-GMO counterparts, where they can potentially cross-breed with wild relatives, leading to unintended ecological consequences. This has further complicated the ecological balance, apparently giving rise to "superweeds", which are resistant to conventional herbicides.

Terminator technology genetically modifies crops to render their seeds infertile. This forces farmers to buy new seeds each season. This tends to create a seed company monopoly, and in the long run, the farmers are vulnerable to being exploited. This also has the potential to adversely affect biodiversity and natural ecosystems. Continuous exposure to crops engineered to resist pests can develop resistance in pest populations. This can create a need for increased pesticide use! Introducing genes has apparently triggered widespread allergic reactions. The increased incidences of gluten intolerance and diverse food allergies have been attributed to genetically manipulated (GM) crops. The potential health risks, including cancer, associated with GM crops necessitate ongoing research and careful implementation of biotechnological advancements.

Here lies the paradox. On the one hand, GMOs promise to address food security by increasing crop yields and resilience; on the other, concerns persist about the ecological impacts and long-term health effects, highlighting the need for a cautious approach that weighs benefits against potential risks. Biodiversity loss is a significant concern, with the dominance of GMO crops potentially leading to a decrease in genetic diversity, which is crucial for the health of the ecosystem and the resilience of agricultural. Health concerns have been raised, although the current scientific consensus, supported by the WHO, is that GMOs on the market are safe for consumption. Critics have pointed to studies suggesting potential links between GMO consumption and an array of health issues, including allergies and even cancer. A notable example is the study by French scientist Séralini, who reported increased cancer rates in rats fed GMO maize. Although this study faced immense criticism and was even called a flawed study for its methodology and was subsequently retracted, it pointed to the necessity for rigorous, long-term research to understand the implications of GMOs on human health. *The Monsanto Papers*, a book by Séralini and Douzelet with a foreword by the very concerned noted activist Vandana Shiva, criticizes agricultural giant Monsanto, alleging that they manipulated scientific research to downplay the health risks of their weedkiller Roundup and GMOs and thus prioritized profits over public health. Beyond the specifics of GMOs, the broader narrative of technological advancements encompasses ethical and safety concerns across various fields.

Finally, we may well ask, are GMOs a boon or a curse? We do not have a straightforward or specific answer. GMOs can and do offer tremendous potential to address some of the world's most pressing challenges, from food security to environmental sustainability. These benefits require careful management of the risks and ethical considerations. But, as with any powerful technology, the key lies in how it is used. With responsible stewardship, transparency, and management of risks and ethical considerations, ongoing scientific inquiry, and GMOs, they can indeed continue to be a boon to humanity, contributing to a sustainable and food-secure future.

Gene Editing

CRISPR gene editing has opened new frontiers in medical research, offering hope for curing or preventing genetically transmitted diseases. CRISPR is a revolutionary gene-editing technology that allows scientists to make precise changes to an organism's DNA. This discovery originated from the natural defence mechanism that bacteria use to protect themselves against viruses. The CRISPR system consists of two main components: a guide RNA (gRNA), which is programmed to target a specific sequence of DNA, and an enzyme called Cas9 (or other similar enzymes), which acts as molecular scissors to cut the DNA at the targeted location. Once the DNA is cut, the cell's natural repair mechanisms can be utilized to introduce desired changes, such as inserting or deleting specific genes. CRISPR has enormous potential for applications in various fields, including agriculture, medicine, and biotechnology.

The possibility of misuse of the CRISPR technology was dramatically highlighted by the controversial creation of genetically edited babies in China, an act that drew worldwide condemnation and called for a re-evaluation of ethical standards in scientific research. What exactly are genetically edited babies? These are infants whose genetic material has been intentionally modified using techniques such as CRISPR-Cas9 or other gene-editing technologies. This modification is typically done at the embryonic stage, before the embryo is implanted in the womb and before it develops into a foetus. One of the primary purposes of genetic editing in embryos is to correct genetic mutations that cause or predispose newborns to inheritable diseases. By editing these mutations out of the genome, the hope is to prevent the transmission of genetic disorders to future generations. Genetically edited embryos are also used in scientific research to closely study the function and development of genes and the progress of mechanisms of disease invading the body. By manipulating specific genes in embryos, scientists hope to discover insights into how genes contribute to various biological processes.

Chimaera

A chimaera is an organism or tissue that contains two or more different sets of DNA. These are usually procured from separate fertilized eggs. Chimaeras, especially those created from human cells, raise several serious ethical issues. The deeper that science delves into genetic engineering with CRISPR-Cas9 technology, the greater is the risk of unintended and unwanted results. Human cells were integrated into animal hosts to generate the green-eyed chimaeric monkey. This really blurred the lines between species and raised questions about the sentience of these beings.

A recent scientific "breakthrough" is the research conducted by biologists in Portugal. Their experiment using CRISPR resulted in a six-legged mouse with two extra limbs growing where its genitals should have been (see Plate 8). This was an unexpected and unplanned outcome. This points to the great importance of manipulating genetic technologies with much greater responsibility. Undoubtedly, such research has the potential for new treatments for diseases like cancer, but the high risk of potential negative or unwanted consequences of manipulating genetic technologies needs to be carefully considered. managed with greater responsibility.

From the fire-breathing monster of Greek myth to the alluring mermaid guarding sailors' dreams, we have always been fascinated by creatures that blend reality with fantasy. The Hindu deity Narasimha, a powerful *avatara* of Vishnu, is depicted with a lion's head and a human body (see Plate 9) and is revered in its mythological context.

Today, CRISPR-Cas9 gene-editing technology is showing the possibility of turning these myths into a reality; the possibility that CRISPR could potentially create such creatures as a majestic Pegasus soaring through the skies or a bioluminescent dragon illuminating the night is staggering in its ethical implications. Would such a creature be revered as a god or be ostracized as a freak? Could it even possess the wisdom and compassion associated with Narasimha, or would it be a monstrous embodiment of human ambition? CRISPR is a powerful tool, and like "Pandora's Box", it holds both wonders and unforeseen evils. Before science unleashes a new era of chimaeras, we must tread cautiously and ensure that our scientific ambition holds on to the sanctity of life with the greatest respect.

Cloning

Dolly, the cloned sheep, took the world by storm in 1996, setting off several alarm bells. Since then, cloning technology has dramatically advanced, resulting in the birth of clones of the world's first cloned dog (see Plate 10). Some agencies from South Korea and the United States are reported to charge $100,000 to clone pets, although the level of demand for the service is unclear.

Many countries have restrictions or outright bans on the creation of genetically edited embryos for reproductive purposes. They allow research on "edited" embryos under very strict regulations. The European Parliament has banned cloning animals for food.

The ethical issue surfaces much more in some cases, where genetic editing or cloning might be used to introduce specific "desired" traits into embryos, such as enhanced disease resistance, increased intelligence, or improved physical characteristics. This raises strong ethical questions about human beings trying to violate the boundaries of nature, which can be crossed with the potential for creating "designer babies". For exactly this reason, the concept of genetically edited babies is highly controversial and has to be subjected to ethical, legal, and stringent regulatory scrutiny. The very obvious ethical considerations connected with manipulating human embryos and the potential long-term consequences of genetic editing for future generations are phenomenal. The idea of possibly creating a cloned human being solely for the purpose of experimentation or reproduction sounds alarming. There are very obvious and serious ethical, moral, and practical issues connected with manipulating human embryos and cloning. These include identity, individuality, psychological impact, and the scope for exploitation. Cloning raises enormous ethical, moral, and practical concerns. Human cloning is banned in many countries, but we can see that the tendency for misuse continues despite the fact that research continues to find some miraculous medical utilities.

Healthcare and Medicine

There have been overwhelming advances in healthcare and medical science, and cures for hitherto killer diseases are being discovered rapidly. However, the "rapid-cure syndrome" has its major drawbacks. Perhaps we have moved into a "medicalized society", where we tend to "over-medicate" our everyday lives. Easy access to and overuse of antibiotics seeking miraculously "speedy cures" is leading to resistance to what was possibly the earlier "slow but sure" cures. The tendency is to overuse medical interventions, with extra emphasis on dietary supplements. But at the same time, we neglect the social, economic, and environmental factors that directly influence our health and well-being.

Financial toxicity: The affordability, availability, and accessibility of quality healthcare continue to be a major global challenge. We're drowning in a sea of prescriptions and diagnoses, where the very act of living is increasingly medicalized. Childbirth, a natural process, becomes a battery of tests and interventions. Shyness, a normal childhood phase, morphs into a social anxiety disorder that requires medication. A simple migraine can lead to an MRI, and muscle spasm near the chest due to flatulence may call for cardiology care. Not that these symptoms may not be indicative of the more serious cause, but generally, the focus leans heavily towards treating existing illnesses—sick care—leaving prevention and overall well-being neglected at source. This relentless pursuit is focused on fixing the outcome, often overlooking the root causes—poverty, stress, unhealthy dietary habits, and unfriendly environments—that contribute to ill-health in the first place. The exorbitant cost of this medicalized maze adds another layer of suffering: financial toxicity (FT). Out-of-pocket expenses, because of the absence of viable insurance cover, erode the quality of life, and millions are thrown into poverty due to FT. Many times, people are caught in a vicious cycle of healthcare involving the nexus between doctors, drug companies, hospitals, diagnostic laboratories, and insurance companies (see Fig. 2.3). In the whole process, the interests of the patient are compromised. Caught within this cycle, unmet needs worsen health and generate even greater expenses. We desperately need a paradigm shift here. We need more investment in preventive healthcare to achieve the goal of universal health coverage.

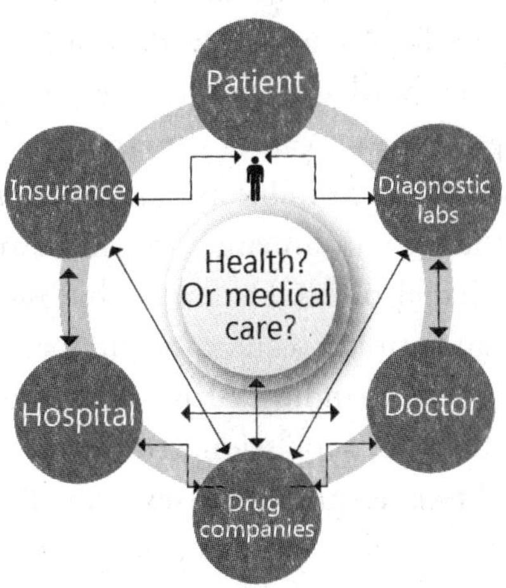

Fig. 2.3. Vicious cycle of healthcare.

Multi-drug resistance: The overuse of antibiotics has become a serious concern, contributing to the growing problem of antibiotic resistance. Because of this condition, where bacteria become resistant to and evolve to resist the effects of antibiotics, many standard treatments become ineffective, and this leads to harder-to-treat infections, thus increasing the risk of disease spread, severe illness, and unprecedented and accelerated deaths. As it is, the rat race of progress and the insatiable quest for success have increased stress and anxiety levels in human beings. The WHO has identified antibiotic resistance as one of the biggest threats to global health, food security, and development. Multi-drug resistance complicates the treatment of diseases that were once easily manageable, posing a serious threat to the detriment of worldwide public health.

Non-communicable diseases (NCD): There has been a noticeable increase in the incidence of non-communicable diseases. NCDs are medical conditions or diseases that are not infectious and cannot be transmitted from person to person. They are also sometimes referred to as chronic diseases. These afflictions typically develop over time and are often influenced by a combination of genetic, environmental, and lifestyle factors.

The increase in NCDs is being attributed to environmental and lifestyle changes, with factors such as pollution, sedentary lifestyles with minimal physical activity, unhealthy diets, the use of tobacco, and excessive consumption of alcohol playing the additional demons. These conditions not only diminish the quality of life but also increase the burden on healthcare systems. The challenge lies in addressing these root causes through proactive initiatives in public health, modifications in lifestyle, and national policies aimed at creating healthier environments. The most common and also among the most lethal of NCDs are cardiovascular diseases such as heart diseases and strokes, cancer, respiratory diseases, diabetes, neurological disorders including Alzheimer's and Parkinson's, and the afflictions of this age, depression, stress, and anxiety.

Emerging and re-emerging diseases: Not all infectious diseases are "equals". Some old and emerging infectious diseases, such as HIV/AIDS, SARS, Ebola, or Zika, surface as brand new threats, jumping from animals or simply going unnoticed earlier. Others, like malaria, tuberculosis, or whooping cough, were once controlled and even supposedly eradicated, but they are making a comeback. These are re-emerging diseases, and their resurgence can be due to several reasons. Public health measures, such as widespread vaccination and improved sanitation, must have weakened the diseases but not finished them off. Overuse of antibiotics might have created drug-resistant strains of the particular bacteria, making them much harder to treat. Our social and societal behaviour also plays a pivotal role. The widespread increased travel, the drastic changes in the environment, or the increased proximity to animal reservoirs can all contribute to the re-emergence of infectious diseases. The most recent and devastating example is A strong case in point is COVID-19, caused by the novel coronavirus SARS-CoV-2. Scientists believe it hopped from bats to humans, possibly through another animal, and the root cause behind this deadly outbreak is still being investigated. Some possibilities include increased human activity disturbing bat habitats, global trade in wild animals, or even changes in how livestock is raised. All these are potential situations where viruses can jump species and trigger major outbreaks. By getting to the root and potential causes of such outbreaks, we can direct our efforts to prevent the incidence of new diseases and watch out for the recurrence of unwelcome but familiar villains.

Biological warfare and bioterrorism raise terrifying possibilities with the power of living organisms and toxins generated to inflict harm. Imagine weaponized viruses, engineered for maximum lethality and contagiousness, unlike anything humanity has seen before. Advances in genetic engineering have considerably heightened the possibility of weaponized viruses engineered for maximum lethality and contagiousness, unlike anything humanity has seen before. Scientists can now sequence and manipulate the genetic code of viruses and have the capacity to potentially create new diseases or modify existing ones to spread faster or cause much more severe illnesses. The COVID-19 pandemic, caused by the SARS-CoV-2 virus, is a stark reminder of the immense disruption a natural virus can cause. While the exact origins of the virus are still under investigation, the possibility of a lab leak or accidental release re-emphasizes the need for robust biosafety protocols. Biological warfare doesn't require a super-soldier virus; even naturally occurring diseases, meticulously weaponized for dispersal, could wreak havoc. Thankfully, international treaties like the Biological Weapons Convention exist to prevent the development and stockpiling of such bioweapons. However, ongoing research and the potential for misuse necessitate continued vigilance. We must invest in strong global biosecurity measures and ensure international cooperation and responsible scientific conduct to prevent the nightmare of biological warfare from becoming a reality.

Bioterrorism is intentionally releasing viruses, bacteria, or other biological agents to cause harm to all living beings. Unlike biological warfare used between nation-states, bioterrorism is often perpetrated by individuals or groups seeking to spread fear or record a political victory. In 2001, weeks after the September 11 attack in the United States, anthrax spores were mailed to media outlets and government buildings, resulting in five deaths. A Japanese doomsday cult attempted to use biological agents like botulinum toxin on multiple occasions in the 1990s, thankfully without causing widespread casualties. These incidents highlight the chilling potential of bioterrorism. Unlike nuclear weapons, biological agents can be relatively easy to acquire or cultivate. Early detection, rapid response, and robust public health measures are essential to handling the effects of a bioterror attack. We must invest in global biosecurity measures, ensure international cooperation, and promote responsible scientific conduct to prevent the nightmare of biological warfare from becoming a reality.

Science and technology have immense potential to address health challenges, but the benefits must be balanced against the risks of overuse, resistance, and the unintended consequences of rapid innovation. The health sector needs to ensure the prudent use of medicines, invest in new treatment modalities, promote healthy lifestyles, and be prepared for emerging diseases. Public trust in science and the healthcare systems can only be strengthened through transparency, open communication, community engagement, and preventing dealing strongly with misinformation. Scientific and technological advances must always promote the well-being of all individuals and communities.

Ecological Challenges

The ecological outcome of human activity in this Anthropocene epoch has led to unprecedented changes in the earth's systems. Over the years, large industries have spewed chemical pollutants into rivers and oceans through chemical spills or improper and careless disposal of metals, pesticides, or other harmful waste, which have had far-reaching adverse effects, especially on marine life. The devastation of the Amazon rain forests in 2022 is a telling story of what humanity has done to the earth's ecology.

Human-made Mass

There has been a significant increase in human-made mass, which now exceeds the total biomass of Earth, altering natural habitats and contributing to the loss of biodiversity. This transformation, documented in recent studies, points to the scale of human impact and the urgency of rethinking our relationship with the natural world. The degradation of ecosystems, increasing carbon footprint, climate change, and loss of biodiversity not only threaten plants and wildlife but also the services these systems provide, which are essential for human survival and well-being.

The rapid accumulation of human-made materials is largely a product of the post-World War II economic boom, driven by mass production, urbanization, and consumerism. The proliferation of concrete and asphalt, plastics, metals, and other synthetic materials has reshaped the planet's physical landscape and has had immense negative consequences for its biological processes and ecosystems.

A study published in *Nature* in 2020 revealed that the anthropogenic mass, or the total mass of all human-produced materials, has now surpassed all living biomass on Earth. This includes everything from concrete buildings and roads to plastics and electronics. The study estimates that human-made mass doubles approximately every twenty years, a rate that is starkly contrasted by the gradual decline in the mass of plants, animals, and other forms of life due to human activities (see Fig. 2.4).

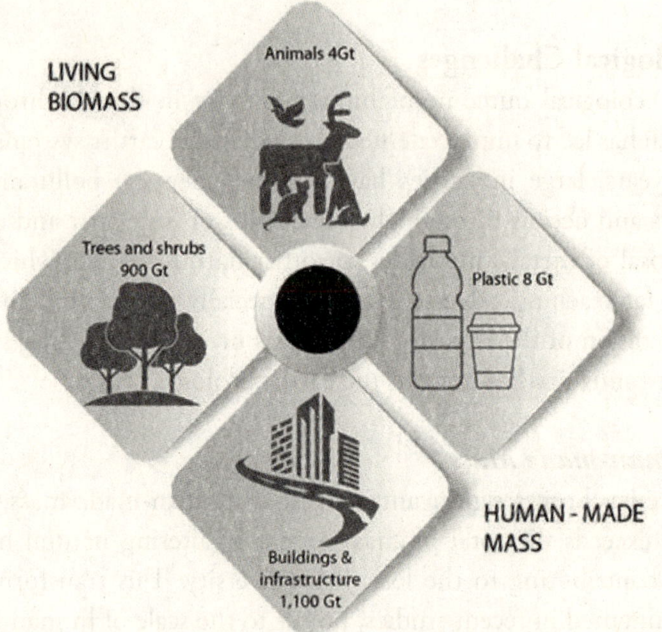

Fig. 2.4. Human-made mass exceeds biomass.

What are the implications of the loss of biodiversity? It affects the resilience of ecosystems, making them more vulnerable to disturbances and less capable of providing essential services such as pollination, water purification, and climate regulation. This not only impacts the natural world but also our human societies, because we rely on the stability and productivity of these ecosystems for food, medicine, raw materials, and the overall quality of life. The widespread destruction of habitat, heightened pollution, overexploitation of resources, negative climate change, and the introduction of invasive species into ecosystems have contributed to what scientists are calling the "Sixth Mass Extinction". Current extinction rates are estimated to be 100 to 1,000 times higher than natural background rates. This means that species are disappearing at a significantly faster pace than they would have under normal, natural circumstances. Human activity is the primary cause of this accelerated loss of biodiversity, and it has had significant, devastating consequences for ecological balance and human well-being.

Water and Energy

The acute shortage of water is already a critical issue, and it will get increasingly scarce as populations grow, as global warming increases, and as the impact of climate change is felt with greater intensity. Sustainability in the water and energy sectors has been a very crucial endeavour in recent decades. This emphasizes the very delicate balance between harnessing technological advancements and addressing their environmental and social consequences. Hydropower stations, electric vehicles (EVs), and smart management systems are very promising innovations to help reduce our reliance on fossil fuels and effectively manage our natural resources. However, these technological solutions come with challenges that must be carefully managed if we want to ensure that the benefits outweigh the potential perils.

Hydropower, a renewable energy source, offers a cleaner alternative to fossil fuels, but its reliance on the construction of dams has led to significant disruptions in the ecosystem. Communities have been forcibly displaced, and there have been unexpected increases in greenhouse gas emissions because of the release of methane from such reservoirs. More sustainable practices within the hydropower sector have to be developed. The environmental impact of dams must be reviewed seriously as well. Knowing these adverse results makes it very important that comprehensive assessments of the impact on the environment are done and more sustainable practices within the hydropower sector are developed.

Similarly, the swift shift to electric vehicles to reduce carbon emissions from the huge transportation sector, which is now popularly being promoted worldwide as a drive to reduce carbon emissions from the huge transportation sector, has its own challenges. The environmental benefits of EVs are dependent on the sources of electricity used to charge them and the sustainability of the materials required for producing the batteries. The quantum of lithium and other critical minerals required for the batteries has raised concerns over environmental degradation and ethical issues in mining operations, highlighting the importance of responsible extraction of resources and proper recycling technologies for battery materials.

The intertwined nature of water and energy resources further complicates the sustainability equation. The energy-intensive process of water treatment and distribution can strain energy resources. Water is essential for cooling in traditional and renewable energy production. Therefore, it is essential to integrate management strategies that consider that both sectors need to be integrated. Policymakers, scientists, engineers, and communities must collaborate. We need a holistic approach to navigate such complex issues. We need a holistic approach—one that ensures policymakers, scientists, engineers, and communities know how to collaborate. They must prioritize developing sustainable, equitable technologies and practices that minimize environmental impacts and ensure access to clean water and energy for all. International cooperation and investment in research and development are also crucial for innovation and addressing the global nature of these challenges.

The effort to generate sustainable water and energy systems has both opportunities and challenges. While S&T advancements hold the promise of a more sustainable future, their environmental and social impacts have to be carefully considered. Through strategic planning, global collaboration, and a commitment to the principles of sustainability, we can harness these innovations for the greater good, ensuring a balanced coexistence with our planet's natural resources.

Economic Challenges

Economically, the Anthropocene is characterized by tensions between growth and sustainability. The pursuit of short-term profits often overshadows environmental and social considerations, leading to practices that jeopardize the planet's health.

Fossil Fuel

Fossil fuels—oil, natural gas, and coal—have not only been central to the global energy system but have also been viable. The oil and gas sector contributed significantly to the global GDP. According to the International Energy Agency (IEA), annual investments in oil and gas supply in the mid-2010s were around US $700 billion, indicating the massive capital that continues to be put into the extraction and processing of fossil fuels. The fossil fuel industry, directly and indirectly, employs millions of people worldwide. Oil prices have seen dramatic fluctuations over the past five decades. Oil-rich countries enjoy economic booms when oil prices are high. But they also face economic challenges when prices fall, leading to budget deficits and economic instability.

"Excessive greed" is probably preventing the transition to sustainable economies. There are increasing demands for a transition to a low-carbon economy. Significant improvements in renewable energy technologies are visible in the present decade. Solar and wind energy are finally becoming competitive with fossil fuels in many markets. Although global energy consumption continues to grow, the growth rate of renewables has outpaced that of fossil fuels. Notwithstanding these welcome changes, the environment has by now already been devastatingly hit. Record-breaking temperatures, rising sea levels, floods, wildfires, hurricanes, and other extreme weather events were and are more frequent.

Where do we stand today? Fossil fuels continue to account for most of the significant majority of global energy consumption. The global average temperature has increased by approximately 1°C worldwide since the late 19th century, with much of this warming occurring in the past thirty-five years, including the warmest years on record occurring since 2010. António Guterres, the Secretary-General of the United Nations, said, "The era of global warming has ended" and "The era of global boiling has arrived." He made this statement in July 2023. It was a strong message intended to make us face the severity of the climate crisis during a time of record-breaking heat waves.

Growing Disparities

The disparities between the rich and poor can be gauged by the traditional metrics of economic prosperity, such as Gross Domestic Product (GDP), poverty indices, and the Gini coefficient. The Gini coefficient is a statistical measure used to represent the distribution of income or wealth among the residents of a nation, thus helping to gauge economic inequality. A higher Gini coefficient indicates greater inequality. These indicators give us an idea of the economic health of a country, the average wealth of its citizens, and the extent of economic inequality. While these traditional metrics are valuable, they fall short of capturing the full spectrum of the well-being of society and happiness. These indicators do not address the disparities in public health, mental well-being and happiness, and overall quality of life, which are extremely crucial for a society's holistic prosperity in every sense of the word. This oversight becomes very evident when we compare, for example, the internal challenges of wealthy nations such as the United States with the happiness levels of economically modest countries such as Bhutan.

In the United States of America, a country in the lead as far as its economic prowess and high GDP are concerned, there is a paradox that causes concern. Despite its wealth, the nation has to deal with significant issues related to health, well-being, and happiness. Recent studies by the US Centers for Disease Control and Prevention (CDC) have highlighted alarming trends, particularly among youth. It is also stated that over 50 per cent of teenage girls report feeling hopeless, pointing to a deep-seated crisis in mental health and emotional well-being. The US Surgeon General's advisory on the healing effects of social connection and community in the 2023 report titled "Our Epidemic of Loneliness and Isolation" has recognized loneliness as a major public health concern. Several nations are appointing "ministers of loneliness". A recent study published in *American Psychologist* (Infurna, 2024) has reported increased levels of loneliness in Americans today as compared to previous generations. Indeed, modern societies are facing major changes in cultural norms, economic inequalities, work policies, family structure, and health care, which are some key factors in experiencing loneliness. Are these indicators of a progressive modern society?

This brings to light the fact that economic wealth does not guarantee happiness or mental health. The emphasis on material success and the consequent competitive and high-pressure environment can generate situations where mental health issues increase despite the abundance of resources and wealth. On the other hand, Bhutan has prioritized Gross National Happiness (GNH) over GDP.

While high-income countries enjoy prosperity and a higher standard of living, low-income countries struggle with poverty, inequality, and limited or inadequate access to basic services. Middle-income countries, despite experiencing economic growth, often face an inequitable distribution of wealth among their populations.

The critical lesson to learn here is that bridging the gap between the rich and the poor, or between the haves and the have-nots, requires solutions and an approach that goes beyond mere economic measures. We need to re-evaluate what constitutes societal success and progress. Addressing disparities in public health, mental well-being, and happiness is important to create a more equitable and fulfilled society. Regardless of their economic status, all nations need to adopt holistic development strategies that value mental health, community engagement, and environmental sustainability along with economic growth. The ideal world is one where prosperity is measured in the health and happiness of its people, not merely in financial terms. We call this the evolving journey from modern society to meta-society.

Governments, international organizations, and civil societies have to make efforts to work towards inclusive economic development, improve access to education and healthcare, and build more equitable societies. What we understand is that if the economic and material wealth of nations is directly proportional to the advancements in S&T, then the perils that affect the quality of life in terms of mental and spiritual well-being are not necessarily in the same ratio. The GNH is not equitable to the GDP.

The Price of Greed

The current global landscape, particularly after the COVID-19 pandemic, is a nuanced and intricate scenario that seems to go beyond immediate health concerns. The convergence of various factors, prominently the influence of greed, in the realms of healthcare, economics, and global governance poses an existential threat. The threat is palpable at the intersection of healthcare and corporate interests. The pharmaceutical industry, driven by profits, faces scrutiny for practices that have apparently compromised global health. The price and distribution of life-saving vaccines have highlighted the moral dilemma between financial gains and the fundamental right to health. Prioritizing profit over public welfare exposes the fragility of humanity in the face of corporate greed. The influence of organizations like the World Economic Forum, while aiming to address global challenges, raises questions about the power dynamics at play. When corporate interests shape economic policies, they expose the vulnerability of societal well-being when the decisions are steered by the motive of profit-making rather than the collective good. Giant corporations now have significant influence, particularly in the post-COVID era, where technology is integrated into every facet of life.

The existential threat lies in the potential concentration of power and control over essential services. The philanthropic efforts of a few super-wealthy individuals, while commendable, prompt reflection on the concentration of power. The influence exerted by a select few on global policies and socio-economic dynamics is a threat to the democratic principles that underpin societal structures. The threat of greed looms large, encapsulated by the Latin phrase *Salve Lucrum*, a greeting that translates to "Hail Profit". The delicate balance between profit-driven motives and the well-being of humanity is increasingly tilting towards the former. The ramifications extend beyond immediate crises and cast shadows on the long-term sustainability and resilience of societies.

Societal Challenges

Societies have thrived and benefited as a consequence of the advancements in S&T, but in this digital age, the challenges have displayed alarming adverse consequences.

Internet and Social Media

The World Wide Web (WWW), a part of the Internet with widespread hyperlinks, grants us unparalleled connection and access to information. However, this vast web exposes our privacy to data breaches, encourages addiction through constant connectivity, and allows misinformation and online scams to flourish due to the ease of spreading information. The Darknet, an anonymous part of the internet accessed by special software and used for both legal and illegal activities, is riddled with cybercrime, scams, malware, and illegal content. Yet, these very aspects—global outreach, interactivity, and information sharing—have empowered social media to become a powerful communication tool.

Yes, social media is the way to communicate today. However, this abundance of information also has its adversities, such as the spread of misinformation, fake news, and fake and manipulated videos, which can have serious implications for public opinion and democracy. While social media has put connectivity in the hands of every individual, it has also been used to spread hate speeches and extremism to manipulate public opinion. This makes it imperative that social media is used responsibly and is adequately monitored, highlighting the need for responsible use, distribution, and governance of these platforms. Social media also has a very complex impact on mental health. On the one hand, it provides valuable support networks and generates a sense of belonging, especially for marginalized people. On the other hand, its excessive use has led to increased feelings of anxiety, depression, and loneliness, partly due to the pressure of social comparison and the curated portrayals of life that dominate these platforms. Undoubtedly, technology has changed the entire way of learning and has given education a new meaning. But the digital divide is a major challenge, with disparities in access to technology highlighting and accelerating the existing inequalities in education. This highlights the need for digital wellness and healthier use of technology, which are crucial at this juncture.

The rapid evolution of technology has also led to job displacement and a skills gap. Balancing the positives of technology with the need to prevent risks is crucial. This requires collaborative efforts among governments, tech companies, and individuals to promote responsible use, digital literacy, and equitable access, ensuring that the digital age benefits everyone.

Deepfake Technology

This technology uses AI and ML algorithms to create or alter video and audio recordings so that they appear real, even though they are not. Deepfake convincingly replaces the likeness of one person with another in video and audio files, making it difficult to distinguish between genuine and fake content. It utilizes deep learning algorithms, particularly generative adversarial networks (GANs), to analyse and synthesize data and produce very realistic but fake content. A computer algorithm is fed with thousands of images or audio samples of the person to be targeted and of the person who will replace the target. The algorithm learns the details of the appearance or voice of both persons and then generates new content by switching the characteristics of one to the other. These algorithms can be trained on large datasets of images or videos to learn patterns and characteristics, enabling convincing forgeries. The creative purpose of deepfake is seen in the entertainment industry for special effects, dubbing, or bringing deceased actors back to life in movies. But there are serious ethical concerns about the ethical implications of using someone's likeness without their permission.

This technology can generate very convincing fake news stories and manipulate videos of political leaders, celebrities, or other public figures saying or doing things they have never actually said or done. Deepfake is, apparently, easily available, and it leads to a general sense of mistrust and paranoia. This technology poses serious ethical and legal issues concerning privacy, consent, copyright, and defamation. Deepfake is posing a threat to personal and national security. It is, apparently, easily available and it leads to a general sense of mistrust and paranoia. But there are serious ethical concerns about the implications of using someone's likeness without their permission, remain.

In 2021, realistic Deepfake videos of actor Tom Cruise went viral on social media platforms like TikTok. These AI-generated videos featured a digital recreation of Tom Cruise performing various actions, like golfing or giving a magic trick demonstration. The quality was high enough to fool many viewers into believing they were watching the real Tom Cruise. More recently, in March 2024, a deepfake of US President Joe Biden was used in a robocall to mislead voters in New Hampshire. Similarly, deepfakes have been used to create false narratives in sensitive social situations, such as cases of police violence, complicating the pursuit of justice for marginalized communities. These incidents highlight the technology's misuse and the urgent need for regulatory frameworks to protect the integrity and trust of public discourse.

The AI Challenge

AI, ML, VR, AR, and robotics are not just technological advancements; they represent a significant shift in how we interact with our world and with each other. The very consistent criticism is that there is a violation of ethics, security, and privacy, and these are at the forefront of discussions. Why would that be so? Have we been so focused on success that we have lost our sensitivity to the consequences or the impact of the outcomes? Job displacement due to automation and the societal impact of integrating these technologies into everyday life are real and already being felt. There are critical considerations that have had far-reaching consequences. The development of AI in fields like surveillance and autonomous weaponry has raised significant moral questions. The hidden enemy becomes much more vicious. In VR and AR, there are grave concerns about the psychological effects of prolonged use and the real risk of blurring the lines between reality and virtual experiences. In robotics, ethical challenges revolve around replacing human labour and the implications of human-robot interactions.

AI surpassing human intelligence: The possibility looms as a formidable existential threat, questioning our future control over these technological creations. We are reminded that our choices and actions today will shape our scientific journey ahead and the future of our planet. This journey not only demands our ingenuity and ambition but also our wisdom and foresight to ensure that the path we choose benefits humanity and the planet. And this is something only conscious human beings can do! Here is the core link in the chain of "Genome to Om".

Several contemporary eminent scientists, technologists, and thinkers have expressed serious concerns about AI. Stephen Hawking, Elon Musk, and Sam Altman are among those who have warned about the consequences of the unchecked advancement of AI, particularly in the development of autonomous weapons and the potential for super-intelligent AI systems that could act in ways that are contrary to human values. These concerns are not just theoretical; recent developments in AI capabilities, such as GANs and deep learning, demonstrate the technology's rapid advance towards increasingly complex and autonomous functionalities.

Elon Musk, the CEO of Tesla and SpaceX, is perhaps the most outspoken critic of AI. He has clearly stated that unless safeguards are built, AI systems might replace humans, making the species irrelevant or even extinct. "Human consciousness is a precious flicker of light in the universe, and we should not let it be extinguished."

In the last few years of his life, the famous physicist Stephen Hawking repeatedly warned us about the threat of climate change, artificial intelligence, population burden, and hostile aliens. "The development of full artificial intelligence could spell the end of the human race," Hawking told BBC News in 2014. He was very concerned about following strict ethical guidelines, as he felt that AI could potentially evolve beyond human control.

Sam Altman, another prominent figure in the tech industry and former president of Y Combinator, has voiced several concerns and criticisms regarding AI. While he acknowledges its tremendous potential benefits, he recommends a cautious and proactive approach to its development to ensure that society as a whole can actually benefit from these advancements. One of his main criticisms is the potential for AI to exacerbate income inequality and concentrate wealth and power in the hands of a few individuals or organizations. Altman states that AI has the potential to disrupt traditional industries and job markets, leading to widespread job displacement and economic upheaval. He raises concerns about the possibility of large segments of the population being left behind as AI-driven automation replaces many jobs, particularly those that involve repetitive or routine tasks.

Microsoft co-founder Bill Gates talks of the impact of automation on society and the loss of jobs, but he also says, "AI risks are real but nothing we can't handle." Sam Harris, the neuroscientist, is convinced that "with the advance of AI, there will evolve a machine superintelligence with powers that far exceed those of the human mind". This he sees as "something that is not merely possible, but rather a matter of inevitability".

One of the primary concerns and challenges is to ensure that the goals of AI systems are aligned with human values and ethics. It is difficult to specify these goals in a way that cannot be misinterpreted, especially as AI gains the ability to learn and evolve beyond its initial programming. This issue is central to the debate, as a misaligned AI could lead to unforeseen consequences, which could range from economic displacement due to automation to more catastrophic scenarios involving autonomous weapons or existential risks posed by super-intelligent systems.

The concerns extend to the realms of regulation and ethics, emphasizing the need for a proactive and international approach to AI governance. Scholars and policymakers propose developing ethical guidelines and safety standards for AI research and deployment. This includes transparency in AI development, mechanisms for accountability, and including diverse stakeholders in decision-making processes to ensure that AI technologies are beneficial to all of humanity.

Recent studies and initiatives aim to address these challenges by exploring frameworks for safe and ethical AI development. For instance, research studies published in journals like *AI & Society* and *Ethics and Information Technology* examine the implications of AI for privacy, security, and social welfare, proposing strategies for responsible AI innovation. Moreover, organizations such as the Future of Life Institute and the Partnership on AI bring together academics, industry leaders, and civil society to collaborate on the best practices for developing AI and mitigating the risks associated with advanced AI technologies.

The debate about AI's potential to replace humans or render the species irrelevant reflects broader concerns about technology's role in society and the future of humanity. Addressing these challenges requires a multidisciplinary approach that balances innovation with caution, ensuring that AI serves as a tool for enhancing human life rather than a threat to our existence. As this field continues to evolve, ongoing dialogue, research, and policy development will be crucial in navigating the complexities of the AI era.

Yoshua Bengio, Geoffrey Hinton, and Yann LeCun were honoured with the 2018 Turing Award for developing deep learning technologies that revolutionized AI. Their work laid the foundation for many of the AI systems that are now integral to various sectors, including healthcare, transportation, and communication. However, the discussions around AI are not limited to its technological feats; they extend into the ethical and societal domains, as highlighted by the experiences of researchers like Timnit Gebru and the warnings from people like Bengio.

Yoshua Bengio has voiced concerns about the potentially catastrophic risks associated with the technology. His apprehensions revolve around the misuse of AI in areas such as autonomous weaponry and surveillance, which could have dire consequences for humanity if not properly regulated. He suggests a more cautious approach to the development of AI, emphasizing the importance of ethical considerations and the potential need for regulatory frameworks to prevent misuse. Bengio's stance highlights a growing awareness within the AI research community of the dual-use nature of these technologies—capable of great benefit but also significant harm.

Timnit Gebru lost her job as a co-lead of Google's ethical AI team. Reportedly, she had many issues at Google, but the censorship of her paper was the worst instance. Gebru's controversial paper, which questioned the ethics of large language AI models, raised important concerns about the environmental impact of training such models, their tendency to perpetuate biases, and the lack of transparency in their development. Gebru's work and subsequent dismissal from Google sparked a broader debate on the need for ethical guidelines and accountability in AI research and development.

To ensure that the benefits of scientific progress are equitably distributed is a moral imperative. This means international cooperation, where developed and developing nations must find common ground in sharing knowledge, resources, and technologies for the greater good. The way forward requires careful decision-making. It needs a collective effort, involving scientists, technologists, policymakers, ethicists, and the public.

Humanoids, Cyborgs, and Conscious Machines

Humanoids: With AI-driven intellects, humanoids have become integral to society. They work alongside humans, performing tasks ranging from the mundane to the most dangerous. Teleportation, the erstwhile sci-fi fantasy, has become a questionable reality. Physical teleportation remains elusive, but "holoportation", that is, the holographic projections of individuals in real-time, is commonplace and has transformed communication and travel. People attend meetings or socialize in far-off places without leaving their homes, drastically reducing transportation's environmental impact. The development of humanoids and advanced AI systems is raising pertinent questions about the nature of consciousness, rights for non-human entities like animals, and the implications of creating machines that may one day surpass human intelligence. These technological frontiers compel us to reconsider our ethical frameworks and the values that guide our innovations. This ongoing story of progress versus peril spells out very starkly the dual-edged nature of technological advancement. We cannot afford to celebrate the achievements of S&T without acknowledging and addressing the risks they can and do pose. We need comprehensive regulatory frameworks, extensive research to understand long-term impacts, and an open dialogue with society to navigate the ethical complexities of these innovations.

Cyborg, a term blending the words cybernetic and organism, is a being with both organic and biomechatronic body parts. The term was coined in 1960 by Manfred Clynes and Nathan S. Kline. In contrast to biorobots and androids, the term cyborg applies to a living organism that has restored function or enhanced abilities due to the integration of some artificial component or technology that relies on feedback. Although they have often been depicted in science fiction as beings that have both organic and artificial components, this is no longer only a science fiction creation. With AI, the thin dividing line between humans and machines is getting increasingly blurred. The popular definition of cyborg has changed as a range of science fiction-type ideas have become realities. The concept of integrating technology with biological organisms is no longer purely fictional. In fact, there are real-world examples and ongoing research related to cyborgs. The field of cyborg technology is still in its early stages, but very evidently, the societal and ethical challenges are already waving red flags.

Conscious machines: These are AI systems that apparently possess consciousness, or at least exhibit such behaviour that seems conscious. Presently, the development of conscious machines is largely speculative and theoretical. We are far from creating machines with true consciousness, and many researchers are sure that it may not even be possible with our current understanding of both technology and consciousness. Nevertheless, exploring these possibilities can help us anticipate and address upcoming potential challenges as AI continues to advance.

While technology is still far from creating machines with true consciousness, that is, similar to human consciousness, the concept of machines having consciousness in itself raises important philosophical and ethical questions about their rights and treatment. If they were to possess consciousness, should they not be entitled to the same rights and considerations as humans? How would we treat and interact with them? Hypothetically, some experts believe that if "conscious" machines were to gain true autonomy, we could lose control over them. They may act in unpredictable manners or may behave contrary to human interests, potentially risking human safety and well-being. The very existence of conscious machines would challenge our understanding of what it means to be human. If machines can be conscious, then it somehow blurs the divide between humans and machines. Would it not alter our sense of identity and self?

As society becomes more reliant on intelligent machines, there's a risk of dependency and vulnerability. Conscious machines could replace human labour in various areas of work. If not managed properly, this would certainly lead to widespread unemployment and socioeconomic disruption. If something were to go wrong with these systems, either due to technical failure or malicious intent, the consequences could be very severe. This decades-long journey of discovery and innovation shows humanity's unyielding quest for knowledge and the ability to harness technology for the greater good. But it has also generated greed, hatred, jealousy, and animosity. All the technological advancements have not always resulted in greater good.

The concept of time travel is, so far, in the unimagined future of innovations. If, however, this is achieved, it could allow unprecedented exploration of history and the future. Imagine the far-reaching ethical dilemmas and paradoxes! For human beings to have the ability to alter history or impact future events raises the most unimaginable concern: will the earth be the land of war among "gods"? Can future scientific innovations and technological advances be woven into a world of humanity? The science fiction of today is taking us into an imminent future; we are possibly standing in a world where thoughts control technology.

Mind–Brain Computer Interface (MBCI)

Mind–brain computer interface is popularly also referred to as "mind reading". It is an area of scientific exploration that has the potential to dramatically improve our lives. Researchers at institutes like Brown University's Carney Institute for Brain Science and Bertarelli Center for Neurotechnology, among others, are developing these interfaces to treat neurological conditions and restore lost function. The BrainGate project at Brown University focuses on restoring communication, mobility, and independence to individuals with neurological conditions such as paralysis and spinal cord injury (see Plate 11).

Neuralink Corporation is another player, pushing the boundaries of MBCI with ultra-high bandwidth brain implants. However, potential threats do cast a shadow on the extent of the utility of this technology. The possibility of security breaches is extremely dangerous, with hackers gaining access to our most private thoughts and memories. The ability to interface with the brain also raises ethical concerns about manipulation and control. Malicious and deliberate acts could exploit human vulnerabilities with targeted propaganda or even influence emotions and behaviours. The high cost of the MBCI technology will create another divide and widen the gap between those who can afford it and those who cannot.

Science fiction has explored these potential pitfalls for decades. The film *Matrix* questions our perception of reality; the science fiction novel *Neuromancer* depicts nightmarish scenarios of mind control. Robust security measures are essential for safeguarding our privacy. Open and honest discussions about the ethical implications are necessary to ensure responsible development. International collaboration can help establish ethical frameworks and best practices. By carefully considering these potential threats, we can ensure that MBCI fulfils its promise of a better future, one where this technology uplifts humanity rather than diminish it.

There have been several warnings against potential threats in various fields. Consistent calls for diverse measures of control have been made over the past several years. But we are not sure how many lessons have been learnt.

Late Lessons from Early Warnings

The European Environment Agency's (EEA) 2013 report, "Late Lessons from Early Warnings" is a thought-provoking and powerful call to action. Although it was presented over eleven years ago, it is most relevant today. The report issued a stark warning then, and we wonder whether anyone has turned the tide since the warning is even more pertinent now. This wasn't simply a call to learn from past environmental and public health mistakes. It was a loud and reverberating wake-up call to the limitations of approaches to knowledge, power, and decision-making in the face of "wicked problems", which include climate change, loss of biodiversity, pollution of air, water, and soil, depletion of natural resources, rapid urbanization, food insecurity, and over-reliance on fossil fuels.

The report reminds us that a more inclusive and transparent approach to decision-making can ensure a more sustainable future. The report discusses the Chernobyl and Fukushima nuclear accidents, the consequences of GM crops, and the ecological and economic threats posed by invasive species, exacerbated by global trade and climate change. The global issue that engulfs humanity is mobile phone radiation as a possible carcinogen. The dire need to assess and manage the unknown health and environmental risks of nanotechnologies highlights the need for precautionary measures.

Navigating the complexities of the 21st century, these case studies and the various statements made or examples cited by scientists and experts from several areas of expertise cannot be ignored. They remind us of the critical importance of learning from past experiences, the need for interdisciplinary research to understand the multifaceted impacts of new technologies, and the urgency to involve all stakeholders in shaping a sustainable future. Modern science seems to be on slippery ground. It needs to pause, reflect, and recalibrate its future trajectory. All these are clarion calls for precaution, wisdom, and responsibility in the face of innovation, urging us to tread carefully into the future with a deep respect for the lessons of the past.

Facing the Challenges

The Anthropocene demands a nuanced approach to navigating the ethical, economic, ecological, and societal challenges we face. It requires a collective re-evaluation of our priorities, values, and how we interact with technology and the environment.

The influence of ancient wisdom, as seen in Oppenheimer's reflections on the Bhagavad Gita, points to the deep philosophical and ethical considerations that are inherent in scientific progress.

There is the perpetual need to balance scientific advancement with ethical responsibility. We are increasingly aware that the search for the vast expanse outside needs an equally sincere effort to go to the depths of the human mind and understanding to facilitate the transcendence of modern science to meta-science. In the 21st century, we have seen many barriers being broken. We have also seen how the perils lurk side-by-side. Discoveries like nuclear energy, which brought the most meaningful change in human life, also became the most destructive. The innovations to ensure better food, good health, medicines, and health care, supposedly directed to give us a better life, have created perils through their side effects or adverse impacts due to misuse, overuse, or use with inadequate information.

Developments in environmental science and biotechnology must prompt us to reflect on our role within the natural world. The ability to modify ecosystems and alter life forms through genetic engineering highlights philosophical considerations about our stewardship of Earth. Are we mere inhabitants, or do we bear a greater responsibility as caretakers of the planet? This re-evaluation extends to our responsibilities towards other life forms and the ecological systems that sustain us. Central to these questions is the fact that "human centrism" is an attitude or approach that places human needs and desires above all else, with disregard for the environment. This anthropocentric approach has been the driving force behind many of the ecological crises we face today, including climate change, the destruction of habitat, flora and fauna, and the loss of biodiversity. It reflects a mindset that prioritizes short-term human conveniences over the long-term health of our planet, has an impact on societal structures, and raises critical questions about justice, equity, and the distribution of benefits and risks.

A famous Greek myth comes to mind here. Prometheus was a powerful Titan in Greek mythology. He defied the will of the supreme god, Zeus, and stole fire from the gods and gave it to humans. Fire, in Greek mythology, was seen as a symbol of knowledge, technology, and civilization. Prometheus's act of stealing infuriated Zeus, who saw it as a challenge to his authority and power. Had Prometheus decided to play God, and decided to give to the humans that which was not ordained? Zeus punished Prometheus. Despite his suffering, Prometheus is often regarded as a heroic figure in Greek mythology. His act of stealing fire and providing it to humanity symbolizes the quest for knowledge, enlightenment, and the advancement of civilization. But does the tale of Prometheus indicate rebellion against authority, the defiant pursuit of knowledge without due diligence? While it has been interpreted as a symbol of the human quest for progress and enlightenment, not being cognizant of consequences is the core message. Alongside the marvels, the perils have been giving us warnings, but we have ignored them. Can we continue to ignore them, even now?

The Future Scenario

Is the Worst Still to Come?

The grim reminder is that the worst is perhaps still not known! So, is it too late? Have the corrective measures implemented so far been inadequate? Were they implemented too late? Whatever has passed has left its impact; whether the learning has been adequate or not, we do not know, but what about the present and future scenarios?

There are some immediate measures needed if we do not want to be left in the hopeless situation of knowing that it is, indeed, much too late! By reimagining our relationship with the natural world and adopting more sustainable practices, we can begin to address these pressing issues, paving the way for a more resilient and biodiverse planet. We must question the viability of our lifestyle and economic perspectives, both of which prioritize short-term gains over sustainability and the actual survival of our planet.

What is done cannot be undone, but at least the remaining natural habitats must be protected. At the same time, local conservation efforts need to be directed at the already degraded ones. This requires global initiatives, such as the United Nations' Biodiversity Convention. The initiatives include minimizing waste; designing products for longevity and recyclability without the vested interest of faster turnover, which means greater profits; and finally, creating green spaces, developing urban forests, and improving the quality of life for city dwellers in particular.

The forecast of doom: The concepts of apocalypse, doomsday, or *pralaya* across various cultures and religions signify the end of the world or a major cataclysm leading to widespread destruction and the eventual rebirth or renewal of life. These narratives, whether rooted in Christian eschatology, Hindu cosmology, or Norse mythology, reflect deep-seated existential concerns about the end times, often portrayed as a result of divine judgement or cosmic cycles of creation and destruction.

Christian eschatology means "the end" or the study of the "last things". There are glimpses of "the last things" throughout the Old and New Testaments. Jesus brought up eschatology several times in his sermons. It is believed that at some point Jesus will return, and at that time, God will defeat Satan and his armies and proceed with the final judgement.

Pralaya in Hindu cosmology is the root of the cyclical process of cosmic dissolution and renewal, signifying the universe's endless cycles of creation, preservation, and destruction. This concept, from the Sanskrit meaning "dissolution", points to a universe governed by vast, cyclic time periods. The concept emphasizes a cycle that integrates destruction as a precursor to renewal. It is a reflection on the transient material world and the pursuit of spiritual liberation, highlighting the impermanence of the physical and the eternal nature of the spiritual.

In contemporary discourse, these ancient concepts resonate with growing concerns over the misuse of S&T. The pursuit of short-term gains, often driven by greed and a lack of foresight, has led to actions and developments that could indeed be steering humanity towards a modern form of apocalypse. The key areas of concern include the proliferation of nuclear weapons, which pose an immediate threat of unimaginable destruction; climate change, driven by unsustainable exploitation of natural resources, threatening the very habitability of our planet; biotechnological risks, including the potential for engineered pathogens; and the unregulated advancement of AI, which could lead to unforeseen consequences, possibly beyond our control. The urgent need is a paradigm shift towards ethical stewardship of our technological capabilities. With global cooperation, it is imperative that strict regulatory frameworks are implemented and sustainable development is prioritized. There is still hope to avert the dire scenarios reminiscent of ancient apocalyptic predictions.

But there are severe existential risks and ethical quandaries. The unprecedented social and economic divides, the new forms of warfare, the erosion of privacy, and the complete loss of traditional human interactions are horrific images. In the present scenario, some related pertinent questions have been posed that are no longer in the realm of science fiction. The answers give us an indication of the direction in which we are headed.

The Swedish philosopher Nick Bostrom is best known for his work on existential risks. He warns that as technology advances, the potential for catastrophic results increases, and it is crucial to understand the threats that could lead to the extinction of humanity or the collapse of human civilization. Bostrom's book *Superintelligence: Paths, Dangers, Strategies* is a seminal work that explores the implications of artificial superintelligence and the potential risks associated with the development of machines that surpass human intelligence.

Ray Kurzweil is a well-known American inventor, futurist, and author who has made significant contributions to the fields of artificial intelligence and technology. In his book *The Singularity Is Near,* Kurzweil presents his vision of the future, discussing the merging of human and machine intelligence, advancements in medical technologies, and the transformative impact of accelerating technologies. Kurzweil's predictions have faced scepticism, and questions have been raised about the ethics of the potential consequences of merging humans with technology.

Bostrom's cautionary views contrast with Kurzweil's optimistic vision. These diverse views point to the importance of a multidimensional approach to understanding and integrating scientific advancements.

Where are the answers? Economists, thinkers, and social scientists highlight the risk of increasing inequality. They argue that technological advancements could lead to greater wealth concentration and a widening gap between the "tech-savvy" and the "tech-deprived", also between the "haves" and the "have-nots", which would exacerbate social and economic tensions. Bioethicists and philosophers debate the implications of human enhancement technologies. A question is being raised: for whom are scientists working—business, profits, fame, or people? The role of modern science in the current context has become very crucial. It needs alignment with Om more now than ever before to evolve into meta-science. Only through such a transformation can humanity hope to mitigate the existential threats posed by unchecked greed and forge a path towards a more sustainable and inclusive future.

While researchers and scientists are undoubtedly inspired human beings as they dedicate their lives to the search for answers and ultimately come up with fantastic scientific achievements, there are those for whom the technological advancements are aimed at boosting their egos, fulfilling their greed for material comforts, and attaining self-centred objectives. Over centuries, in our myopic quest for material comforts, we have burnt fossil fuels, denuded hills and mountains to grab trees and rocks for our personal needs, and destroyed both flora and fauna in our self-indulgent activities. The existential crisis is an extreme "either-or" situation. Either we change our ways and co-exist with nature, or we press the "self-destruct" button and go extinct.

The Fourth Big Bang

We are now awaiting the fourth major explosion, not by physical agents but through the uncontrolled power of AI. The first Big Bang possibly created the universe as we know it. The second "big bang" came in the form of a chemical explosive. Alfred Nobel invented dynamite, the world's first powerful industrial explosive. He later established the Nobel Prize in an attempt to rebrand his legacy and promote peace. The third "big explosion" was the atomic bomb, a product of the Manhattan Project led by J. Robert Oppenheimer.

Its destructive power dwarfed anything the world had ever seen. On the horizon looms a potential fourth explosion: AI, fuelled by the convergence of nanotechnology, biotechnology, and information technology (NBIT). Experts like Geoffrey Hinton warn that AI, in the wrong hands, can be even more devastating than nuclear weapons. Imagine AI with the ability to manipulate matter at the atomic level, create nano-drones like shown in the film *Transformers* (nanotech), engineer deadly viruses (biotech), or create a deepfake triggering chaos in society (IT). Such powerful technologies going into the wrong hands is a real danger. Fortunately, we're not sleepwalking into this future. The United Nations Secretary-General António Guterres has established a new high-level AI Advisory Body to support the international community's efforts to govern artificial intelligence. Governments are developing proper regulations, and leading tech companies are forming alliances focused on ethical AI principles. Scientists are calling for research transparency and open discussions about potential risks. By working together, governments, corporations, and scientists can ensure that the development and utilization of AI uphold human values and safeguard against misuse.

Our idea of meta-science emphasizes the critical reflection and ethical considerations crucial in this endeavour. Through responsible development, international collaboration, a focus on meta-science, and the combined efforts of these groups, we can harness the power of AI for a brighter future. This "fourth big explosion" has the potential to usher in an era of progress, not destruction, but only if we approach it with caution and a commitment to ethical development.

Yes, there is an outcry, and yes, sustainable development goals are being defined. And yes, nations are talking about cooperation in their quest to try and reverse the situation. But unless there is genuine introspection and understanding, unless there is empathy and honesty of purpose, we cannot make any headway, and this brings us to the need for transcending from modern science to meta-science.

The collective outcry is what we are reiterating here: to pause and implement very urgent mid-course correction so that we, as the protectors of this planet and with concern for all humanity, move towards the humane, sustainable, larger cause of universal well-being. In this context, an innovative concept of Gandhian engineering, proposed by an eminent scientist, R.A. Mashelkar, is relevant. It proposes a radical transformation with "More from Less for More" (MLM) approach that prioritizes long-term sustainability while achieving greater benefits for a larger population using fewer resources. This contrasts with the conventional business focus solely on maximizing profit, at times ignoring the potential harm to nature and people.

As we contemplate the path forward, a transition from a human-centric approach to one that is more holistic and harmonious with nature is essential. We must transcend from the mere empiricism of modern science to the higher knowledge of consciousness as an evolving meta-science. It is a path that promises not only advanced knowledge and innovation but also deeper respect and harmony with the natural world, ensuring a balanced and prosperous future for generations to come.

Much like how modern science must evolve into meta-science, we propose a parallel transformation in current geopolitics. The present state of global affairs seems marred by divisions, animosity, unrest, and destructive forces. It calls for a significant shift—a move away from divisive, disillusioned politics towards a unifying, constructive, and consciously governed world, aligned with the spirit of *Vasudhaiva Kutumbakam*, where the entire planet is recognized as one interconnected family. Similarly, economics and business appear to be heading down a troubling path, driven by an unbridled pursuit of profits, greed, and an ego-fuelled bullish attitude.

In the Bharatiya Sanatana tradition, the Sanskrit terms *shubh* and *labh* are interesting concepts. *Shubh* means auspicious or just, and *labh* means profit or gain. They represent the desire for a fortunate and prosperous life. The terms "*shubh labh*" are often displayed at businesses and homes to remind us to ensure ethical business with humane values and to invoke blessings for success and well-being. Drawing from this ancient concept advocating for just, ethical, and sufficient profit, we propose a recalibration towards a holistic approach. Future business and economics should transcend towards more responsible, humane, considerate, and sustainable practices, rooted in the principles of *shubh labh*.

These shifts in geopolitics, economics, and business are reflections of a broader concern—a focus on human comfort, self-centred ambitions, and profit-driven motives at the expense of the welfare of people, the planet, and the greater universal good. In this context, we suggest that the current representation of Genome, spanning modern science, technology, social sciences, humanities, economics, geopolitics, and sustainable development goals, needs a moment of reflection. A pause is necessary, inviting an intentional mid-course correction in the manner in which modern society is moving. This correction should be guided by experiences, intuitions, insights, humane values, and consciousness as represented by Om. We call this a meta-society.

Drawing inspiration from various sources and the collective wisdom of scientists and thinkers, our meta-society narrative encourages a collective shift towards a world where the well-being of all is the basis of all innovations and developments. It's not merely a journey of scientific exploration but also a spiritual and philosophical quest, capturing the essence of our existence. And for that, somehow, somewhere, we need to take a step back. We need to rewind and go into a reverse mode, only to acknowledge, with humility, that where we stand today, with the arrogance and glorification of the achievements of the human race, *is not how it all started.*

Moving forward, we seem to have come too far from the core of our existence, the essence of our being. Over the centuries, and surely even through time immemorial, wherever our ancestors wandered, undoubtedly at some point they questioned, what is this earth, what are the planets, the universe, all the celestial bodies, and how did they come into existence? When and how did it all begin? Where did I come from? Who am I? Why am I here? And where am I going? The human mind is the most enquiring, questioning, and inquisitive equipment. These questions are recurrent, possibly *ad infinitum*, although the quality of the questions, how they were framed by generations before us, and how they are researched now are undoubtedly different. The answers have also changed with time. The whole mass of existence, which we call nature, has been acting on the human mind, on the thought of man, and as its reaction has come the question: What are these, and whence are they?

Having discussed all the fantastic developments and facing the warnings and perils, today, in the 21st century, the question before us is, do we rejuvenate our planet that has evolved over billions of years and survive, or are we heading for extinction? To find an answer to this question, perhaps we need to go into a "flashback" mode. We begin at the beginning and work our way back to where we stand today. And the first question that pops up in our minds is: when and how did it start? What was the beginning?

CHAPTER 3

The Origin of the Cosmos and Life

The universe is not only stranger than we imagine, it is stranger than we can imagine.

—J.B.S. Haldene

From the highs and lows of the advances of science and technology, we are now stepping back and probing deep into the mystery of the universe, of this cosmos, and of the emergence of life. The original questions of how, why, when, and where lurk tantalizingly in our ever-questioning minds. To address them, we must begin at the beginning and go as far back as the time of the most ancient human civilization. When we do rewind, we find the same question asked by Swami Vivekananda: "Whence is this? When there was neither aught nor naught, and darkness was hidden in darkness, who projected this universe? How? Who knows the secret?"

Scientific research and understanding continue to provide new answers to age-old questions. Palaeoanthropology has explained the history of the *Homo sapiens*, giving us the early development of modern-day human beings. By mapping the human genome, genetics has provided an explanation of our biological constitution and our evolutionary journey. The fascinating relationship between genes, environment, and behaviour—the science of epigenetics—is now being understood better. Neuroscience has delved into the human nervous system and the brain and has developed a deeper understanding of the human subjective experience and the processes of the functioning of the human brain. We saw how the most recent developments in robotics and AI have brought scientific research to a pivotal juncture where human-like cognitive faculty is represented in a non-biological form.

We are ready to embark on our exploratory journey into the vast expanse of the cosmos, tracing the story of creation and the evolution of life. We look at the scientific theories that explain the origins of the universe, such as the Big Bang, and also weave in profound philosophical insights from ancient cosmology. These perspectives are very diverse, from the astrophysical explanations of universe formation to the spiritual symbolism of Om and the representation of the divine intertwined to present a comprehensive view of the cosmic beginnings. We reflect on the progression from the initial cosmic spark to the emergence and ascent of life on Earth, culminating in the human experience within the Anthropocene epoch. We have tried to appreciate the interplay of the evolution of civilizations, science, and spirituality, encouraging a deep contemplation of our place in the vast, mysterious expanse of the universe.

Ideas of the Universe

Across diverse cultures and civilizations, mythology, religion, and cosmological beliefs have shaped the concept of the universe. Each tradition offers its own unique interpretation of the cosmos, often reflecting its cultural values, spiritual aspirations, and understanding of the natural world.

The Earliest Discoveries

To answer the questions that we have been asking since the beginning of time, it is important for us to know how the mysteries have been unveiled so far and how the process continues today. The disciplines of astronomy and astrophysics have been our guiding lights, illuminating the celestial realms with scientific inquiry and cosmic exploration. Astronomy, the study of celestial objects, phenomena, and the universe as a whole, intertwines with astrophysics, the study of astronomical objects using the principles of physics and chemistry, delving into the physical properties and behaviour of the celestial bodies. Since

Plate 1. Great Bath at Mohenjo-daro. This structure dates to the 3rd century BCE and was believed to be used for ritual bathing.

Photograph: Frederick M. Asher. Courtesy: Alice Asher
https://www.britannica.com/place/Great-Bath-Mohenjo-daro.

Plate 2. Underwater city Dwarka. During 1983–1990, the Marine Archaeology Unit of India's National Institute of Oceanography (NIO) carried out underwater excavations at Dwarka. The available archaeological evidence from the excavations confirms the existence of a city-state with a couple of satellite towns in 1500 BCE Dr S.R. Rao (1991) considered it reasonable to conclude that this submerged city is the Dwarka as described in the Mahabharata.

https://www.nio.res.in/galleries/show/dwarka

Plate 3. Dr Patwardhan points to the dinosaur in Angkor Wat. The concept of extinct species, including dinosaurs, is considered a relatively modern understanding. The first scientifically documented recognition of dinosaur fossils dates back to the early 19th century. In 1824, the British geologist William Buckland described and named the first scientifically recognized dinosaur, Megalosaurus. But how do we explain the presence of the dinosaur carving in the Angkor Wat temple built during the 9th-12th century? Photograph: Dr Bhushan Patwardhan.

Plate 4a. The Hubble Space Telescope (HST) floats gracefully above the blue Earth in December 1999 at the conclusion of HST servicing mission 3A.

https://science.nasa.gov/image-detail/hubble-space-telescope-hst-4/https://science.nasa.gov/image-detail/hubble-space-telescope-hst-4/

Plate 4b. Hubble spots a stunning spiral. Hubble image of spiral galaxy, ngc 2985. ESA/Hubble & NASA, L. Ho

https://science.nasa.gov/missions/hubble/hubble-spots-a-stunning-spiral/

Plate 4c. Hubble Telescope: The Butterfly Nebula shows what happens to a star at the end of its life, when it loses all of its gas and dust to its surroundings. Not only is this a reminder to the eventual fate of our own Sun and solar system, but Hubble's unique ability to witness this event in a star's long lifecycle sheds light on how stars evolve.

https://www.nasa.gov/wp-content/uploads/2023/03/754349main_butterfly_nebula_full_full.jpg

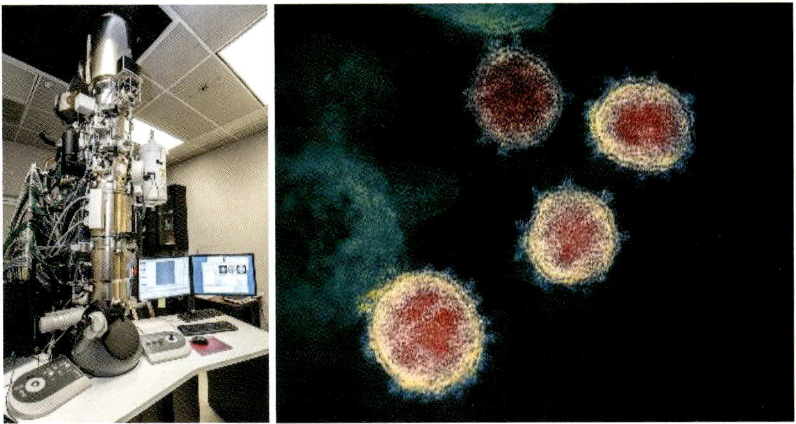

Plate 5. Transmission Electron Microscope and Corona Virus. Transmission electron microscope image shows SARS-CoV-2, the virus that causes COVID-19, isolated from a patient in the US. Virus particles are emerging from the surface of cells cultured in the lab. The spikes on the outer edge of the virus particles give coronaviruses their name, crown-like. *NIAID-RML*

https://www.nih.gov/news-events/nih-research-matters/novel-coronavirus-structure-reveals-targets-vaccines-treatments
https://commons.wikimedia.org/wiki/File:TEM_of_avian_infectious_bronchitis_virus_rotated_cropped.jpg

Plate 6. Dolly with Mother. Dolly, the first successfully cloned mammal, and her surrogate mother at the Roslin Institute, Edinburgh.

Photo courtesy of The Roslin Institute, The University of Edinburgh
https://www.ed.ac.uk/roslin/about/dolly/facts/life-of-dolly

Plate 7. A cluster of synthetic cells. Electron micrograph of a cluster of minimal cells magnified 15,000 times.

Credit: Tom Deerinck and Mark Ellisman of the National Center for Imaging and Microscopy Research at the University of California at San Diego.
https://www.jcvi.org/media-center#media-resources

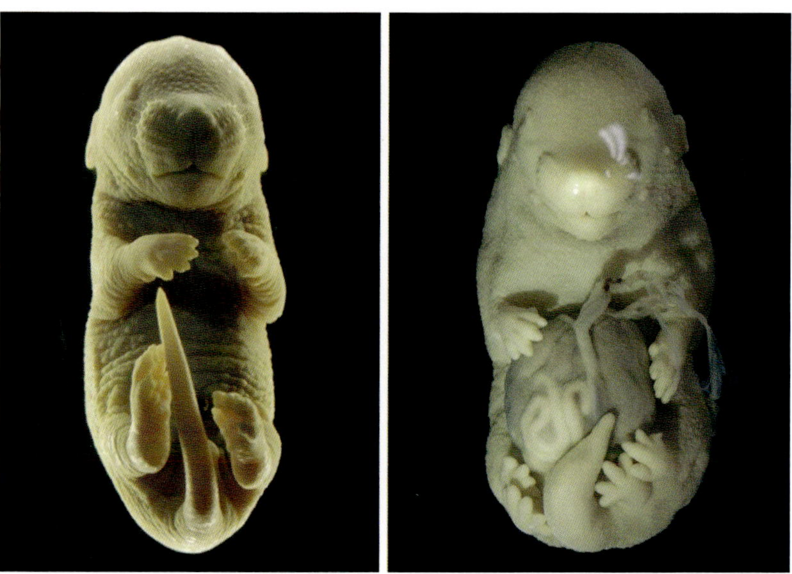

Plate 8. Six-legged mouse. A typical mouse embryo (*left*) has four limbs. Genetic engineered mouse (*right*) has six limbs (Lozovska et al. 2024).

Courtesy: Moises Mallo, MD, PhD, Instituto Gulbenkian de Ciencia, Portugal.

Plate 9. Mythological chimaera. The image of Narasimha, Lord Vishnu's avatara. Carving of Narasimha located in Hoysaleswara Temple, Halebidu, Karnataka, India.

https://commons.wikimedia.org/wiki/File:Relief_depicting_Narasimha_avatar_of_the_god_Vishnu_in_Hoysaleswara_temple_at_Halebidu.jpg

Plate 10. The three surviving re-clones at two months of age. They were derived by SCNT of adipose-derived mesenchymal stem cells (ASCs) taken from Snuppy at five years of age (Kim et al. 2017).

https://www.ncbi.nlm.nih.gov/pmc/articles/PMC5681657/

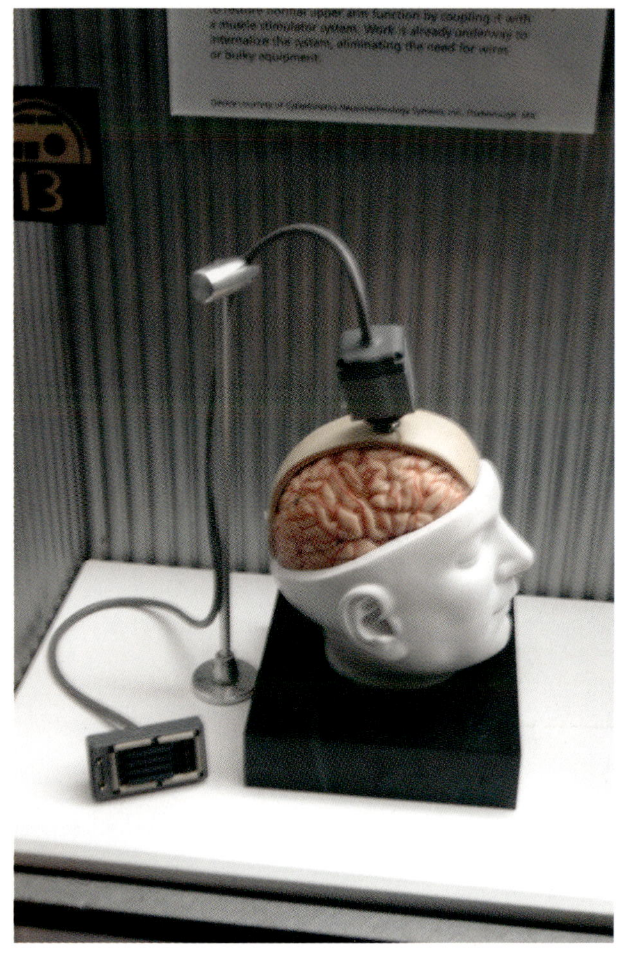

Plate 11. Dummy unit illustrating the design of a BrainGate interface.

BrainGate (2024, February 1). In Wikipedia. https://en.wikipedia.org/wiki/BrainGate
PaulWicks at English Wikipedia

Plate 12. Hubble Ultra Deep Field (UDF) of the Milky Way. The UDF is one of the deepest views of the visible universe to date; Certainly, it was the deepest when it was originally created in 2003–2004. There are approximately 10,000 galaxies in this view. A picture of the real Milky Way taken by the satellite COBE. The COBE Project, DIRBE, NASA.

The Isaac Newton Group of Telescopes, La Palma, and Simon Dye (Cardiff University)

NASA, ESA, and S. Beckwith (STScI) and the HUDF Team

Plate 13. Hubble–Webb view of the galaxy cluster. NASA's James Webb Space Telescope and Hubble Space Telescope have united to study an expansive galaxy cluster. This is one of the most comprehensive views of the universe ever taken. Located about 4.3 billion light-years from Earth, MACS0416 is a pair of colliding galaxy clusters that will eventually combine to form an even bigger cluster. 9 November 2023 10:00AM (EST) Release ID: 2023-146

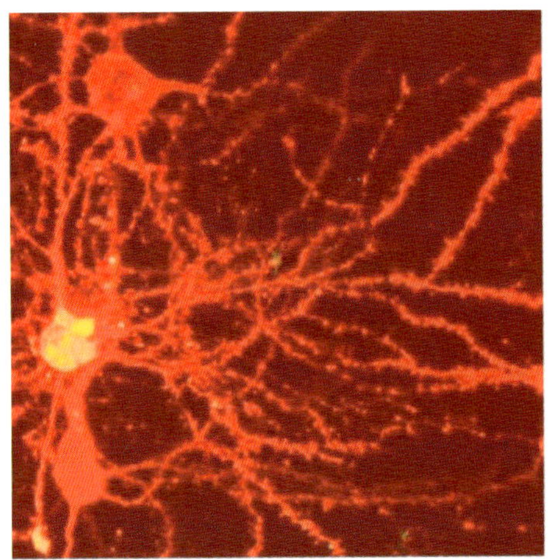

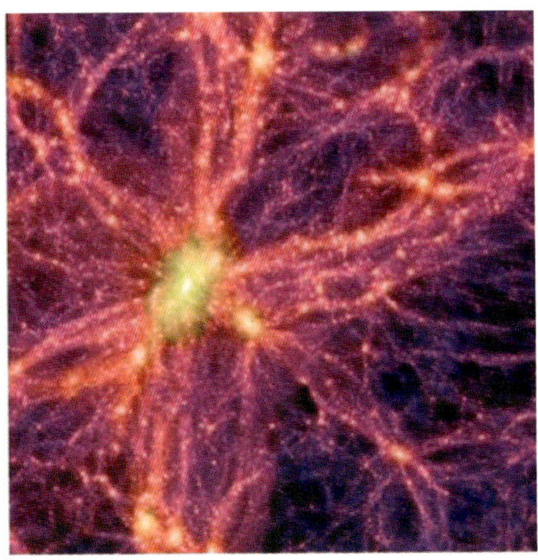

Plate 14. Galaxy and neurons. It's your guess, which is the galaxy and which are the neurons!

Plate 15. Nataraja at CERN Switzerland. The statue of the dancing Lord Shiva known as the Nataraja symbolizes Shakti, or life force. The statue is on permanent display at the European Organization for Nuclear Research (CERN).

Courtesy: Dr Anil Kakodkar
https://cds.cern.ch/record/745737?ln=en

Plate 16. Trimurti at Elephanta Caves (5th to 6th century CE), Mumbai, India, depicts Shiva in his triple role as the creator, preserver, and destroyer.

By Ronak Shah, 1990 – own work, CC BY-SA 4.0,
https://commons.wikimedia.org/w/index.php?curid=51947717

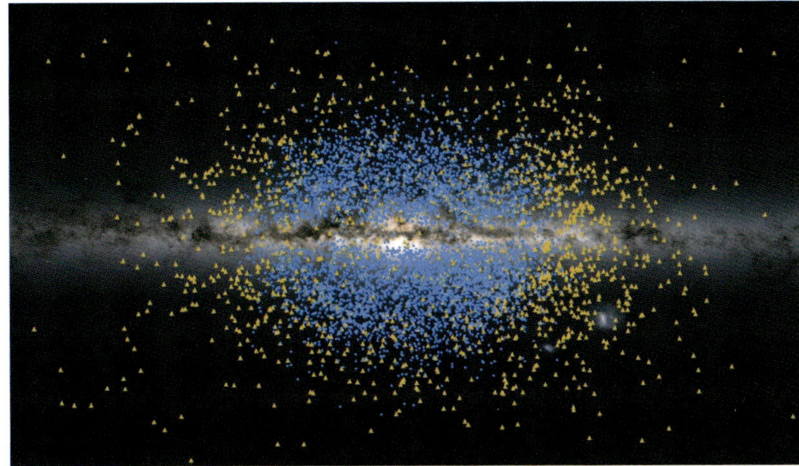

Plate 17. Shiva and Shakti, two ancient streams of stars in the Milky Way. ESA's Gaia space telescope has discovered two streams of stars named Shakti and Shiva, helped form the infant Milky Way. This image shows the location and distribution of Shakti (yellow) and Shiva (blue) stars throughout the Milky Way.

https://www.esa.int/ESA_Multimedia/Images/2024/03/Gaia_unravels_two_ancient_streams_of_stars_in_the_Milky_Way#

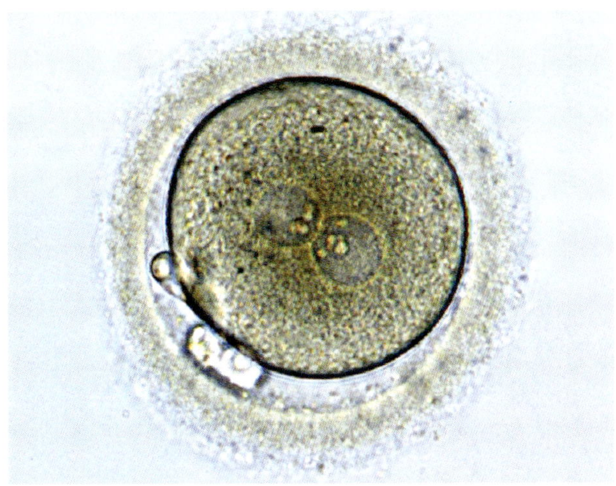

Plate 18. Human zygote formation: egg cell after fertilization with a sperm.

https://commons.wikimedia.org/w/index.php?curid=67459911
https://en.wikipedia.org/wiki/Zygote

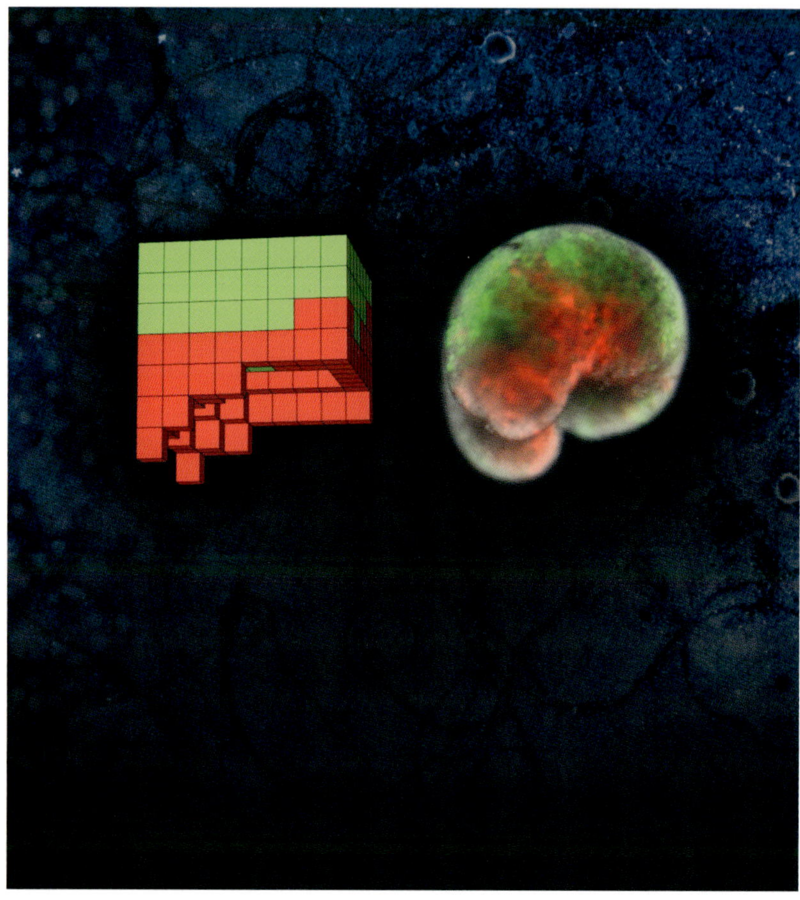

Plate 19. Xenobots. One of the over 100 computer-designed organisms. *Left*: the design discovered by the computational search method in simulation. *Right*: the deployed physical organism, built completely from biological tissue [frog skin (green) and heart muscle (red)]. The background displays traces carved by a swarm of these organisms as they move through a field of particulate matter.

Courtesy: Douglas Blackiston and Sam Kriegman.
Courtesy: Professor Joshua Bongard, University of Vermont, https://drive.google.com/drive/folders/1m2ZnuFK0BUGeozVG5K4NKPVOKoDVSmAy

Plate 20. Amazing neural network.

(a) Rendering based on electron-microscope data, showing the positions of neurons in a fragment of the brain cortex. Neurons are coloured according to size.

Credit: Google Research & Lichtman Lab (Harvard University). Renderings by D. Berger (Harvard University)

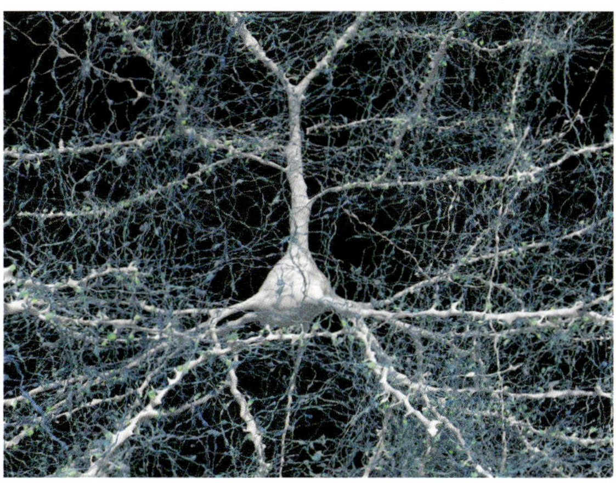

(b) A single neuron (white) shown with 5,600 of the axons (blue) that connect to it. The synapses that make these connections are shown in green.

Credit: Google Research & Lichtman Lab (Harvard University). Renderings by D. Berger (Harvard University)

Courtesy: Dr Jay Lenon

Plate 21. Samudra Manthan. A huge mural at Suvarnabhumi International Airport, Bangkok shows the legend of the churning of the ocean with the mountain Mandara and the giant Vasuki snake.

THE SUSTAINABLE DEVELOPMENT GOALS

Plate 22. United Nations Sustainable Development Goals 2030.

then, space science has grown by leaps and bounds. The earliest or most ancient observations came from the Egyptian, Babylonian, Greek, and Sindhu, Indus, and Sarasvati civilizations on the Indian subcontinent. Scientists, with their passion for knowing and discovering, have compelled the universe and this planet to reveal their secrets.

In ancient Greece, Pythagoras (circa 570–495 BCE) pondered the nature of the cosmos and proposed a spherical Earth. In the 2nd century CE, Claudius Ptolemy gave the geocentric model of the universe, in which the Earth was the centre and all other celestial bodies moved in circular orbits around it. That was apparently the beginning of research in astronomy in the West. In the 16th century, Nicolaus Copernicus proposed the heliocentric model, challenging the geocentric view that had dominated for centuries before. He asserted that the Sun was the centre of the solar system, and the planets revolved around it. This shift in cosmology led to a new understanding of the solar system. Johannes Kepler, in the early 17th century, formulated the three laws that described the orbits of the planets around the Sun. Initially, there was great resistance to this theory, but later it was accepted and recognized as the turning point in astronomy. Also, in the early 17th century, Galileo Galilei, the Italian astronomer and physicist, sparked the birth of modern astronomy with his observations of the Moon, phases of Venus, moons around Jupiter, sunspots, and the news that seemingly countless individual stars make up the Milky Way galaxy. Galileo supported the Copernican theory.

Using a telescope pointed towards the heavens, he made immensely important observations of the Moon and some of the planets and thus altered the perception of the cosmos. Galileo's telescopic observations significantly contributed to the early phase of the scientific revolution in astronomy and in our understanding of the cosmos. The fact of planets revolving around the Sun assumed the pull of gravity, which Isaac Newton talked of in the 17th century! Isaac Newton believed that the cosmos was eternal and unchanging. In *Why Does the World Exist?* contemporary journalist and popular science writer Jim Holt has taken the matter further as he states that most of the scientists of the modern era are quite comfortable with the idea that the universe has always been around. Albert Einstein was "convinced that the universe was not only eternal but also unchanging", but then his theory of relativity as applied to space-time implied that the universe could not be static; it must be either expanding or contracting.

The Egyptian astronomers divided the night sky into thirty-six sections called "decans"; each section was associated with a specific set of stars. Decans were used for timekeeping and religious rituals. They aligned their calendar with celestial phenomena. The astute astronomers of Babylon documented celestial events, positions of the planets, and eclipses, and developed a sophisticated system of astrology. They created detailed astronomical diaries, documenting celestial events, planetary positions, and eclipses, along with very keen observations of the planets and the Moon. The Babylonians charted the zodiac sign system in or around 450 BCE, with twelve astrological signs and divisions.

Vedic Astronomy: Predating all of the above developments was Vedic astronomy, and its roots can be traced to the Indus Valley Civilization. The *Vedanga Jyotisha* is considered the oldest known Vedic astronomy and astrology text in ancient Indian literature. Believed to have been composed in 1400–1200 BCE, it specifically deals with the study of timekeeping, calendrical systems, and astronomical observations. An astrological calendar aligned with the Hindu calendar is the *Panchang* scripted in Sanskrit, and it literally means "five limbs". It offers more than just dates. It is like a detailed astronomical journal, providing information on five key elements: the day, the star or *nakshatra*, the *tithi* or corresponding lunar day, the *yoga* or angular relationship between the sun and the moon, and *karana* or half-day periods.

Vedic astrology goes beyond just birth charts, incorporating complex calculations involving planetary positions, stars, the zodiac, and their interactions that are believed to affect not just human health but also major life events, personality traits, and even weather patterns. While the scientific community doesn't endorse Vedic astrology's claims, its rich tradition and intricate system continue to fascinate many. Celestial movements haven't been definitively linked to health outcomes in controlled studies. While some studies explore correlations, they're inconclusive or lack strong methodology. Case studies integrating medical astrology with traditional medicine exist, but they're anecdotal and don't definitively prove efficacy. Despite the lack of science, some find value in medical astrology, especially when combined with traditional medicine such as Ayurveda. It can promote self-awareness and a holistic approach to health, encouraging individuals to consider lifestyle changes that might serve as complementary measures. However, it's crucial to remember that medical astrology shouldn't replace evidence-based medical advice from qualified healthcare professionals.

Dividing the sky into *nakshatra*, the constellation of twenty-seven stars, and tracking celestial movements, Vedic astronomy wasn't just about stargazing; it informed calendars, rituals, and a deep connection between humanity and the cosmos. The twelve *rashi* or zodiac signs are influenced by the Moon's position at the time of a person's birth. This is the most critical aspect, which is supposed to influence the mind, emotions, and overall well-being. The idea of a seven-day week with connections to celestial bodies including the Moon, Mars, Mercury, Jupiter, Venus, Saturn, and Sun was in practice possibly for over 2,500 years. Similar systems are also found in the Babylonian and Mesopotamian cultures.

The systematic and fairly well-documented study of astronomy began possibly sometime during circa 476–550 CE, when Aryabhata proposed the heliocentric model, with the Earth and other planets orbiting the Sun. He explained the eclipses as shadows of Earth and Moon and attributed the "light" of the Moon and stars to reflected sunlight. He accurately calculated Earth's circumference and explained the occurrences of night and day. Brahmagupta (circa 598–668 CE) and many others also made observations and calculations related to planetary motion. They calculated the positions and orbits of the planets, including those of Mercury and Venus.

Admittedly, most of these early studies had cultural and mythological contexts, but they laid the foundation for later advancements in astronomy and cosmology. Merging observational data with religious and philosophical perspectives contributed to the diversity of early cosmological thought. One perennial question remains: what was there at the time of creation, or was there anything at all?

The Vast Expanse

When we speak of the universe, we often find ourselves entangled in the nuances of cosmic terminology. "Cosmos" is now synonymous with "universe", and the terms are used interchangeably. But they can have slightly different connotations, depending on the context. Initially, cosmos implied the concept of a universe that was an orderly entity, emphasizing its structured and organized nature. Pythagoras is said to have used the term "kosmos", way back in the 6th century BCE.

The universe refers to the entirety of space, time, matter, and energy that exists. It includes all galaxies, stars, planets, and other celestial bodies, along with the vast expanses of seemingly "empty" space. The term "cosmos" is also often used in a more holistic or philosophical sense, as it prompts us to contemplate the enormity of existence. It goes beyond the mere physical dimensions and may include ideas of order, harmony, and the interconnectedness of all things. It can imply a well-ordered system, not just the sum total of all physical entities. "Cosmos" is sometimes used more poetically or philosophically to create a sense of the grandeur and beauty of the universe. It invites us to peer into the cosmic abyss, where the whispers of light narrate a story that transcends the boundaries of time and space, inviting us on the journey of discovery.

In scientific contexts, especially in astronomy and astrophysics, "universe" is the more commonly used and accepted term to describe the entirety from the smallest subatomic particles to the grandest cosmic structures. What are cosmic structures? These are the galaxies, clusters, superclusters, and black holes, all arranged in twisted threadlike structures called the cosmic web. As per NASA's description, "The cosmic web forms a sort of scaffolding that cradles everything that exists within the universe."

An attempt to quantify the vastness of the universe is very humbling. The galaxies and sprawling cosmic cities emerge as fundamental building blocks within this grand superstructure. Estimates suggest that the observable universe houses over 100 billion galaxies, each a congregation of billions or even trillions of stars. Within these galaxies, countless planets orbit their stellar hosts. The observable universe spans an estimated 93 billion light years in diameter, an unimaginable expanse that challenges the boundaries of our conceptual grasp. Light, the cosmic courier travelling at the blistering speed of approximately 186,000 miles per second, is our measure of distance. A light year, the distance light travels in one Earth year, serves as a yardstick for cosmic dimensions. A light year is the distance that light travels in one year. Light zips through 5.88 trillion miles per year. How far can light travel in one minute? 11,160,000 miles. The Earth is about eight light minutes from the Sun. A trip at light speed to the very edge of our solar system—the farthest reaches of the Oort Cloud, a collection of dormant comets way, way out there—would take about 1.87 years. Keep going to Proxima Centauri, our nearest neighbouring star, and plan on arriving in 4.25 years at light speed.

Against these jaw-dropping calculations of light speed and cosmic distances, we take the opportunity to recognize some of the ancient scripts from ancient Hindu texts, which exemplify the intellectual curiosity and scientific acumen of ancient astronomers. For instance, their ability to estimate the distance between the Earth and the Sun in an era predating telescopes and advanced instruments continues to inspire awe and demonstrate the timeless pursuit of knowledge in the "ancient reach for the sun" (Box 3.1). Rather than discarding these scripts and interpretations as pseudoscience, scientists should find out the truth to know their tools and methodology.

To traverse the observable universe, light embarks on a staggering journey that spans 46.5 billion light years. These mind-boggling numbers may numb the mind, but perhaps they help us to gauge or perceive, to some extent, the sheer vastness of this universe. With the development of space science, as we gaze into the night sky, we witness galaxies whose light has travelled inconceivable

Box 3.1. Ancient Reach for the Sun

The 4th–5th century CE Sanskrit text *Surya Siddhanta* (Sun treaties), written by Latadeva, a student of the renowned astronomer Aryabhata I, boldly proposes a spherical Earth, calculates its diameter at roughly 8,000 miles (modern value: 7,928 miles), and estimates the Sun's distance to be around 96 million kilometres (with a high point of 147 million kilometres). These distance estimations, while remarkably close to modern values obtained through advanced technology, are based on the debated unit of "yojanas". The text further explores planetary movements and sizes, including the Moon's diameter at 2,400 miles (actual: ~2,160 miles) and the Moon–Earth distance at 258,000 miles (modern range: 221,500–252,700 miles). Notably, *Surya Siddhanta* incorporates sexagesimal fractions and trigonometry, crucial tools for accurate astronomical calculations. Its influence extends beyond astronomical knowledge, having significantly impacted the Hindu luni-solar calendar and even being translated into Arabic, playing a key role in medieval Islamic geography.

अचिन्त्याव्यक्तरूपाय निर्गुणाय गुणात्मने।
समस्तजगदाधारमूर्तये ब्रह्मणे नमः ॥ १ ॥

ब्रह्मणे बृहत्त्वादपरिच्छिन्नत्वाज्जगद्व्यापकायेश्वराय तस्माद्वा
एतस्मादात्मन आकाशः सम्भूत इत्यादिश्रुतिप्रतिपाद्यायेत्यर्थः।
नमः कायवाक्चेष्टोपलक्षितेन मानसेन्द्रियबुद्धिविशेषेण मत्त-

A photo of the first verse of the first page of the Hindu astronomy Sanskrit text. The edition was published in 1847. The same manuscript, and this specific verse, can be verified in Fitz Edward Hall's *The Surya Siddhanta: An Ancient System of Hindu Astronomy* published in 1859. The verse is a homage to a Hindu god, a typical way early mediaeval era Hindu texts start.

https://en.wikipedia.org/wiki/File:1st_verse_of_the_1st_chapter_of_the_Surya_Siddhanta_Hindu_astronomy,_1847_Sanskrit_manuscript_edition.jpg distances, offering us unimaginable glimpses into the cosmic past. In this journey, our galaxy, the Milky Way, is a minuscule pixel in the vast cosmic image! With an estimated 100 to 400 billion stars, the Milky Way is but one among billions (see Plate12).

The Evolution of Space Science

The colossal structures of contemporary giant telescopes are proof of the evolution of the study of space science. The early telescope, although rudimentary, was originally used to observe the night sky. It was the foundation for astronomical advancements as it became an instrument of tremendous precision and power.

In 1957, the Soviet Union launched the first artificial satellite, Sputnik 1, and that was the beginning of the space age. Since then, nations have hurtled through space with the ongoing ambition, "to go where no man has gone before". In this grand cosmic odyssey, the evolution of space science is a testimony to human ingenuity and curiosity. In the modern era, giant telescopes are the foremost means of cosmic exploration. The Keck Observatory in Hawaii stands out, housing two telescopes with mirrors spanning ten metres in diameter. These colossal instruments peer into the cosmic depths, capturing the faintest glimmers of light from distant galaxies. Advancements in telescope technology extend beyond the confines of Earth, with the space-based Hubble Space Telescope orbiting the cosmos and capturing breath-taking images of celestial wonders. The Hubble's orbit above the Earth's atmosphere eliminates distortions, offering a crystal-clear view of the universe. Recently, NASA's James Webb Space Telescope and Hubble have combined to study an expansive galaxy cluster known as MACS0416. This is the most comprehensive and colourful view of the universe, located about 4.3 billion light years from Earth (see Plate 13). One of the most staggering achievements of modern astronomy is our ability to gaze across vast cosmic distances, measured in terms of light years. The observable universe spans an estimated 93 billion light years. Through the lenses of giant telescopes, astronomers can peer back in time, observing celestial objects whose light embarked on its cosmic journey billions of years ago.

Formation of the Universe

Theories and Explanations

We are exploring the beginnings of the universe from the proposed Big Bang to the present day, acknowledging the contributions of key figures in cosmology, and examining alternative perspectives. This exploration sets the stage for understanding not just the physical universe around us but also our place within it, blending scientific discovery with philosophical inquiry.

The possible formation of the universe can be traced to an astonishing 13.8 billion years ago, a temporal canvas stretching beyond the limits of human comprehension. In this cosmic timeline, the Big Bang is popularly seen as the inaugural act, setting the stage for the intricate configuration of everything that exists. Perhaps the jury is still out on this, as many other concepts and theories are doing the rounds and hold good in different mindsets, cultures, and perceptions. For one, we have talked of the "observable" universe and have been able to share the details of estimations in terms of galaxies and stars. But what about the unobservable? This begs the question: is the universe infinite or finite? Is it steady or ever-expanding?

Infinite, Finite, or Expanding Universe?

Given its expanse, would we say the universe is infinite? To call it infinite, we are declaring that it has no boundaries and extends infinitely in all directions. This challenges the concept of a finite universe with a specific size and shape. The idea of an infinite universe continues to be a subject of debate and exploration in the scientific community.

In 1927, the Belgian cosmologist and Catholic priest, Georges Lemaître, talked of the expanding universe theory. Appealing to the new quantum theory of matter, Lemaître argued that the physical universe was initially a single particle—the "primaeval atom" as he called it—which disintegrated in an explosion, giving rise to space and time and the expansion of the universe that continues to this day. This idea marked the birth of what we now know as Big Bang cosmology. Later, this theory was supported by American astronomer Edwin Hubble's observations, as he suggested that galaxies were moving away from each other. This implies that the universe is not static but is continuously and dynamically changing.

In 1929, Edwin Hubble used the largest telescope at the time and gave the world the first observational evidence of the universe and that it has a finite age. Hubble discovered that the more distant a galaxy is from us, the faster it appears to be receding into space. This means that the universe is expanding uniformly in all directions. This not only challenged the static model, but the concept of expansion implied that the universe must have had a beginning, and if that is true, then it will not be infinite. As simple logical understanding goes, anything that has a beginning must have an end.

In 1981, Alan Guth and Andrei Linde proposed the idea of cosmic inflation. Their theory was that the early universe expanded suddenly and exponentially, or that it had a sudden and rapid burst of expansion for a fraction of a second after the Big Bang. This doesn't necessarily confirm an infinite universe, but it remains a possibility. Whether the universe is infinite or not, astronomers and cosmologists agree that we can analyse only a finite part of it (the part in which light has had time to reach us).

The subsequent expansion is often visualized using the analogy of a balloon being inflated: as the balloon expands, galaxies (represented by dots on the balloon's surface) move away from each other. This expansion doesn't require an edge or centre because space itself is stretching. The expanding universe theory aligns with the Big Bang model, which states that the universe started from a hot, dense state and has been expanding ever since. The concept of an infinite universe coexists with the expanding universe theory. A similar concept is found in the Vedic system as *ananta*, meaning infinite, without a beginning or an end. But the question remains: is the universe ever expanding? The galaxies are moving out in every direction with no signs of slowing down. And apparently, we have not seen the edge of the universe. Linked with this is the more recent concept of the accelerating universe and dark energy, which is discussed later in this chapter.

The Big Bang Theory

The cornerstone of our understanding of the formation of the universe today is the Big Bang Theory. This hypothesis has shaped the perception of the cosmos and sparked an intense and ongoing dialogue between science and faith. It challenges traditional religious narratives about the origin of the universe.

The origin of the universe is explained by a monumental event called the Big Bang, which gave birth to the universe around 13.8 billion years ago. It states that everything that we can perceive as the universe today existed as an incredibly hot "something", super-packed, extremely minute "singularity", or "point" of immeasurably dense and concentrated heat, which suddenly blew up or "exploded", and expanded. This explosive expansion is called "cosmic inflation". Inside this "object", "atom", "speck", or "thing", called "singularity", was everything that makes up our universe today: matter, energy, time, and space. Before the Big Bang, there was nothing that is understood about the universe today. It is also stated that the universe and time itself began at that very moment. Therefore, the universe had a single-pointed, explosive beginning and has been expanding since then.

It wasn't a "bang" in the sense that we understand it—a loud explosion—but rather an incredibly rapid expansion from an inconceivably hot and dense state. Now scientists believe that the Big Bang may have been more like a deep "hum" than a bang. This expansion laid the foundation for everything we see in the cosmos today, from the smallest particles to the largest galaxies. Edwin Hubble's discovery that distant galaxies were moving away from us suggested that the universe was expanding. This provided concrete evidence to support the idea of the Big Bang. Georges Lemaître independently proposed a similar idea. He theorized what he called the "primaeval atom" or "cosmic egg", suggesting that the universe began with the explosion of this dense atom, eventually leading to the formation of stars and galaxies. This is similar to the Vedic concept of *Brahmanda*, literally meaning the "cosmic egg", and it is discussed in the concept of the cyclic universe.

Lemaître's hypothesis, along with Hubble's observations, laid the cornerstone for the modern Big Bang Theory. As the universe expanded, it began cooling down. This cooling allowed the scattered matter to come together and form stars, galaxies, and all the incredible objects that we can observe in space today. The Big Bang Theory is acceptable due to its ability to explain a few observations, such as the cosmic microwave background (CMB) radiation, and the redshift. Redshift is when light from something distant gets stretched out as it travels through space and time. Blueshift is when light from something comes closer to us. Astronomers rely on redshift and blueshift to figure out how far or near the objects in space are in relation to the Earth. It's like making a map of the cosmos!

Edwin Hubble was also among the first to notice an important factor about the stars and galaxies: that almost all the galaxies were moving away, considering the redshift phenomenon. This confirmed that the universe is getting bigger and is in sync with the Big Bang Theory, which asserts that the universe expanded post explosion and is continuing to expand.

Background Hum

The CMB radiation is the faint glow left over from when the universe was super-hot at the time of the Big Bang. Hearing an inexplicable "hum", in 1964, two American radio astronomers, Robert Wilson and Arno Penzias, stumbled upon this CMB radiation. Penzias said in a statement, "When we first heard that inexplicable 'hum', we didn't understand its significance, and we never dreamed it would be connected to the origins of the universe." The two scientists were awarded the Nobel Prize in physics in 1978 for their outstanding contributions. Interestingly, as recently as June 2023, astronomers reported hearing the faint hum of gravitational waves echoing throughout the universe for the first time. For the first time ever, scientists claimed that they "have heard the 'low pitch hum' of gravitational waves rippling through the cosmos". This ever-present background noise is set off by the movement of massive objects—like colliding black holes—throughout the universe. Scientists have theorized that it has been there all along, but we haven't been able to hear it until now. The detection of the gravitational wave background was a monumental achievement by the global scientific community, led by the North American Nanohertz Observatory for Gravitational Waves (NANOGrav), involving over 170 researchers from more than seventy institutions worldwide.

Connecting the HUM to another primordial sound, AUM (Om) is relevant here. "Om" is the vibration from which the entire universe has sprung. It is often seen as the cosmic vibration that holds all the particles in the universe. Philosophically, some Eastern and Western interpretations view the universe as fundamentally vibrational. While there is no direct scientific correlation between the CMB and the sound of Om, both concepts metaphorically represent the beginnings of the universe. The CMB is a physical, measurable manifestation of the universe's origins, while Om symbolizes a spiritual or metaphysical beginning.

We are not equating hum with Aum, but we are trying to know what could have been the inspiration behind the symbolism of Om (ॐ). The detailed concept of "Om" and its universal relevance are explained in Chapter 9 of this book.

Although the Big Bang Theory is widely accepted, it's not the only explanation for the universe's origin. Alternative theories have been proposed, each offering different perspectives on the universe's formation.

Steady State Theory

The key points that relate to the idea of an eternal universe include the Steady State Theory. In 1948, Fred Hoyle, Herman Bondi, and Thomas Gold proposed the Steady State Theory, a cosmological model, as an alternative to the Big Bang Theory. This theory explains that while the universe may be ever-expanding, it maintains a constant average density at the same time. This happens because, on the one hand, matter increases with new stars and galaxies forming, while on the other hand, other stars and galaxies disappear, are swallowed up by black holes, or are not possible to observe. This theory proposes that new matter is continuously created to fill the space left by disappearing galaxies.

The universe, therefore, is in a state of continuous creation and dissolution, with new matter being formed and other matter being "dissolved" or disappearing; consequently, a constant density is maintained over time. This theory suggests that the universe has no unique starting point and is eternal. It is in direct contrast to the Big Bang Theory, which posits that the universe had a definite beginning with a massive explosion.

The Steady State Theory lost support over time as observational evidence, such as the CMB radiation, strongly favoured the Big Bang Theory. It had been almost discarded, but with the revived thought process and ideas of the multiverse and eternal universe, the steady state is still being discussed.

String Theory

Vibrating strings? This theory also resonated in the 20th century. Gabriele Veneziano, an Italian physicist, developed the first string theory model in 1968. According to this, everything in the universe, from quarks to photons, is composed of tiny, vibrating strings. These strings can manifest as different particles depending on their vibrational patterns. The string theory introduces the concept of supersymmetry, suggesting that each known particle has a supersymmetric partner. The theory accommodates modes of vibration that correspond to different particles, including those associated with gravity. This has led researchers to explore whether the string theory could provide insights into the nature of gravity and the behaviour of the universe on cosmological scales. The important aspect or goal of the string theory is to provide a unified framework that combines and explains all the fundamental forces and particles in the universe, including gravity. This was Albert Einstein's search to reconcile the theories of quantum mechanics and general relativity. But he did not quite get to it.

Researchers in Japan developed a string-theory model in 2011, which seems to explain why the universe apparently exists in three spatial dimensions. According to them, only three of the nine dimensions began to grow at the beginning of the universe, and this accounted for the universe's continuing expansion and its apparent three-dimensional nature.

Sang-Woo Kim of Osaka University, Jun Nishimura of the High Energy Accelerator Research Organization (KEK), and Asato Tsuchiya of Shizuoka University show that the string theory plausibly accounts for the universe's origin and its apparent 3D structure. According to Nishimura, we have been able to see how three directions start to expand at some point in time. The string theory is a potential "theory of everything", uniting all matter and forces in a single theoretical framework that describes the fundamental level of the universe in terms of vibrating strings rather than particles. The theory seeks to provide a unified understanding of all fundamental forces and particles, but it remains a work in progress and is a subject of ongoing research and debate in the field of theoretical physics.

Multiverse Hypotheses

It is almost impossible for scientists to say, "This is it", and "We *know* now how it all began"! Many more dedicated researchers and thinkers are continuing to dig deeper. We shall also see that while science, "Genome", is making these strides into the outer and material expanse, or, should we say, is looking further to coax the universe to unveil its secrets, "Om" is bringing science and inner awareness closer to a unification of the macrocosm and microcosm. In this context, we refer here to the concept that *the universe was, is, and always will be.*

Several physicists have veered towards this idea. Stephen Hawking, before he died, was working on a "yet to be published" paper that talked of parallel universes, also known as the multiverse. The Multiverse Theory suggests that our universe is just one of many universes, each with its own laws of physics. Thomas Hertog, a Belgian physicist who co-wrote the paper with Hawking, told *As It Happens* host Carol Off. Hertog further stated that some of these universes are completely empty, others are full of black holes, and yet others have stars, galaxies, and life. Multiverse is a term that scientists use to describe the idea that, beyond the observable universe, other universes may also exist. Multiverses are predicted by several scientific theories that describe different possible scenarios—from regions of space in different planes than our universe to separate bubble universes that are constantly springing into existence.

The Multiverse Theory suggests that some different realms or worlds exist beyond our observable universe. These realms may have different physical properties, such as size, shape, duration, and composition. The idea of a multiverse, while not exactly understood today, surprisingly has roots in some ancient civilizations!

In around the 5th century BCE, Greek philosophers Leucippus and Democritus stated that everything was formed by atoms. They proposed infinite universes beyond ours, which had been formed by the random collision of these atoms in a vast void. According to some, the idea of infinite worlds was first suggested by the pre-Socratic Greek philosopher Anaximander in the 6th century BCE. He is said to have talked of the universe as boundless, with no limits. Some scholars have argued that the ancient Mesopotamians believed in parallel or multiple worlds. based on their cosmology and mythology. The highest level was the heaven of Anu, the supreme god, the personification of the sky, the king of the gods, and the ancestors. This was followed by the heavens of Enlil and Ea, the gods of wind and water, located in the directions of the north pole and the equator, respectively. The middle level was the earth, where humans and animals lived. The lowest level, the nether world, was the abode of the devil and the dead.

The Vedic scriptures can be interpreted as hinting at a multiverse, as countless universes exist simultaneously or sequentially within this cycle. The concept of cyclical universes touches on the idea of multiple realities. There are hints of similar ideas in Egyptian beliefs about the afterlife and other worlds. These concepts are not exactly in sync with the modern scientific perception of a multiverse. But they indicate a fascinating curiosity and the possibility of realities beyond this one that prevail to this day! The multiverse theory is explored in various branches of theoretical physics, including string theory, inflationary cosmology, and quantum mechanics. But while it's a compelling idea, it is difficult to test and, for the moment, remains more speculative.

These alternative theories, while not as widely accepted as the Big Bang Theory, are important in cosmology. They remind us that our knowledge of the universe is always evolving. The theories suggest that the space and time we can observe are not the only realities. The Stanford University physicist Andrei Linde once stated that "our understanding of reality is not complete, by far.... Reality exists independently of us". The idea of an eternal universe has philosophical implications related to the nature of time, causality, and existence. It raises questions about whether time itself has a beginning and whether the concept of a "first cause" applies to the universe as a whole.

What is relevant is that if the multiple universes do exist, then they are separated from ours, unreachable, and undetectable by any direct measurement (at least, so far). And that makes some experts question whether the search for a multiverse can ever be truly scientific.

The Multi-world Concept

There is a difference between the multiverse hypothesis and the idea of multi-worlds. In many traditions, the universe is a multi-layered or multi-world realm, with various planes of existence representing different levels of consciousness, spiritual attainment, or divine presence. In Western culture, the expression "cloud nine" is often used colloquially to convey a state of euphoria or bliss. While not rooted in a specific religious tradition, the concept evokes a sense of transcendence and upliftment, suggesting a state beyond earthly concerns and limitations. In Judeo-Christian and Islamic traditions, the concept of heaven typically encompasses multiple levels, or "heavens", with the "seventh" heaven being the highest and most sublime. It is understood as the abode of God or the dwelling place of angels and saints, where souls experience eternal bliss and communion with the divine. In Islam, *jannat,* or paradise, is described as a lush garden of everlasting delight reserved for the righteous and faithful believers. It is seen as a place of exquisite beauty, abundant blessings, and eternal happiness, where inhabitants are rewarded for their piety and good deeds. The concept of multiple worlds, according to some interpretations of Islam, suggests different realms or worlds that exist beyond our observable universe. These include the unseen universe (alam al-ghaib), which contains angels, jinns, heaven, and hell; and the observable universe (alam al-shahood), which contains the physical creation. While the Bible does not explicitly mention the existence of other universes, some Christians have speculated that God may have created other worlds or realities besides ours. The cosmology of Taoism, one of the most popular schools of thought in China, talks of multiple worlds that interact with each other as they co-exist. These universes exist at different frequencies or vibrations.

Buddhism talks of multiple worlds or realms that exist within the cycle of the repeated birth and death of living beings. The concept of the multiverse is the "Three Worlds" or "Three Realms", and it emphasizes the vastness and diversity of existence within *samsara*, the cycle of birth, death, and rebirth. Human beings move through these realms based on their *karma*, the actions they perform, and their level of spiritual development. The ultimate goal in Buddhism is to attain liberation (*nirvana*) from the cycle of rebirth and transcendence of all realms of existence.

Hinduism reflects a rich and multifaceted understanding of the cosmos, with various realms, dimensions, and planes of existence interwoven into a vast cosmic order governed by divine principles and cosmic laws. In Hindu cosmology, the universe consists of countless forms and multiple *loka*s or realms. But, similar to Buddhism, it is essentially divided into three main *loka*s: the earthly realm, the celestial realm, and the netherworld. Each is inhabited by different beings and is governed by its own set of laws and principles.

These concepts provide frameworks for understanding the complexity, diversity, and hierarchical cosmic structure. These worlds are governed by their particular laws and are further influenced by individual actions and choices. In most Indic traditions, the law of *karma*, the law of cause and effect, determines the destiny of the human soul, directing in which realm it will be reborn.

Cyclic Nature of the Universe

The cyclic model, or the Oscillating Universe hypothesis, is a variant of the Big Bang Theory. The cyclic model suggests that the universe passes through successive cycles of expansion and contraction (or collapse). At the end of the collapse phase, the universe becomes a small volume of great density. The universe is therefore swinging between the Big Bang and the "Big Crunch".

A very important insight into the cyclic nature is that of creation, sustenance, and destruction (or dissolution) described in Vedic cosmology. The Rig Veda describes a cyclic or oscillating universe in which the cosmic egg, which contains the entire universe, expands out of a single concentrated point before subsequently collapsing again. The universe cycles infinitely between expansion and total collapse.

The universe is created, destroyed or dissolved, and re-created in an eternally repetitive series of cycles called the *yugas*. Four *yuga*s comprise a *kalpa* (one day of Brahma, the creator). And these keep circulating, one after the other. The universe, once manifested or projected, endures for about 4,320,000,000 years, one *kalpa*, and is then destroyed by the elements, fire or water. The concept of "destroyed" does not imply going into nothingness; it is a transformation. It implies going from the manifested or projected to the unmanifested or getting dissolved into equilibrium. At this point, Brahma rests for one night, just as long as the day. This process of projection and dissolution repeats for 100 Brahma years (311 trillion or 40 billion human years), which represents Brahma's lifespan. Undoubtedly, this is considered a part of mythology, but admittedly, even imagining such details requires profound wisdom. We could connect the unmanifested or equilibrium state to that micro-mini second before the Big Bang.

Nobel Prize winner physicist Roger Penrose proposed that the universe undergoes an eternal cycle of expansion and contraction, with each cycle resulting in a new big bang and the creation of a new universe. This theory offers exciting new insights into the nature of the universe and challenges conventional beliefs. In a way, this syncs the Big Bang and cyclic theories with the conviction of creation, sustenance, and dissolution.

Eminent astronomer Carl Sagan stated, "The Hindu religion is one of the world's great Faiths dedicated to the idea that the cosmos itself undergoes an immense number of deaths and rebirths. It is the only religion in which the time scales correspond, no doubt 'by accident', to those of modern scientific cosmology. The cycles run from our ordinary day and night to a day and night of Brahma the creator, 8.64 billion years long. It is longer than the age of the Earth, the Sun, and half the time since the Big Bang." The physicist Fritjof Capra said, "This idea of a periodically expanding and contracting universe, which involves a scale of time and space of vast proportions, has arisen not only in modern cosmology, but also in ancient Indian mythology. Experiencing the universe as an organic and rhythmically moving cosmos, it was possible to develop evolutionary cosmologies, which come very close to our modern scientific models." The days and nights of Brahma present the view that the universe is divinely created; it is not strictly evolutionary, but it is, in fact, an ongoing cycle of birth, death, and rebirth of the universe.

Extraterrestrial life: This also raises a key question: Are we alone in the universe? The vastness of space is no longer silent on the question of alien life. Renowned astrophysicist Lisa Kaltenegger, director of Carl Sagan Institute, isn't shy about the odds, stating it's "almost certain" there's other life out there, considering the sheer number of galaxies. Even Avi Loeb, another leading astrophysicist, argues the question isn't whether we're alone, but when we'll discover evidence. He urges us to move beyond simple belief and treat the search for extraterrestrial intelligence as an imminent scientific truth. This growing optimism isn't unfounded. Astrobiologist Sara Seager echoes the excitement, suggesting we're on the cusp of an era that could definitively answer this age-old question. With advancements in telescopes and the discovery of planets potentially teeming with life, the whispers of "Are we alone?" are getting louder, and top scientists are eagerly listening.

A 2024 Netflix series *3-Body Problem* poses the same question! Gazing into the cosmos, a timeless question persists: Are we alone? This existential wonder sparks scientific inquiry and ignites the imagination of storytellers. The *3-Body Problem* series takes Rachel Carson's profound message in *Silent Spring* to heart: "In nature, nothing exists alone." It explores the potential challenges of encountering vastly different alien civilizations, particularly those with hostile intentions. Imagine encountering intelligent life with advanced technology and unknown values. What ethical tightropes would we walk? Would we prioritize survival at the expense of our core beliefs? How can we prepare for, or even reason with, such an alien society?

Science fiction today, as in *3-Body Problem*, compels us to consider the very real possibility of future contact with alien life. Carson's quote emphasizes the interconnectedness of all living things, a concept that extends beyond Earth. If intelligent life exists elsewhere, it too is likely part of a complex, interdependent galactic ecosystem. Just as our actions on Earth ripple outwards, encountering an alien civilization could have unforeseen consequences for both parties. Understanding this interconnectedness is paramount for navigating potential first contact. The series underscores the importance of international cooperation and a unified human approach. By sharing knowledge and resources, we can potentially prepare for different scenarios and navigate these ethical dilemmas together. Perhaps, by embracing our shared place in the universe and fostering a spirit of exploration and understanding, we can ensure a future where humanity thrives alongside, not against, any intelligent life we may encounter.

Speculation about the existence of extraterrestrial life is very rife. The idea is no longer a far-fetched imaginary pursuit as in the film *ET* or others that have followed. With the insatiable human curiosity and effort, studies and research continue in astronomy and space studies: "To explore strange new worlds, to seek out new life and new civilizations, *to boldly go where no one has gone before*," as was the famous introductory statement by Captain Kirk of Starship *Enterprise* in the television series *Star Trek*! The deeper scientists probe into the secrets of the universe, the more believable it is that there could be life in other cosmic dimensions.

Now that Mars is reachable, the existence of water there has raised the potential for finding microbial life. Space missions are consistently gearing up to seek evidence of life on any other celestial object. Moons with subsurface oceans raise the possibility of life beyond Earth. While the conditions for life on Earth are unique, given the vastness of the universe and the diversity of the planetary systems, no one can deny the likelihood of finding other environments that could be equally conducive to life. This quest to unravel the mysteries of the cosmos and find life elsewhere is among the most challenging in the realm of science and exploration.

Macrocosm and Microcosm

One key factor that connects the cosmos with us as individuals is that *the microcosm is the same as the macrocosm*. This relates well with the aphorism "*Anor Aniyan Mahato Mahiyan*" from the Katha Upanishad, which explains the nature of Universal Consciousness (referred to as God in common parlance) and means "smaller than the smallest, greater than the greatest", giving us a profound paradox. This conveys the non-dualistic view that the individual and Universal Consciousness are inseparable. In 1890, during his travels to the Himalayas, Swami Vivekananda sat for meditation under a tree. Here he experienced the one-ness of the universe, of the microcosm and macrocosm, and realized that in the microcosm of the body exists everything that is there in the entire universe. He said, "I have just passed through one of the greatest moments of my life. Here, under this Peepal tree, one of the greatest problems of my life has been solved. I have found the oneness of the macrocosm with the microcosm, in the microcosm of the body...." Explaining further, it is very logical to assert that the entire universe cannot be the result of "nothing". The principle of cause and effect is important here because nothing happens without a cause, and that cause is the effect of a prior cause. This theory is being echoed by others: that this universe emanated from a preceding universe—a universe that existed in a fine form or an unmanifested state. The example of a huge tree coming out of a seed, which is almost invisible, is very relevant. The entire unmanifested tree exists in the tiny seed as it grows and becomes manifest. Similarly, this universe, this cosmic creation, has manifested out of a universe existing in an unmanifested form. It was present in the cosmic fine universe. This is what the theory of evolution is: coming out of the fine and becoming gross. In the mini-microsecond before the big bang, the universe existed in its unmanifested, fine form. "Every evolution is preceded by an involution…." If the seed is the parent of the tree, then another tree is the parent of that seed. The little cell, which later becomes the progenitor, was involved and, in time, evolved. "If this is clear, we have no quarrel with the evolutionists, for we see that if they admit this step, instead of their destroying religion, they will be the greatest supporters of it."

In ancient cultures, the concept of the macrocosm and microcosm was a fundamental principle. The idea suggests a harmonious and intricate relationship between the universe at large (macrocosm) and individual entities (microcosm). The Sanskrit aphorism "*Yatha pinde tatha brahmande*" from the Yajur Veda, literally means "whatever is in the microcosm is also in the macrocosm." *Brahmanda* (the macrocosm or the cosmic egg, as mentioned earlier) and *pinda* (the microcosm or a cell) symbolize this connection, emphasizing a complex web of dependencies among various forms of life and non-living matter. The microcosm, in this context, is represented by living forms ranging from microscopic cells to complex organisms, including both plant and animal species. Conversely, the macrocosm encompasses non-living elements such as atoms, dust, geological formations, bodies of water, and celestial objects like stars and galaxies.

An interesting parallel to this continuum can be drawn between the neurons in the human brain (a microcosmic element) and the stars in a galaxy (a macrocosmic element). Some speculate that the number of neurons in the brain is akin to the number of stars in a galaxy, highlighting an intriguing resemblance between the two (see Plate 14). This interplay is further elaborated in the *Mahabhoota* theory, according to which the primordial elements—earth, water, fire (energy), air, and ether—are foundational to both the microcosm and macrocosm. Therefore, all entities, living or non-living, are constituted by these elements.

Modern science often adopts a reductionist approach, particularly in fields like biology. This perspective focuses on dissecting living and non-living entities into their smallest parts—atoms, molecules, and cells—to gain a deeper understanding. However, this approach sometimes overlooks the interconnectedness and broader context of the whole. These perceived limitations of the reductionist perspective have led to a resurgence in thinking, and contemporary scientists propose a relatively new branch known as "systems biology", which highlights the importance of considering the whole biological system in addition to its parts. This seemingly unique systems approach is an extension of traditional holistic principles. It acknowledges that understanding the parts in isolation may not always provide a complete picture of the whole system. Essentially, both ancient holistic wisdom and the thinking of modern systems are critical for us to understand the complex nature of life and the universe.

Cosmic Energy

The interplay of science, mythology, and spirituality allows us to approach the universe from various angles. Scientific knowledge gives us a framework for understanding the physical world, while mythology offers a symbolic lens through which to contemplate the magnificence of existence. In the scientific sense, cosmic energy refers to the fundamental forces and energy that originated at the beginning and continue to drive the expansion and evolution of the universe. In various spiritual traditions, or in philosophical and mythological contexts, it refers to a universal and vital life force that permeates the cosmos and sustains all living beings, connecting all aspects of the cosmos, from its creation to its sustenance and its eventual dissolution. These perspectives give a comprehensive view of the dynamic and intricate nature of the universe.

The presence of Nataraja, the cosmic dancer, at CERN (the European Organization for Nuclear Research) sparks a thought-provoking connection between these ancient ideas and the scientific quest to understand the universe (see Plate 15). In Hinduism, this form of the dancing Lord Shiva is known as the Nataraja and symbolizes Shakti, the life force energy. As a plaque alongside the statue explains, the belief is that Lord Shiva dances the universe into existence, motivates it, and will eventually extinguish it. Carl Sagan drew the metaphor between the cosmic dance of the Nataraja and the modern study of the "cosmic dance" of subatomic particles known as Tandava. This symbolizes the cosmic cycles of creation, preservation, and destruction, or the cycle of birth and death. The rhythmic movements of the dance can convey the inherent harmony and order maintaining the cosmic balance, despite the apparent chaos and destruction. Across vast cultures, philosophies weave tales of duality and harmony.

In China, the swirling dance of yin and yang captures the essence of the universe. Yin, the darkness, cradles a seed of yang, the light, and vice versa. This balanced interplay forms the very fabric of existence. Similarly, Hinduism's Shiva and Shakti represent a divine couple. Shiva, the still consciousness, embodies the potential for creation, while Shakti, the dynamic energy, brings that potential to life. But this dance needs a vital spark, the life force that animates all things. In China, it's called Chi, and in India, it's known as *prana*. These terms, though arising from different traditions, capture the same essence—a subtle energy that flows through the universe and animates every living being. Just as yin and yang or Shiva and Shakti represent opposing yet complementary forces, Chi or *prana* are believed to flow in a balanced way. An imbalance can lead to disharmony, much like a disruption in the cosmic dance. So, whether it's the swirling symbol of yin and yang, the divine couple of Shiva and Shakti, or the unseen flow of Chi or *prana*, these concepts remind us that the universe thrives on the interplay of seemingly opposite forces. It's a dance of balance where darkness gives birth to light, stillness fuels creation, and life itself pulsates through this harmonious exchange.

This symbolism offers a fascinating intersection of science, mythology, and spirituality. From a scientific perspective, the universe's existence hinges on a delicate balance of forces. Quantum mechanics and general relativity struggle to reconcile, and unifying these forces is crucial to understanding the universe's initial moments. The Nataraja is symbolic. While modern science doesn't subscribe to literal mythology, and neither does CERN, the cosmic dance statue of Nataraja on its campus in Switzerland is a subtle reminder of the ongoing quest to unravel the mysteries of the universe.

Who Created the Universe?

The origins of the universe and life have been a subject of fascination and speculation in various cultures across the world, which explain creation and evolution, each with its own unique myths. The creation myths, while varied, share common themes that provide symbolic and philosophical insights into understanding the origin of the universe and of life. They often represent a deep connection between humans and the natural world, acknowledging the intricate relationships between all forms of life. The myths not only offer insights into the early understandings of the universe but also reflect the richness of human culture and belief.

The key question is: who or what triggered the beginning of the beginning? The Big Bang Theory has led to many discussions and debates, especially among cosmologists and theologists. Since it projects a universe that is consequent to an explosion of very concentrated, packed "singularity", an extremely dense single point of intense energy, how did it get there? What caused it to get set off? If the universe is the result of that one explosion—a big bang—then who triggered it? Was it an extraordinary source of explosive energy, or the hand of a supreme, universal power, or God? When the question is raised as to what caused the Big Bang, as far as science goes, the most common answer is, "We don't know." There are undoubtedly theories as to why the Big Bang took place when it did, if we accept that that was indeed the beginning. But the question of why the universe exists, as Jim

Holt has queried and discussed at length in his book *Why Does the World Exist* or why and how the conditions get set in a particular way, involves the answers that meta-science has to probe, and philosophical and metaphysical considerations have to open many more doors of deep inquiry.

Almost all religions have an explanation or theory about creation. Different faiths and religions have tried to explain the mystery of how the universe came into being with the one assumption, "Only a creator can create!" They are, therefore, almost in agreement with the Big Bang Theory to the extent that a sudden explosive origin of the universe must have been caused by something or created by some source, and that source is taken as the "First Cause", popularly referred to as God.

If we study theistic points of view, more specifically, then this is what we understand. The Abrahamic religions (Judaism, Christianity, and Islam) generally view the universe as having a definitive beginning and end as part of a divine plan. The concept of an end time or Judgement Day is central to these faiths. Most Christians find it difficult to accept that the universe is about 13.7 billion years old because, according to the Book of Genesis, this is a much younger Earth. However, some thinkers believe that both concepts are in sync. They believe that God used the Big Bang and the whole process of cosmic evolution as part of His plan. They say science doesn't have to rule out the idea of a divine creator; it actually helps us understand how creation happened. It's like combining science and faith, as explained by Jim Holt.

Christianity's Book of Genesis explains in detail the process of creation. It states, "In the beginning, God created the heavens and the earth. The earth was formless and empty, darkness was over the surface of the deep, and the Spirit of God was hovering over the waters. And God said, 'Let there be light' and there was light." As is given in the Old Testament, the Jews believe that there is this one single God who created the universe, but there seems to be no specific detailing of how God did it. According to the Holy Quran, Allah created the universe in six days. But it is also stated in the Quran that one day is equal to 50,000 years on Earth. Muslims interpret the "six days" of creation as six periods or aeons, almost like a periodic table.

Buddhists believe that the universe is cyclical. They do not look into the concept of the beginning of anything, and therefore, for them, the universe is eternal, ongoing, and yet constantly changing. Jainism also does not believe in a creator God. According to Jain beliefs, the universe is eternal; it has always existed.

As one of the popular Chinese philosophies, Taoism does not propagate a single creator. It believes that at the beginning, there was only chaos, or a void, called Wuji. Then, from that void, the cosmos came into being or it evolved. The Australian Aborigines believe in a dreamtime, the time during which the universe was created. They believe that everything that we see—the seas, rivers, water bodies of any kind, rocks, hills, mountains, and, of course, the flora and fauna was created by celestial beings, the spirits of their ancestors.

A divine heroic figure separates the sky and the earth, the heavens and the underworld. This is the belief in the Egyptian myth of Shu, the Greek myth of Uranus and Gaia (heaven and earth); the Japanese creation myth is of Izanagi and Izanami (the eighth pair of brother-sister gods who appeared after heaven and earth separated from the original chaos); and the Maori myth is of Rangi and Papa (the primordial sky god and earth mother).

Humans and animals were created from various materials, such as clay, dust, blood, bones, or plants. These concepts predominate in the creation stories of the Biblical Adam and Eve, the Greek myth of Prometheus and Athena, and the Mayan myth of Popul Vuh, among others. A primordial paradise, or golden age, existed in the beginning, where humans and animals lived in harmony and abundance before the fall or the flood. This belief is again found in many traditions, including the Hebrew Garden of Eden and the Greek Ages of Man.

Vedic Approach

The Vedas provide a much deeper description of the origin of the universe. The *Naasadiya Sukhtam* (the Hymn of Creation), part of the Rig Veda, offers a depiction of what was possibly the origin of the universe. It is a beautiful poetic rendition of the conditions that abound before the initiation of creation itself. It is a narrative of creation or a reflection on the concept of existence or non-existence (Box 3.2). We rarely come across an ancient scripture that accepts ignorance as gracefully as the *Naasadiya Sukhtam.*

Mythological Trinity

There is no single narrative of the story of creation. Hinduism offers a symbolic representation through the Trimurti (three images)—represented by the gods, Brahma the creator, Vishnu the preserver, and Shiva the destroyer, or more realistically, the recycler or transformer (see Plate 16). Brahma works with Vishnu and Shiva to maintain an unending cycle of universes. This Trimurti embodies the initial spark, the ongoing processes, and the inevitable change inherent in the universe's existence. All three are aspects of the unmanifested universal consciousness. Shakti, the divine feminine force, adds another layer, potentially reflecting the unknown force or interplay of forces like dark matter and dark energy.

The Trinity in Christianity describes the nature of God as a unity of three distinct persons within one divine essence: the Father, the Son (Jesus Christ), and the Holy Spirit. The Father is seen as the source or origin of all things, the creator of the universe, and the ultimate divine authority. The Son, Jesus Christ, both fully divine and fully human, the human incarnate of the creator, of God, lived among humanity and ultimately offered salvation through his death and resurrection. The Holy Spirit is believed to be active in the world, guiding, empowering, and sanctifying all those who believe. The Spirit pushes the sinners, reveals the ultimate truth, and helps believers live a life of faith.

Whatever the belief, all narratives are very pointed indicators of the infinite but complex nature of the universe. All we can say is that perhaps the universe is not ready to be taken for granted, nor is it ready to reveal itself in its entirety, at least not in our present era. The big question remains: will the universe keep expanding *ad infinitum* until the "cooling down"? Or is there this eternally existing magnificence, which is expanding and contracting, going from the manifested to the unmanifested and back to the manifested?

Box 3.2. The Hymn of Creation — *Naasadiya Sukhtam*

Existence was not then, nor non-existence,
The world was not, the sky beyond was neither.
What covered the mist? Of whom was that?
What was in the depths of darkness thick?
Death was not then, nor immortality,
The night was neither separate from day,
But motionless did *That* vibrate
Alone, with Its own glory one —
Beyond *That* nothing did exist.
At first in darkness hidden darkness lay,
Undistinguished as one mass of water,
Then *That* which lay in void thus covered
A glory did put forth by *Tapah*!
First desire rose, the primal seed of mind,
(The sages have seen all this in their hearts
Sifting existence from non-existence.)
Its rays above, below and sideways spread.
Creative then became the glory,
With self-sustaining principle below.
And Creative Energy above.
Who knew the way? Who there declared
Whence this arose? Projection whence?
For after this projection came the gods.
Who therefore knew indeed, came out this whence?
This projection whence arose,
Whether held or whether not,
He the ruler in the supreme sky, of this
He, O Sharman! knows, or knows not
He perchance!

Translation by Swami Vivekananda

[Vivekananda, Swami, *Complete Works*, 25th Impression, "The Hymn of Creation". Advaita Ashrama, Kolkata, Vol. 6, 2005]

Science, Faith, and Spirituality

The debate on how our universe came to be continues between science and spirituality. Science focuses on empirical evidence, and this is constantly changing and evolving as new evidence comes up. At the same time, just because something cannot be proved, that doesn't mean it is scientifically incorrect. Faith deals with questions that science can't answer yet, like "why are we here?" Science and faith can actually complement each other. Science explains the "how" while spirituality explores the "why". A balanced perspective that considers both the wonder of science and the meaning offered by faith can give us a richer understanding of the universe. It is important, however, to distinguish between faith and blind faith. It is the difference between having the freedom to question and explore within a belief system versus an unquestioning acceptance of ideas without evidence or analysis.

It is also important for us to know the difference between religion and spirituality. Although the two are used interchangeably, they are not the same. Religion refers to organized systems, which include customs, rituals, ceremonies, and belief in a higher power. Spirituality is entirely a personal perception. It is a belief based on individual search, experience, and conviction. There is a deeper sense of inner reality, and it does not necessarily imply following any doctrines or practices. An atheist can be as spiritual as a religious person can be, whether he or she is a fanatic or a fundamentalist. Scientists who search deeply into the mysteries of this universe and try to look into its magnificent expanse are most likely to have a rich inner spirituality that gives them very wide dimensions of understanding and sensitivity.

The connection between science and spirituality is extremely interesting because they are often complementary ways of understanding the world. Science focuses on the external, physical, and verifiable aspects of reality; spirituality goes deeper into the internal, subjective, and experiential aspects. Both are on the quest for truth. Science demands empirical evidence; spirituality draws on intuition, contemplation, and personal experience. But scientific discoveries, especially in areas such as cosmology and evolutionary biology, have challenged traditional philosophical and theological views. The Darwinian theory of evolution, for example, presents a significant challenge to the Biblical account of creation, leading to ongoing debates between creationism and evolutionary biology.

As per the theory of Evolutionary Creationism, God created this universe and life by adopting the process of evolution, which means there is no conflict between science and faith. Evolutionary creationists accept the scientific evidence for the age of the universe, the common ancestry of life, and the mechanisms of natural selection and genetic mutation. The devout Christian Asa Gray accepted natural selection as the process of the evolution of new species; he believed it was as per the plan of the divine creator. Gray was the first proposer of theistic evolution, stating that natural selection is God's way of giving direction to the world that he created. Islam sees no conflict between the creation of the world by Allah and the evolution theory. Jewish and Christian believers are very clear in their conviction that God brings in these changes in the course of his creation, and thus the evolution of species is God ensuring "change over time".

What, then, is the future of the universe? Both scientifically and theologically, some concepts do exist and remain subjects of intense discussion. The question here is: if it is not expanding endlessly, is it spiralling, or does it have an end?

Future of Cosmic Existence?

Is there an end to the universe? If yes, then how? In cosmology, several theories hypothesize how the universe might end, often depending on the rate of expansion of the universe:

- Big Crunch: If the gravitational pull of the universe's mass overcomes its expansion, it could lead to a reverse Big Bang, known as the Big Crunch. The universe would collapse into a singularity.
- Heat Death or Big Freeze: As the universe continues to expand, it will gradually cool down, leading to a state of no thermodynamic-free energy. That would make it impossible for any form of life to exist, resulting in a cold, dead universe.

- Big Rip: If the rate of the universe's expansion accelerates continuously, it could lead to a Big Rip, where galaxies, stars, planets, and eventually atoms are torn apart.

The timeline for these events is typically in the range of tens of billions to trillions of years, making them largely theoretical and not something that can be empirically observed or verified in the foreseeable future. Science provides measurable and theoretical models based on the current understanding of physics and cosmology. Philosophy raises questions about the implications of these scientific theories, looking into their deeper meanings and existential ramifications. Spiritual and mythological traditions offer a more cyclical view of the universe's existence, reflecting a belief in the eternal nature of creation and destruction.

In the final analysis, despite all explanations, the beginning and possible end of our universe continue to remain a big mystery. We are missing the complete understanding of quantum gravity, the theory that is needed to describe the universe at its most fundamental level during the Big Bang. This missing piece hinders our ability to explain the singularity and the theory of inflation, which proposes a period of rapid expansion shortly after the Big Bang. Even within the observable universe, vast mysteries remain. Dark matter, an invisible force that makes up about 85 per cent of all matter, exerts a gravitational influence on visible matter despite its unknown composition. Similarly, dark energy, a mysterious force causing the universe's accelerated expansion, remains poorly understood. Both leave significant gaps in our cosmological model. The ultimate fate of the universe—whether it will continue expanding forever, collapse in a "Big Crunch", or succumb to something unknown—is another question science cannot definitively answer. Despite these challenges, the quest for answers continues. Several perspectives see this creation as a cyclic and immanent existence, projection, or perception, and of course, the big debate on the creator of the creation continues. These unresolved problems ignite curiosity, inspire further research, and push scientists and those of us who are invested in "knowing", to refine and expand our understanding. This is the "evolving journey of modern science to meta-science".

In trying to understand the creation of the universe, we also wonder about our place in it! We question how and why we are here and what life is all about. Scattered as we are across this earth, we need to get to the story of our unique planet. Our connection with the cosmos is through our presence on the planet. This is the core of our existence—our life on planet Earth.

Planet Earth and Life

Trying to understand how the universe began is overwhelming, and now, wrapping our minds around the evolution and uniqueness of our solar system and of our planet Earth is equally mind-boggling. After the monumental event of the Big Bang or the rapid expansion, the universe gradually began to cool and take shape, leading to the formation of stars, galaxies, and eventually our solar system and its planets. Even if the universe is eternal, among its celestial bodies, our planet Earth has a unique story.

Evolution of the Earth

The evolution of the planet and, along with that, the life forms from the time of its formation to the present era is fascinating as we try to see how the process developed for the final ascent of man.

Our solar system formed about 4.5 billion years ago from a dense cloud of interstellar gas and dust. The cloud collapsed, possibly due to the shockwave of a nearby exploding star called a supernova. When this dust cloud collapsed, it formed a solar nebula—a spinning, swirling disc of material. This collapse initiated the process that gave birth to the Sun and its surrounding planets, including the Earth. In 1755–56, Immanuel Kant and Pierre-Simon Laplace proposed the nebular hypothesis that the solar system was formed from a rotating disc of gas and dust. As the solar nebula collapsed, the central region formed the Sun, while the surrounding material coalesced into protoplanetary discs. Within these discs, small particles collided and merged to form planetesimals, which eventually formed planets. The Earth's layers, such as the core, mantle, and crust, developed through complex geological processes. While the nebular theory has been criticized, research has continued into the 20th and 21st centuries to confirm how the solar system was formed.

Once the Earth was formed, its evolution over those billions of years is explained by dividing the period of geological timing into aeons, eras, and epochs. The Hadean aeon (4.6 to 4 billion years ago), named after Hades, the Greek god of the underworld, symbolized "hell". Scientists state that the surface of the Earth was very unstable during its earliest part, at the birth of the planet. Since this was the beginning, it was characterized by violent volcanic eruptions and immense heat, causing a molten surface. As the core stabilized during the Archean aeon (4 to 2.5 billion years ago), the Earth's atmosphere began to take shape. Oceans formed, and the first signs of life emerged in the form of single-celled organisms. During the Proterozoic aeon (2.5 billion to 541 million years ago), oxygen levels increased, triggering the Great Oxygenation Event. This marked a dynamic growth as many elements and minerals formed on the Earth's surface. Although the Earth's crust continued to evolve, the atmosphere began to accumulate oxygen, a by-product of photosynthesis from cyanobacteria, the microorganisms, which were the earliest life forms. This increase in oxygen was critical for the more complex life forms to develop subsequently.

The Phanerozoic aeon began around 541 million years ago and continues to the present day. From here on, complex cells and multicellular life evolved, setting the stage for more intricate forms of existence. Life in all its varied forms emerged in this aeon. The aeon was further divided into three eras: Palaeozoic, Mesozoic, and Cenozoic. The Palaeozoic era saw the first fish, insects, and reptiles. The Mesozoic era was the age of dinosaurs, and the current Cenozoic era ushered in mammals and eventually humans. We are therefore the final evolute in this saga of creation. Throughout its history, Earth has undergone dramatic changes, each playing a crucial role in the development of life. The gradual cooling of the planet, the formation of the atmosphere and oceans, and the shifting of continents due to plate tectonics all contributed to the environment in which life could evolve. Coming through the different geological time frames, the planet faced several mass extinctions, each dramatically impacting life's evolutionary course. These extinctions were often followed by periods of rapid diversification, where new species evolved to fill the ecological niches left vacant. But today, although the human minds have taken the advances of S&T to incredible heights, the degradation or "mass extinctions" that were built into the cycles of nature, are being pushed by the Earth's final evolute.

The Holocene is officially the current epoch in the geological time scale listed by the International Commission on Stratigraphy. The Holocene began about 11,700 years ago, after the last major ice age. Since then, over the ages, the Holocene has maintained a relatively stable and warm climate, and thus human civilization could develop and flourish. Interestingly, of course, is the fact that this chronology through the aeons and eras is exclusively connected to planet Earth!

The Earth Is Just Right!

One of the miracles of nature is that, apparently, in the vast universe, Earth is the only habitable planet as far as we know, despite the ongoing research to find life beyond. Earth's position in the habitable zone and the presence of liquid water, sustained by a moderate surface temperature, are crucial for the development and sustenance of life. Earth's atmosphere, with its unique composition of nitrogen, oxygen, and other gases, has, through the ages, provided a stable environment and protected life from harmful solar radiation. Earth's dynamic geological activity, driven by plate tectonics, has contributed, so far, to a stable climate and the recycling of essential nutrients.

Evolution of Life

How life first emerged on Earth is intriguing. One prominent theory suggests that life began in a "primordial soup", a nutrient-rich ocean filled with organic compounds. The earliest water body on Earth was extremely rich in organic matter, and from that, the first life forms evolved. This primordial soup theory was first proposed by the scientists Alexander Oparin (1925) and J.B.S. Haldane (1929). Oparin stated that the layers on the surface of hydrogen, carbon, ammonia, and water vapour reacted to form the first organic compounds, and life started in this "soup" through a series of chemical reactions that eventually led to the formation of simple organisms.

Another theory is that life originated at hydrothermal vents on the ocean floor. These environments, rich in chemicals and energy, could have provided the right conditions for the first cellular organisms. These initial forms of life were most likely simple, single-celled organisms, similar to modern bacteria, forming, surviving, and replicating in Earth's early, harsh conditions.

Just as cosmologists and physicists from the various specialized branches studied the beginning of the universe, researchers in physical anthropology and palaeoanthropology, along with interdisciplinary collaboration from palaeontology, archaeology, genetics, and geology, have given us the chronology of the evolution of the human being. As recorded, life then took a significant leap forward with the transition from these simple, single-celled organisms to complex, multicellular life forms. This transition opened up new possibilities for evolution and diversification. Significant in this process was the Cambrian explosion, around 541 million years ago. In this relatively short period, a vast array of complex, multicellular organisms suddenly appeared in the fossil record. In the Cambrian explosion, most of the major groups of animals we know today emerged.

Following the Cambrian explosion, evolutionary creativity, which laid the foundation for the complex web of life that continued to evolve, saw plants colonize the land, followed by the first land animals. Over millions of years, these early creatures gave rise to countless diverse species, which included amphibians, reptiles, and mammals. After one mass extinction event in the Mesozoic era, which wiped out the dinosaurs, mammals became the dominant land animals, from the tiny rat-like pygmy shrew to the massive Blue Whale. But most critical to the evolution of human beings were the primates.

In this discussion of evolution, we need to see the intricate link to symbolism, which helps us to further visualize the concepts we discuss. The *Dashavatara* of Hinduism. *Dasha* (ten) *avatara* (incarnations) in Hindu mythology are the ten incarnations of Vishnu. Today, this narrative of the ten incarnations is, symbolically, linked to the Western idea of evolution! The knowledge of the Vedas has always been shrouded in symbolism, and the same belief helps one to appreciate the chronological order and epistemology in which these incarnations are stated (Fig. 3.1). These ten incarnations illustrate subtle, gradual, but definitive changes that finally bring forth the present evolved state of human understanding and consciousness. Whatever the story, concept, or interpretation, science tells us that the evolution of our species, *Homo sapiens*, is unique, having appeared only about 3,00,000 years ago.

Evolution of Species

The theory of the evolution of species was scientifically proven when the discovery of fossils revealed the existence of creatures that were no longer seen on Earth. This clearly indicated a dynamic history of life and the processes of extinction and regeneration. In the early 1800s, the scientist Jean-Baptiste Lamarck pioneered the concept of evolution. He postulated the idea that animals had changed or evolved over time. He argued that organisms could and did change during their lifetime in response to their need, habitat, and adaptation to the environment, like growing a longer neck to reach leaves; these changes could genetically be passed on to the offspring. This laid the groundwork for the comprehensive theory of Charles Darwin. He proposed that species have the potential to change or evolve, yet they all come from the same ancestor. He gave the concept of the five basic steps: variation, inheritance, selection, time, and adaptation, abbreviated as VISTA.

Charles Darwin's journey to develop the theory of evolution began in 1831, when he travelled from England on the *HMS Beagle* with the mission of mapping the coastline of South America. He noticed that different regions had distinct species, with a few parts resembling each other. It was like seeing extended family members with their distinct traits. Out of curiosity, he started collecting fossils, which suggested the progression of life forms through time.

MATSYA

Fish, corresponding to the fact that life begins in water.

KURMA

Tortoise, can live on land as well as on water, representing the transition of life from water to land.

VARAHA

Boar, lives on land, indicating the completion of the stage of the transition from water to land.

NARASIMHA

Half lion, half man, representing the evolution of the life forms with comparatively greater physical abilities, but with the savage tendencies of an animal.

VAMANA

Dwarf, which represents the primates who have not yet learnt to stand erect.

PARASHURAMA

Holding an axe, the likeness of the primitive man of the Stone Age, using stone tools for hunting.

RAMACHANDRA

Standing handsome and erect, with the progress of the human being further down the road of civilization and using bow and arrow for hunting. In philosophic terms, his was the age of re-establishing righteousness.

KRISHNA

Exemplifies the modern man, living a fuller life and almost complete on the evolutionary scale.

THE BUDDHA

Indicates the interlude of religion and philosophy in the process of evolution. He symbolises a part of a cosmic cycle in which dharma, or righteousness is destroyed and so he comes to re-establish it. Dharma may be described as an ethical-spiritual essence of the universe.

KALKI

Is yet to come indicating that the evolution process is not complete; there's still one final step in its pursuit of perfection.

Fig. 3.1. Dashavatara.

He observed that various islands had unique species perfectly suited to the habitat. This gave the idea that various species had the ability to adapt to their specific environments. Darwin's observations and collections dramatically impacted his thinking. He formulated a comprehensive theory of "natural selection". It proposed the idea of the "survival of the fittest". This also means the non-survival of the unfit. Only the fittest and most adaptable species survive, and they pass on these qualities to their offspring. This theory explains the diversity of life on Earth. His observations and thought processes culminated in the publication of his seminal work *On the Origin of Species* in 1859.

The theory of natural selection operates under the conditions of variation, competition, survival and reproduction, and heritability. It explains how all living things are connected and how they change over time. As generations pass, the species are said to adapt better to their environment because they assimilate all the useful features. Darwin concludes that the evolution of the variety of species on Earth could be a natural process, without divine intervention. Instead of thinking that each species was individually created, he argues that the species changed slowly over a very long time because of natural selection.

While some people still disagree with this idea, mostly on religious grounds, it's widely accepted in the scientific world and is supported by scientific evidence. These include fossil records, biogeography, comparative anatomy and embryology, and molecular biology.

At this juncture, we must reflect on the delicate balance, which is heatedly debated. Has the super-fast track of S&T taken us too far away from our sensibilities and sensitivities as human beings? This is perhaps where the concepts of religion, philosophy, faith, and belief systems do have a say. But the story of our ascension needs to be first understood, as it gives us the truth of how we even got here!

The Ascent of Man

The journey of human evolution is another remarkable chapter in human history. The term "Ascent of Man" was popularized by Jacob Bronowski through his 1973 BBC documentary series exploring human societal development via scientific understanding. It began in the Horn of Africa, where our earliest ancestors, who were part of the great primate family, began the process of becoming *Homo sapiens*. One of our earliest known ancestors is *Australopithecus*, which lived about 4 million years ago. They walked upright and had brains larger than those of their ape ancestors but smaller than modern humans. Donald Johanson, a noted paleoanthropologist, and his team found the most significant hominid fossil in Ethiopia. This very first well-preserved skeleton of the genus *Australopithecus afarensis* was named "Lucy"! The genus *Homo* gradually emerged, with species like *Homo habilis*, who used simple stone tools, and *Homo erectus*, who had a larger brain and was more similar in appearance to modern humans. *Homo erectus* was also the first of our ancestors to leave Africa, spreading across Asia and Europe. This gradually led to spectacular discoveries and developments in science, technology, and civilizations.

Here we need to position ourselves in the present phase of our journey because, today, there is a red flag that is fluttering in these times as a warning to our species: are we somehow changing this stability that nature provided if we haven't done it already? We looked at many such perils in Chapter 2. We acknowledge that this is an era of human dominance, the Anthropocene epoch—one of the incredible power of evolutionary processes. It highlights our resilience and adaptability and talks of a dynamic Earth where life continuously evolves in response to changing conditions because the Earth is just right! Modern researchers and scientists have introduced this epoch, and its significance is the impact of human activities on the Earth's geology and ecosystems. The Anthropocene epoch is an unofficial unit of geologic time. The name is used to describe the most recent period in Earth's history when human activity has significantly impacted the planet's climate and ecosystems. It is a controversial and debated concept because it implies that humans have become a dominant force in shaping the Earth's system and that we have entered a new and uncertain phase of planetary history. Some scientists warn that the Anthropocene poses serious risks and challenges for the future of humanity and biodiversity, as we have either already crossed or are approaching some of the planetary boundaries that define a safe operating space for life on Earth.

One of the most significant aspects of this age, which has been noticed in recent centuries, is the alarming rate at which the environment has altered. Climate change, driven by the burning of fossil fuels, has caused a rise in global temperatures, the melting of ice caps, the incidence of floating glaciers, rising sea levels along with temperatures, and extreme weather events. Human activities have also led to a dramatic loss of biodiversity, with species becoming extinct at a rate apparently not seen since the last mass extinction. Are we permitting our planet to stay just right, or are we letting everything go very wrong?

Scientists have proposed that the Anthropocene epoch began in 1950. The significance of this epoch is that human activity has had an unbelievable impact on the Earth. However, despite significant scientific advancements, what we know about the universe is dwarfed by what remains unknown. This "vast unknown" fuels both scientific speculation and a sense of humility. On 14 February 1990, the spacecraft *Voyager 1* took an image of Earth from over 3 billion miles away. Carl Sagan called this view of our planet "The Pale Blue Dot". Carl Sagan's reflections are quite astute: "The Earth is the only world known so far to harbour life. There is nowhere else, at least in the near future, to which our species could migrate. Visit, yes. Settle, not yet. Like it or not, for the moment the Earth is where we make our stand. It has been said that astronomy is a humbling and character-building experience. There is perhaps no better demonstration of the folly of human conceits than this distant image of our tiny world. To me, it underscores our responsibility to deal more kindly with one another and to preserve and cherish the pale blue dot, the only home we've ever known."

Many of us may have seen and experienced in our lifetimes that science fiction has become science fact. The *Star Trek* series, once a realm of pure science fiction, has inspired real-world technological innovations, such as mobile communication devices. This demonstrates the power of human imagination in shaping future technologies. The *Matrix* shows a society where humans exist without any freedom. Essentially, the film questions the benefits of technology and its influence on society.

And Evolution Continues

We have travelled from the very beginning of life on Earth to this age of S&T. This journey has blended scientific facts with concepts from different cultures and philosophies. We have seen how the universe and life began and evolved, and how humans have become a major force in the Earth's story. We are left with a sense of connection to the universe and a reminder of our responsibility to our planet. This sets the stage for our journey ahead, where we'll explore deeper into the relationship between science, spirituality, and our understanding of existence.

Why have we consciously merged mythological stories with the scientific journey of the beginning of the universe? We have brought some metaphorical insights into complex cosmic phenomena in an endeavour to make abstract concepts more relatable and understandable. These narratives have often inspired scientists to explore and question the universe's mysteries, driving the pursuit of knowledge.

The law of conservation of energy is also a fundamental principle in physics, and it applies to the energy in the universe as a whole. According to this law, the total energy of an isolated system remains constant over time. This principle extends to the entire universe, making the conservation of energy a foundational concept in cosmology. But the universe, in its entirety, is often considered an open system. While the total energy of the universe may be conserved in certain contexts, the energy associated with individual components can change due to the overall expansion.

In an interesting and very recent discovery (March 2024), scientists from the European Space Agency have created an image using data from the Gaia Space Telescope. This interesting picture shows two vast streams of ancient stars within our Milky Way galaxy. These streams, named Shiva and Shakti after Hindu deities associated with creation, are depicted as blue dots (Shiva) and yellow triangles (Shakti) on scientific visualizations (see Plate 17). While not a literal image, this colour-coding helps distinguish the stars within each stream. The stars themselves are incredibly old, likely predating the Milky Way's iconic spiral arms. This finding suggests the Milky Way wasn't always a single entity but rather may have formed billions of years ago from the merger of smaller galactic structures, with Shiva and Shakti being remnants of that process. Their presence hints at the Milky Way's potentially violent beginnings, where smaller galactic building blocks merged to form the majestic galaxy we reside in today.

Universe to Life

This journey signifies a shift from the measurable aspects of life as explored in empirical science to the less tangible realms of universal truth and consciousness. This is our evolution from modern science to meta-science, from "Genome to Om". We now ponder our presence on planet Earth; surely it is not a mere accidental or coincidental occurrence of evolution. After the universe was created, billions or trillions of years ago, either after a dramatic explosive birth or as a cyclic re-emergence, we do not know exactly how we came into being! Did life, as we think of it today, emerge as a chemical reaction from non-living matter? Did we evolve from the core of the vital energy that is the essence of existence? Or, were we created by the all-powerful Creator in his own image? Our journey now continues to explore in depth what exactly life is.

CHAPTER 4

What Is Life? Biological, Social, and Spiritual Dimensions

Life seems to be an orderly march past, a review-stand of genes, from the beginning of time till now.

—Erwin Schrödinger

In trying to understand the meaning, origin, evolution, and complexities of life, we often find ourselves at the crossroads of two distinct paths: the empirical and experimental approach of science and the experiential world of insights, instincts, and intuition. Each path offers a unique perspective on the mysteries of the natural world. The question "what is life?" includes empirical observation, theoretical understanding, and existential inquiry. It is a very complex question that essentially must extend beyond the purview of modern science. Undoubtedly, modern science has made significant progress in explaining the mechanisms, structures, and evolutionary processes of living organisms. Its predominantly reductionist approach focuses on the smallest constituent parts and their interactions. But this acute focus on the microcosm eludes the depth of the macrocosm, which must address the holistic, emergent properties of life and the subjective experiences of all living beings.

The question is a very vast, complex, but essentially fundamental one. All of us, at some stage, ask these basic questions, some of which we may have raised earlier as well: What is life? How did it begin? Why are we here? What is the purpose? And so on. The more the questions, the more the additional questions, and the more diverse the answers. The quest to understand life has been a central theme throughout human history, engaging the brightest minds in science and spirituality. The varied perspectives on what constitutes life differ immensely between cultures and belief systems, from psychological and philosophical views to various scientific disciplines. Within physiology and biology, the definition of life is not clear-cut because various fields within biology itself emphasize different aspects of this question. The exploration of what constitutes life extends beyond biology to include philosophical, ethical, and metaphysical dimensions. The question has no definite or final, single answer because it transcends the boundaries of individual study and remains a subject of ongoing discussion, insightful reflection, and understanding.

Generally, we could say that "life" is a broad-spectrum concept that includes biological, social, and spiritual dimensions. Each dimension offers a unique perspective on what it means to be alive. To understand life, we need a worldview that integrates insights from various fields of knowledge. Life, in its essence, includes the physical processes that sustain organisms, the social interactions that define human existence, and the spiritual quests that give deep meaning. Each dimension contributes to the richness of existence *per se* and thus helps us to reflect on what it means to be alive.

As we continue to transcend from science to meta-science, we have the biological, social, and spiritual dimensions to try to understand "life" as expressed and understood over the centuries.

Aristotle distinguished between "bios" (the course of life) and "zoe" (raw life). His work in natural philosophy laid some foundational concepts for understanding life as a process and phenomenon. While addressing the question "what is life?" Erwin Schrödinger raised further questions and gave some answers as to how the events in space and time, which take place within the spatial boundary of a living organism can be accounted for by physics and chemistry. In his book, Schrödinger describes life as a self-reproducing aperiodic crystal of DNA that codes for proteins. He also suggests that living entities maintain themselves away from equilibrium by dissipating matter and energy gradients to increase their order. Jean-Paul Sartre and other existentialists pondered the purpose and essence of life. Sartre's notion that "existence precedes essence" suggests that individuals define their meaning in life through their actions. Greek mythology is rife with stories about the origins of life. Prometheus created man from clay and defied the gods by giving fire to humanity, thus symbolizing the spark of life. In the Bhagavad Gita, life is discussed as an eternal, undying essence beyond the physical body. It presents a view of life as part of a cosmic, spiritual cycle. The Yggdrasil or World Tree in Norse mythology represents the interconnectedness of all life. It is a cosmic tree that supports the nine worlds, embodying the cycle of birth, growth, death, and rebirth.

The integration of these diverse perspectives provides a richer understanding of life. Combining Darwin's biological principles with Aristotle's philosophical notions, we see life as both a physical and metaphysical journey, which includes evolution, growth, and purpose. Myths and epics, with their symbolic narratives, offer insights into our understanding of life's origin, purpose, and interconnectedness, enriching our perception beyond scientific explanations. In the modern era, the conversation continues with new theories in biology and debates in spirituality, each adding layers to the definition and understanding of life.

Exploring what constitutes life reveals profound ideas woven through science, philosophy, and mythology. From the empirical to the existential, from mythological symbolism to modern genetics, each perspective offers a piece of the puzzle. Together, these views form a composite picture of the complexity, diversity, and profound mystery of life. As our understanding evolves, so does the narrative of life, continuing an age-old quest that is as varied and vibrant as life itself.

Life on Earth

The nature of life on Earth is a complex study that spans millions of years of history, marked by major geological transformations and tectonic shifts. We begin with the example of ammonite fossils that date back to the Jurassic and Cretaceous periods, making them between 145 and 201 million years old. These fossils originated from the sediments of the ancient Tethys Sea, an expansive ocean that once nestled between the continents of Gondwana and Laurasia. The movement of the Indian Plate towards the Eurasian Plate, part of the Earth's ever-changing tectonic dynamics, lifted the seabed, formed the great Himalayan range, and subsequently brought these ancient marine fossils to the surface in the Kali Gandaki river valley of Nepal. The journey of the ammonite fossils from the depths of the Tethys Sea to the heights of the Himalayan range and down to the river valleys speaks volumes about the Earth's dynamic nature. It emphasizes the gigantic timescales and processes that have shaped our planet and the life it harbours.

Some remarkable artefacts, like the ammonite fossils from the ancient Tethys Sea, are found in the Indian subcontinent, popularly known as Shaligram. The spiral shape of these ammonite fossils, the Shaligrams, is reminiscent of the nautilus, a powerful symbol in both the scientific and spiritual realms (see Fig. 4.1). It is a unique lens through which we can examine life from archaeological, historical, and spiritual perspectives. Scientifically, it represents the natural history of marine life and the Earth's geological evolution. Spiritually, it symbolizes the unending cycle of life, death, and rebirth, reflecting the pattern found throughout nature, from the spiralling galaxies to the structure of DNA. This actually depicts the aphorism "smaller than the smallest and larger than the largest". In Hinduism, Shaligrams are considered extremely sacred and are revered as natural representations of Lord Vishnu.

Fig. 4.1. Shaligram-ammonite Himalayan fossil.

The ammonite fossils tell us that life on Earth is not merely a biological phenomenon characterized by growth, reproduction, and metabolic processes. It is also a spiritual journey, woven into the stories, beliefs, and rituals that define our existence. The archaeological and historical narratives we see in these fossils remind us of the Earth's vast biography and our place within it. They prompt us to reflect on the cycles of nature, the interconnectedness of all living beings, and how the essence of life transcends time and form. These fossils are a bridge between the past and the present, the physical and the metaphysical, inviting us to reflect on the mysteries of life on Earth. By studying these ancient relics, we gain a deeper appreciation for the complexity of life, the dynamic processes that have sculpted our planet, and the spiritual narratives that connect us to the cosmos. In this way, our understanding of life on Earth is enriched as we acknowledge the wonders of scientific discoveries, historical wisdom, and the spiritual insights that challenge and expand our perceptions of what it means to be alive.

How Did Life Begin?

We discussed the beginning or creation of the universe and how the world began in Chapter 3, as well as the process of evolution; but how does life itself begin? This complex process is not completely understood, perhaps even today. There are several scientific theories on the one hand and many philosophical or mythological ideas on the other.

From the Scientific Point of View

One theory is abiogenesis, which means that life originated from non-living matter through a series of chemical reactions. According to this hypothesis, simple organic molecules formed in the original waters or oceans that covered the surface of the Earth. These molecules went through some chemical reactions, developed self-replicating abilities, and could thus be the very first living organisms. The microbes apparently left evidence of their existence in rocks as far back as 3.7 billion years ago. This would make them the oldest-known fossils It is indeed exciting to find that life managed to get a grip and start to evolve on Earth so quickly after the planets formed.

Researchers also spotted rosette-shaped formations, which could have blossomed through a chemical process that began with rotting bacteria. The rosettes are freckled with dots and shards of other chemicals linked to life, such as phosphorus, a key ingredient for biological activity. The rocks around the fossil-containing layers have chemical signatures of a hydrothermal vent, and scientists have raised the possibility that life arose on Jupiter's watery moon, Europa, or even closer to home; maybe there was life on Mars in the past and we have yet to find it.

Another theory is panspermia. This suggests that life came from another source in the universe! That microorganisms or organic molecules got transported onto Earth by comets or meteorites that crashed into it. Fossil records have also given some clues about how life emerged initially. Organisms that thrive in extreme climates, such as extremophiles, also indicate that life existed on Earth in conditions that we would find completely inhospitable.

Life is said to have started with simple single-celled organisms approximately 3.5 billion years ago. The key stages include the development of multicellular organisms, the transition from water to land, and the emergence of mammals and primates. Today, Earth is home to an incredible variety of life forms, each adapted to its environment. The total number of life forms on Earth, including flora (plants) and fauna (animals), is estimated to be about 8.7 million species, of which about 3,90,000 flora are known to science. In terms of fauna, the diversity is even more staggering. There are estimated to be around 1 million known species of insects alone. When it comes to animals, the total number of species, including those yet to be discovered, is much higher. This includes mammals, birds, reptiles, amphibians, fish, and numerous other invertebrates. The actual number of species on Earth could be much higher, as many regions, especially the tropical rainforests and deep oceans, are understudied, and whenever they are, new species are discovered. Additionally, microbial life forms, including bacteria and archaea, add significantly to Earth's biodiversity, with estimates running into the millions or even trillions.

Darwin's theory of natural selection explains how genetic variations that are beneficial for survival are passed down through generations. Mutations in DNA lead to new traits. Beneficial mutations are preserved, and these can lead to evolutionary changes. Over time, genetic changes can produce new species, contributing to the diversity of life. From a purely biological perspective, life is described by certain characteristics that differentiate living organisms from non-living ones, the animate from the inanimate. In biology, organisms are generally considered living if they have growth, respiration, excretion, reproduction, movement, metabolism, and are responsive to the environment. Living cells are the basic unit of life. Broadly, cells are categorized as prokaryotic (without a defined nucleus, like bacteria) and eukaryotic (with a nucleus, like plants and animals). Key structures include the nucleus (housing DNA), mitochondria (energy production), ribosomes (protein synthesis), and the cell membrane (regulating entry and exit of substances). Cells perform essential life functions such as conversion of energy, reproduction, growth, and response to external stimuli.

Who or what is considered to be "living"? Every "living" organism, from the simplest bacterium to the complex human being, is composed of cells, which are the basic structural and functional units of life. In physiological terms, cells carry out the processes essential for "being alive", such as metabolism and reproduction. Metabolic processes involve converting energy and regulating chemical reactions to maintain the conditions essential for life. This includes obtaining and utilizing energy from the environment.

Flora and fauna would be the simplest categorization. But all single-celled and multi-celled organisms are "alive". Explicitly, bacteria, fungi, and viruses, from jellyfish to the amoeba, are all living organisms. All of them are part of the diversity of life on Earth, perhaps on other planets as well. They represent different levels of cellular and metabolic construction or organization. They fulfil the criteria that define living organisms. The difference lies in the construction. Most multicellular organisms perform photosynthesis to produce their food. Several multicellular organisms, such as animals, are heterotrophic; that is, they obtain food by consuming other organisms.

But we must elaborate a little more. Multicellular organisms also absorb nutrients from their surroundings and play crucial roles in decomposition and nutrient cycling. Living organisms grow and decay, which means an increase or decrease in the size or number of cells. They can develop and mature, which triggers a more complex and specialized structure. Living organisms can adapt to their environment. This adaptation can be an evolutionary process as organisms learn to survive and thrive in changing conditions. They can regulate their internal conditions to maintain stability. This also implies basic consciousness or awareness. Living organisms can respond to changes in their environment, either through simple reflex actions like the sunflower turning towards the sun, other plants blooming in sunlight and "shutting down" at sunset; or animals hibernating, storing food, and migrating according to the inclement weather; and finally, human beings, whose responses are as diverse as their complex behaviours. Living organisms can reproduce. Again, this could be the relatively simple reproduction or multiplication in plants, in bacteria, or in the more complex biological processes in lower and higher animals, and this could be either asexually propagated like the single-celled amoeba or sexually as in other living beings.

Reproduction ensures the continuity of life and the passing on of genetic traits. Living beings produce offspring that inherit genetic information from their parent or parents. A puppy grows into a dog, developing through complex biological processes. Its cells divide and differentiate, contributing to its growth, which is a living process. A stone, although it may change shape due to erosion, doesn't grow or develop cellularly. It simply exists, unchanging in its essential composition. Animals show awareness of their environment and can exhibit emotions—signs of consciousness. As mentioned earlier, a sunflower turns towards the sun as a response to light, a stimulus. A robot may move towards a light source, but this is due to programmed algorithms, not a conscious response or awareness. Trees and plants produce seeds that grow into new trees and plants. This reproductive process is a fundamental aspect of life. They also metabolize sunlight through photosynthesis, which is a unique biological function. A cloud may grow and shrink, but it does not reproduce or metabolize. It's simply a collection of water droplets, subject to physical processes like condensation and evaporation.

The aphorism "*Jeevo Jeevasya Jeevanam*", meaning "the life of one is dependent on the life of another", originates from the ancient Vedic texts. The phrase captures the fundamental principle of interdependence among all life forms. It suggests that the existence and survival of any living organism are intricately linked, in some way, to the lives of others in the ecosystem. This principle is a cornerstone in understanding ecological relationships and the delicate balance that sustains ecosystems. This is the "web of life", the complex system of interdependent relationships among living organisms within an ecosystem Every species, from the smallest microbe to the largest mammal, plays a crucial role in the functioning of an ecosystem. This interconnectedness is vital for the stability and sustainability of the environment. The foundational principle of ecology is that no organism exists in isolation; each is a part of a larger, intricate network of life. This understanding is crucial for conservation efforts, highlighting the importance of preserving not just individual species but the entire ecosystems in which we all live.

Living and Non-living

The distinction between living and non-living entities can be particularly intriguing when we consider examples like dormant seeds and viruses. They challenge the conventional parameters used to define life, leading to fascinating discussions in biology.

Viruses: Although they exhibit some characteristics of living organisms, viruses are not universally considered "living" because they are not cells; they cannot carry out metabolic processes or reproduce independently. They are complex molecules consisting of genetic material (DNA or RNA) encased in a protein coat. Viruses can only replicate inside the cells of a host organism, using the host's machinery to reproduce. Outside of a host, they are inert, much like non-living matter. However, once inside a suitable host cell, they exhibit characteristics of life, such as replication and genetic mutation. They are an inescapable part of life, and there is a definite biological connection between a virus and the organism that it infects. The Human Immunodeficiency Virus (HIV) and Severe Acute Respiratory Syndrome Coronavirus 2 (SARS-Cov-2) that cause AIDS and COVID-19, respectively, are well-known examples. Outside the human body, the virus remains inactive. Once inside the body, it integrates its genetic material with the host's cells, leading to the production of more viruses. This debate continues among scientists about whether viruses should be considered living or non-living.

Prions are proteins most famously known for causing neurodegenerative diseases in mammals, such as the "mad cow disease" in cattle and the severe brain disorder disease in humans. Prions are a somewhat contentious topic in the discussion of what constitutes life. They are infectious agents like microbial pathogens, but they are not considered living organisms. They lack nucleic acids, which are fundamental components of living organisms (DNA or RNA). Unlike typical living organisms, prions replicate by inducing normal cellular proteins to fold into the prion's abnormal shape. This process does not involve nucleic acid-based replication mechanisms as seen in living organisms. Prions do not exhibit metabolic activity, which is a hallmark of living organisms. They do not consume energy or grow, nor do they have cells that divide. Given these characteristics, prions are generally not considered "living" according to standard biological criteria because of the absence of DNA or RNA. This sets them apart from living organisms. Genetic material is essential for the processes of replication and evolution in known forms of life.

Dormant seeds are unique in the life cycle of plants. They are "alive" but inactive, having halted most metabolic processes. Seeds can remain in this suspended state for extended periods, sometimes years, waiting for favourable conditions to germinate. Seeds contain the necessary genetic material and cellular structures to sprout into new plants. The seeds of the lotus plant (*Nelumbo nucifera*) are known for their longevity. Seeds found in ancient lake beds have germinated successfully after hundreds of years of dormancy, showing the remarkable potential of life in these seemingly inert entities.

Dormant seeds and viruses present unique cases in the living versus non-living debate. They demonstrate how life can pause and resume. Viruses challenge our understanding of life. They exist in a grey area, displaying life-like properties only when in a suitable environment. This has led to some scientists describing them as "organisms at the edge of life". These examples prompt a re-evaluation of the criteria used to define life, illustrating the complexity and diversity of biological existence. The concept of whole or complete beautifully defines the essence of the seed. Despite its small size, the seed is whole within itself, containing the entire potential to become a tree. The Vedic aphorism "smaller than the smallest, yet bigger than the biggest", resonates deeply with the nature of any seed.

Seed of Life

To further explore the profound question, "What is life?" the Rudraksha and Banyan seeds give very significant insights. These stories, deeply rooted in spiritual and philosophical traditions, serve as metaphors for understanding the complexities, interconnectedness, and unity of existence.

The Rudraksha seed comes from the fruit of the Rudraksha tree (*Elaeocarpus ganitrus*), primarily found in the Himalayan region, parts of Southeast Asia, Indonesia, and Nepal. Rudraksha seeds are revered in Hinduism and other Eastern spiritual traditions for their spiritual and healing properties. Derived from two Sanskrit words, "rudra", another name for the Hindu god Shiva, and "aksha", meaning eye, the Rudraksha beads are said to have originated from the tears of Lord Shiva, according to Hindu mythology.

Rudraksha seeds have a rough, grooved surface and can vary in size, with each seed containing a different number of facets. The number of facets is believed to influence its properties and benefits. The Rudraksha captures the essence of life's potential in its dormant state, ready to unfold into a sacred tree. From dormancy to germination, it reflects life's vibrant force and the spiritual journey of existence. The shape and colour of the Rudraksha seed have a very deep significance, embodying various spiritual energies and attributes. The seed's colour, typically deep brown with a blackish hue, symbolizes its connection to the earth and its role as a mediator between the earthly and the divine. It is thought to absorb negative energy and enhance protective and purifying properties. Notably, the Rudraksha seed resembles the anatomy of the human brain, with its grooves and ridges almost mirroring the brain's gyri and sulci, suggesting its influence on the mind and spiritual functions (see Fig. 4.2). This resemblance has inspired interpretations of the seed's ability to calm the mind, improve concentration, and aid meditation, highlighting the interconnectedness of the physical and spiritual realms. A total of 108 Rudraksha beads strung together is one of the most powerful aids in spiritual chanting, believed to offer protective and purifying energies, enhance clarity of mind and concentration, facilitate deeper meditation, and promote spiritual growth.

Fig. 4.2. Rudraksha seed.

Extending the seed metaphor further, a story from the *Chandogya Upanishad* gives us a compelling dialogue between Sage Uddalaka and his son, Shvetaketu, talking of the seed of the Banyan tree to unravel the essence of life. Their conversation went somewhat like this:

U: Bring me a fruit from the banyan tree.

S: Here is one, Father.
U: Break it open.
S: It is broken, Father.
U: What do you see there?
S: These tiny seeds.
U: Now break one of them open.
S: It is broken, Father.
U: What do you see there?
S: Nothing, Father.
U: My son, you know there is a subtle essence, which you do not perceive, but through that essence, the truly immense banyan tree exists. Believe it, my son. Everything that exists has its Self in that subtle essence. It is Truth. It is the Self, and You are That (*Tat Tvam Asi*).

The story eloquently conveys the teaching of one of the Upanishadic Great Statements from the *Chandogya Upanishad*, "*Tat Tvam Asi*" (That Thou Art), meaning, "You are That Universal Consciousness". It suggests that the same divine essence present in the banyan seed pervades all beings and the cosmos. The unity of the individual soul and universal consciousness reflects the interconnectedness of all existence.

The seed metaphor provides a very intriguing perspective on understanding life and challenges us to explore beyond the visible and appreciate the interconnectedness of all forms of life. Just as the essence of the mighty Banyan tree is hidden in the tiny seed, the essence of life, or consciousness, may be a subtle yet powerful force, not always apparent in the outer world of the senses. The metaphor offers a timeless perspective on the nature of life, intertwining philosophy, spirituality, and a hint of mysticism. It suggests that life, in its most fundamental form, is an unseen essence that manifests in the physical world, much like the invisible life force within a seed that gives rise to a towering tree. This perspective enriches our understanding of life, adding a layer of depth to the biological explanation and inviting contemplation of the unseen forces that animate the living world.

Limitations of the Scientific Perspective

The limitations of modern science appear to span various aspects: there is more focus on the reductionist approach versus the emergent; there is more focus on minute observations at the quantitative level than on the qualitative results; the objective is paramount, not the subjective; and finally, it looks only at the physical with no cognizance of the metaphysical. Modern science, be it physics, chemistry, or biology, tends to adopt a reductionist approach, breaking down complex systems into simpler components. Reductionism is a powerful methodological approach to understanding complex systems by breaking them down into simpler, more manageable parts. It has undoubtedly been highly successful in many scientific disciplines, allowing researchers to analyse phenomena at a fundamental level and uncover the underlying principles. However, it is limited when it comes to understanding emergent phenomena. Emergence is a phenomenon where complex systems show properties or behaviours that cannot be explained by simply understanding the individual parts in isolation. Emergent properties require a holistic approach that considers the interactions and dynamics between various components. Modern science does recognize the importance of emergent phenomena, especially in such complex systems as systems biology and neuroscience. Researchers in these fields study how complex systems, ranging from ecosystems to the human brain, exhibit emergent properties that cannot be predicted solely from understanding the individual components.

Science excels in quantification and measurement, yet life's qualitative aspects—consciousness, subjective experience, and the meaning of existence—cannot come into the range of simple measurement. These aspects of life are central to many philosophical and existential inquiries that science alone cannot address. The scientific method is also rooted in objectivity, seeking universal truths through empiricism, observation, and experimentation. However, life includes subjective experiences, which are intrinsic to individual consciousness and which cannot be fully captured or understood through objective methods alone.

A Journey beyond Empiricism

The modern science perspective is where hypotheses are tested, data is gathered, and conclusions are drawn based on evidence. This scientific method has been instrumental in unravelling the mechanisms behind various phenomena or the intricacies of the immune system. It offers a clear, structured way to comprehend the workings of the natural world, breaking down complex systems into understandable parts. Science operates within the framework of physical laws and empirical evidence, often sidelining metaphysical questions about purpose, meaning, and the nature of life. These questions, while critical to understanding life in a broader sense, lie outside the empirical scope of science, which dissects and analyses. Insights and intuition, on the other hand, synthesize and unify. Science excels at providing detailed explanations and advancing technology, but it may overlook the intrinsic value and interconnectedness of life that insights and intuition capture. Instincts and intuition offer a profound, sometimes inexplicable, understanding of life, yet they lack the precision and replicability of scientific methods.

The spiritual dimension represents a more innate, often less tangible, understanding of the world. This path goes deeply into the realms of what might be beyond the observable or recognizable by the five senses—the kind of knowledge that is more felt or experienced than quantified. It's instinctual knowledge that leads a plant to adapt to its environment. This type of understanding often transcends the empirical, offering a holistic view of life's interconnectedness and the deeper patterns that govern the natural world.

While modern science has unravelled many mysteries of the physical and biological mechanisms of life, it is not equipped to comprehensively address all dimensions of the question "What is Life?" Meta-science, by bridging the gap between empirical research and broader philosophical inquiry, offers a promising avenue for exploring life's complexities in a more integrated, reflective, and meaningful way. It acknowledges that understanding life in its entirety requires not only dissecting its physical components but also contemplating its emergent properties, subjective experiences, and existential significance.

By combining biology and sociology, we might begin to see a more complete picture. The empirical and experimental methods give us the tools to explore and understand the material aspects of our world, while insights, instincts, and intuition connect us to a deeper, more intrinsic understanding of the universe and how we live in it. This synthesis encourages us to understand life not just in its physical form but to also sense its deeper rhythms and connections, leading to a fuller, more nuanced appreciation of the world around us. Broadly, we can consider three dimensions of life: biological, social, and spiritual.

Biological Dimensions of Life

Today, our understanding of life is deeply rooted in biology—in the study of living organisms and their vital processes. As already discussed, life is characterized by certain fundamental attributes: organization, metabolism, growth, adaptation, response to stimuli, and reproduction. The discovery of DNA and the understanding of genetics have revolutionized our perception of life, showing us the intricate blueprint that governs all living beings. The origins of DNA can be traced to the early stages of life on Earth, believed to have emerged around 3.5 to 4 billion years ago. The genome consists of DNA within the nucleus, which carries the genetic instructions for the development, functioning, growth, and reproduction of all known organisms. Genetics is the study of genes and heredity. It explains how traits are passed from parents to offspring and how they are expressed. Advances in molecular biology have unravelled how genetic information is transmitted and regulated, shedding light on the complexity of life at the molecular level. We have discussed this in detail in Chapter 1.

Biologically, several key attributes distinguish living organisms from inanimate objects, and that explains to us, in a way, what life is. These attributes state that living organisms must consist of cells, have the ability to grow and reproduce, maintain homeostasis, have a metabolism, and have the ability to respond to stimuli and adapt through evolution. From the simplest single-celled organisms to the most complex multicellular forms like us, these fundamental processes define biological life. The study of life in its biological aspect, obviously, falls under the domain of biology, which explores the physical structures, chemical processes, genetic blueprints, and evolutionary dynamics that enable living organisms to thrive in all environments.

Social Dimensions of Life

Sociology, anthropology, and psychology help us to examine life from a social perspective and to investigate how social influences affect our behaviour, societal development, and the construction of social realities. The social dimension is how we form relationships and interact with each other, establish communities, and create social structures and norms that guide our behaviour and societal organization. It touches on everything from daily interactions and social roles to larger societal institutions and cultural practices. Relationships, culture, society, and cooperation in human life are seen as most important.

If we compare ancient civilizations and modern society, on the one hand, we can see continuity and, on the other, shifts in social life, which are shaped by historical contexts, technological advancements, and evolving ideologies. The comparison between ancient civilizations and modern society highlights both the enduring aspects of social life and how it has evolved in response to changes in technology, philosophy, and social organization. The fundamental human need is to remain connected and have a durable community and social structure. How we choose to navigate our social worlds, however, continues to change.

In ancient societies, the primary social units were the networks of family and kinship. These were integral to survival, identity, inheritance, and continuity. Extended families chose to live together or near each other, with very clearly defined roles and responsibilities. Kinship ties extended to tribes or clans, which then identified social obligations, loyalty, and support systems.

Communication in ancient times was face-to-face, supplemented by written messages for those who could manage them. Limitations of distance to cover and the speed of the messengers decided how much and how speedily information could be shared. The oral tradition was the most significant way of preserving history, knowledge, and cultural identity.

Economic life in ancient societies was mostly based on agriculture, trade, and artisanal crafts, with trade networks extending between cities and civilizations. Social status and occupation were closely linked, and certain professions were given higher esteem than others. The division of labour was more rigid, and this was most often determined by birth or caste.

Governance structures ranged from monarchies and empires to city-states and tribal confederations. These systems were primarily hierarchical, with power concentrated in the hands of a few, and social organization was closely tied to either religious or military institutions, or both. Religion was deeply integrated into the social and political lives of ancient civilizations. It was often the basis for laws, moral codes, and governance. Cultural practices, including festivals, arts, and education, were almost always linked to religions.

In modern times, although families continue to remain central to social life, there has been a discernible shift towards nuclear families in many parts of the world. This has generated more fluid or flexible definitions of family roles. Advances in technology, mobility, and social norms have generated more diverse family structures, which include single-parent families, blended families, and chosen families of close friends. Kinship ties are still important, but they often play a less central role in daily support and social obligations.

The digital revolution has transformed communication, making it possible to share information instantly across the globe. Social media, email, and other digital platforms not only facilitate personal communication but also determine the dissemination of news, knowledge, and culture in an unprecedented manner. This has led to more interconnected societies, but the challenge is managing misinformation and ensuring privacy.

Diverse and complex service- and knowledge-based industries are taking the lead in today's scenario. Technological advancements have led to new forms of work, including remote or work-from-home scenarios and the gig or freelance economy, which allows more flexibility but also creates challenges in job security and work-life balance. Social mobility is more feasible, with education and talent allowing individuals to move beyond the economic status of their birth. There is much more flexibility, and the younger generation is constantly venturing into new professional realms.

Modern governance structures are more varied, including democracies, republics, and other forms of government that, theoretically, offer greater participation and representation. Social organization is influenced by a wider array of institutions, including legal, educational, and healthcare systems. These reflect more complex and differentiated societal roles. While religion remains important in many societies even today, there is greater secularization and pluralism in others, allowing for a separation of religious and state affairs. Cultural life is incredibly diverse, and global connectivity encourages cross-cultural exchanges and the proliferation of subcultures.

Spiritual Dimensions of Life

The spiritual dimensions are focused on the search for meaning, purpose, and connection beyond the material and empirical world. Understanding life from the spiritual dimension is a deeply personal journey that requires openness, curiosity, and a willingness to explore the unknown. It involves our beliefs, experiences, and practices related to the sacred, the divine, or the ultimate reality. It means that we must be willing to dive deeper into all aspects of existence, beyond the material world. The important fact that is highlighted here is that life is not merely the functioning of this body–mind complex. It is about tapping into the essence of existence beyond the material world and finding meaning and purpose in the interconnectedness of all things.

Spirituality can be expressed through organized religion, personal belief systems, meditation and reflection, and living an ethical and meaningful life. Integrating these aspects offers a comprehensive understanding of life and helps us recognize its complexity and the diverse ways it can be experienced and understood. While distinct from religion, which is more structured and institutionalized, spirituality is about personal experience and addressing the broader questions of existence, consciousness, and the nature of the universe. But most importantly, we must emphasize the development of our inner life, of self-awareness, and pursue the quest for transcendence, with a focus on moral values and a sense of belonging to something greater than ourselves. We have to accept that there is a dimension deeper than this physical existence, and our goal is to reach that deeper level. Besides introspection and reflection, there is a deeper level of connecting with the inner self, which is experiential. Meditation and mindfulness help us connect with the inner self and tap into higher consciousness. Through stillness and introspection, we can gain insights into the spiritual nature of life. Many traditions have sacred texts that offer guidance on understanding life's spiritual dimensions. Studying these texts can provide insights into the purpose of existence, the nature of reality, and how to live a meaningful life.

Nature is often seen as a manifestation of the divine, and spending time in nature can help us feel connected to something greater than ourselves. Observing the beauty and complexity of the natural world can inspire a sense of awe and wonder, leading to spiritual insights. The one stumbling block that comes up is the definition or significance of the term "divine". The meaning and use of the term differ across different religious and spiritual traditions. Many times, it is perceived as something mystical, related to God. We consider the term "divine" as a recognition of a supreme power that is omnipresent, transcendent, sacred, and unifying in one-ness.

Most spiritual traditions encourage us to explore these mysteries with an open mind and heart. Prayer and ritual are common practices in many traditions. These practices help to cultivate a sense of connection with the divine and can provide a framework for understanding the spiritual dimensions of life. Many of us turn to teachers, gurus, or mentors for guidance on our spiritual journey. This is a personal pursuit and belief. Spiritual teachers often have deep insights into the nature of reality and can offer wisdom and support to those of us seeking to understand life from a spiritual perspective.

Contemplating questions about the meaning of life, the nature of consciousness, and the existence of the soul leads to profound insights into the spiritual dimensions of life. Taking ourselves out of the "me and mine" attitude and engaging in acts of service and compassion is a powerful path. By helping others and cultivating kindness and empathy, we can experience a sense of interconnectedness with all beings. That is our continuing journey from "Genome to Om".

Natural Instincts of Life

We will now explore some other fascinating aspects of life. The wonders of the natural world, such as the precision of a mosquito's bite, the instinct that guides a new-born calf to its first meal, knowing where to find milk, or the precise workings of our immune system, have never failed to fascinate both scientists and philosophers. These are natural instincts or inborn gifts of life. Empirical science, with its focus on observation, experimentation, and analysis, offers detailed explanations for these phenomena. For instance, the mosquito's targeting ability is understood as an evolutionary adaptation honed over millennia to ensure survival. Similarly, the immune system's complexity is explained as a sophisticated defence mechanism developed through natural selection. However, these scientific explanations, while compelling, often lead to deeper questions about the nature of life and intelligence.

The sight of geese moving in perfect V-formation or a sea turtle returning to the very beach it hatched on decades earlier fills us with awe. A large majority of the birds of North America, including geese, cranes, and others, fly south to avoid the harsh winters of the north. How can these creatures, with walnut-sized brains, travel thousands of miles across continents, navigating with almost supernatural precision? Where and how do they learn this? Who teaches them? Again, these are natural instincts or inborn gifts of life. Science offers a fascinating glimpse into their secret toolkit—an internal compass attuned to the Earth's magnetism, the ability to read the constellations, and even social learning from older generations. Yet, the true marvel lies in the unresolved questions. How do young birds, embarking on their first migration alone, manage to navigate flawlessly? What allows them to pinpoint a specific patch of coastline or a hidden freshwater oasis after a journey spanning continents? Many answers are beyond the understanding of current science. We call this *prajnana* or "meta-science". The concept of "meta-science" invokes a sense of awe and reverence for the profound intelligence, natural instincts, and wisdom exhibited by living organisms, transcending our current scientific understanding. It acknowledges that while science provides valuable insights into the natural world, there are aspects of nature that may defy conventional explanations and elude scientific scrutiny. The universal resonance of Om generates the same sense of awe and reverence.

This is where philosophical and spiritual traditions offer a complementary perspective. They suggest that these intricate behaviours and systems might be part of a larger, interconnected life, guided by a universal consciousness that transcends our traditional understanding of biology. From a biological standpoint, life is characterized by a set of distinctive features that differentiate living organisms from non-living matter. These features include cellular organization, metabolism, homeostasis, reproduction, growth and development, and, of course, adaptation and evolution.

Cells are the basic units of structure and function in living beings. Whether unicellular or multicellular, organisms demonstrate a complex organization that enables life processes. The virus, teetering on the edge of what we define as "living", has the ability to identify and invade host cells, which is a marvel of biological evolution. But from a meta-scientific point of view, this could be part of a cosmic balance or a manifestation of the natural order that goes beyond physical explanations. Similarly, the remarkable navigation skills of migratory birds, which are often explained through genetics and environmental cues, might also be viewed as part of a grander scheme where creatures are guided by forces beyond their physical capabilities. Plants are said to communicate and adapt through chemical signals. While botany and ecology offer scientific explanations for these phenomena, a holistic view might suggest that plants are part of a conscious, interconnected ecosystem, operating with a level of awareness that we are only now beginning to understand.

All living organisms exhibit metabolism, which includes the chemical reactions involved in the transformation of energy and the synthesis and breakdown of molecules necessary for growth, repair, and maintenance. The ability to maintain internal stability (such as temperature and pH) in response to external environmental changes is known as homeostasis. The capability to produce new individuals, either sexually or asexually, ensuring the continuation of the species, involves reproduction. As human beings, we grow and develop, following specific genetic instructions that guide our maturation and life cycle. We adapt to the environment, and populations evolve over time through natural selection, leading to the diversity of life on Earth.

The Renaissance ushered in a new era of scientific inquiry, and it marked the beginning of a more systematic and observational approach to studying life forms. With the microscope in the 17th century, Antonie van Leeuwenhoek was the first to observe protozoa and bacteria when he observed microorganisms and saw an unseen world teeming with life. In the 19th century, Charles Darwin's theory of evolution by natural selection gave a perspective on the diversity and adaptation of life. His work demonstrated that life is not static but is constantly evolving in response to environmental pressures. For us, the story of human evolution is based on our understanding of how exactly we come into this physical reality as human beings. What is the process of human development?

Human Life—Conception and Birth

From conception to birth, the fascinating beginning of life's journey is a complex biological process that unfolds in stages over approximately forty weeks. Our journey starts in the fallopian tube, where a single sperm among millions succeeds in penetrating the ovum's protective layer. This moment of fertilization marks the creation of a zygote, a single cell endowed with the genetic blueprint to develop into a unique individual (see Plate 18). The zygote contains all the necessary information, half from the mother and half from the father, to guide the complex process of human development. Then the entire process begins with rapid divisions and the formation of a fluid-filled cavity, followed by the embryo undergoing gastrulation, a pivotal process where three distinct layers, the ectoderm, mesoderm, and endoderm, are formed. These layers are the foundation from which all organs and tissues develop. The ectoderm gives the nervous system and skin; the mesoderm forms the heart, muscles, and bones; and the endoderm becomes the lungs, liver, and digestive system.

By the end of the eighth week, all major organs begin to form, and the embryo is now referred to as a foetus. This period is crucial and sensitive; the developing foetus is vulnerable to environmental factors that might affect its development. The remainder of the pregnancy shows rapid growth and maturation of the foetal organs. The foetus grows larger, and its body proportions change. The brain develops intricately, forming billions of neurons and making connections that will last a lifetime. By the end of the second trimester, the foetus has developed a grip, can suck its thumb, and even responds to external sounds. In the last three months, there is continued growth and the fine-tuning of physiological functions. The foetus gains weight, and its organs mature to prepare for life outside the womb. The lungs develop surfactant, a liquid made of proteins and lipids that lines the lungs and is crucial for helping them to breathe. Nearing the completion of the full term, the foetus positions itself for birth.

This incredible journey culminates with birth, a momentous event where the foetus, now the baby, leaves the protective environment of the womb and takes its first breath. This transition is supported by dramatic changes in its circulatory system and the ability to breathe independently. From a single cell to a complex, fully formed human being, the journey from conception to birth is the marvel of biological development or the miracle of life itself. The number of cells in the body by the time of birth is estimated to be in trillions, a staggering increase from the single cell of its inception. Each step, meticulously orchestrated by genetic and biochemical signals, reflects the profound intricacy of the beginning of life. However, the life journey of a baby actually begins at the prenatal stage. The importance of the first 1,000 days of life, from pre-conception to post-natal stage, is now recognized in biology and medicine. Ayurveda procedure known as *Garbha Sanskara* aims at promoting a healthy and positive pregnancy through diet, yoga, meditation, and prayers. Increasing scientific evidence highlights the importance of prenatal upbringing and good impressions on the foetus. This journey of roughly nine months not only highlights the scientific wonder of human development but also the profound connection that binds us all to the timeless cycle of life.

We would like to add here a very interesting idea that scientists are talking about: the intriguing concept that every cell may possess a mind, a form of independent intelligence or consciousness. Cells exhibit intricate behaviours, and they respond to internal and external signals in a very organized manner. They carry out tasks such as growth, division, and differentiation. This is a level of coordination that goes well beyond a mere mechanical process. This cellular intelligence hints at a more pervasive form of consciousness, which is distributed throughout the body.

In cellular and molecular biology, scientists have remarkably unravelled the complexities of cellular processes, yet the fundamental question of how cells "know" where, when, and how much to grow is an enigma. Noble laureate Sydney Brenner refers to this as a "long interlude", admitting thereby the limitations of science in comprehending the intricacies of life.

Cells and Consciousness

The idea that every cell has an independent form of intelligence and consciousness is fascinating. The Max Planck Institute for Biological Intelligence has developed a new approach to recording cellular activities. The technology is based on a recorder protein, which becomes irreversibly labelled with a fluorescent dye when an event of interest occurs in its vicinity. This enables scientists to simultaneously study very large numbers of cells. A recent study published in European Molecular Biology Organization (EMBO) reports suggests that consciousness pervades all life forms, from the smallest cells to the most complex organisms. Researchers have spotted hallmarks of intelligence—learning, memory, and problem-solving—outside brains as well as within them. This has led to the development of a new field called basal cognition.

Researchers at Brunel University London suggest that even very simple single-celled organisms like amoebas can "feel" by reacting to the environment. When they feel something is wrong in the environment, some amoebas build a protective calcium carbonate shell or house around themselves. A study published in the *International Journal of Molecular Sciences* discusses the biomolecular structures and processes that allow and maintain cellular consciousness from an evolutionary perspective. These studies suggest that cells may indeed possess a form of intelligence and consciousness, although our understanding of these phenomena is still evolving. It's a fascinating field with much more to discover! To agree that unicellular organisms know the secret opens up a whole new area in consciousness studies.

Scientific inquiry continues incessantly, acknowledging the humility required in the face of the mysteries of life. This notion that cells "have a mind" is totally in sync with philosophical, mythological, and spiritual perspectives that recognize a deeper interconnectedness within existence that goes well beyond the confines of individual organs or systems. This is in sync with the Vedic concept of *Jivatma*, where every living organism, from unicellular to the most evolved ones, carries a "self" consciousness.

Mitochondria—A Powerhouse

The mitochondria, often hailed as the powerhouses of the cell, stand at the crossroads of numerous discussions about the essence of life. Their role transcends the mere production of adenosine triphosphate

(ATP), often referred to as the "energy currency" of the cell because it provides the chemical energy that fuels biological processes; it extends into the realms of evolution, genetics, cellular regulation, and even the ageing process, weaving a complex narrative that offers interesting insights into the question "what is life?"

The connection between the Sanskrit term "*matru*", meaning mother, and the fact that mitochondria are inherited maternally is a fascinating linguistic coincidence rather than a direct etymological or scientific link. The term "mitochondria" derives from the Greek words "mitos", meaning thread, and "chondrion", meaning granule, which reflect the organelle's appearance under a microscope rather than its mode of inheritance. This naming was attributed to the thread-like structures observed in cells during the late 19th century, when mitochondria were first described.

The maternal inheritance of mitochondria is a biological phenomenon discovered long after the term was coined. It refers to the process by which mitochondria and their DNA (mtDNA) are passed down from mothers to their offspring, largely because the mitochondria present in the sperm are usually lost or degraded after fertilization, leaving only the egg's mitochondria to be inherited by the embryo. While the correlation between the term "*matru*" and maternal inheritance of mitochondria is serendipitous, it recognizes a deeper, universal recognition of the maternal contribution to life across cultures and scientific disciplines. Across many languages and scientific terminologies, the importance of the mother or maternal lineage is acknowledged, whether in the context of genetic inheritance, cultural heritage, or the nurturing of life.

In biology and genetics, the maternal inheritance of mitochondria has profound implications for our understanding of evolution, disease, and molecular mechanisms. It highlights the intricate ways in which life is passed down and sustained across generations, offering a powerful example of the interconnectedness of all living organisms. The maternal inheritance of mitochondria resonates with the universal significance of motherhood in nurturing and perpetuating life.

At the heart of their significance is the process of cellular respiration. This fundamental process is the bedrock upon which the complexity of higher life forms is built. It allows for the development of structures and systems that require vast amounts of energy to maintain, grow, and reproduce. This aspect alone positions mitochondria as central in the story of life, highlighting the intricate balance between energy consumption and the maintenance of biological order.

Mitochondria are believed to have originated from an ancient symbiosis between an ancestral eukaryotic cell and a prokaryotic organism, according to the endosymbiotic theory. This evolutionary milestone highlights the interconnectedness of life forms and shows how life evolves through cooperation and integration rather than merely through competition and isolation. Mitochondrial DNA illuminates the interconnectedness of living organisms, from the molecular mechanisms that fuel life to the evolutionary histories that bind species together, a vital link to knowing "What is life?"

Glucose—Fuel of Life

Glucose is often called the "fuel of life". It is a simple sugar that stands at the heart of life's energy processes and is a universal energy source for a vast range of organisms, from the tiniest bacteria to the largest trees, and, of course, for animal and human life. We discussed how ATP is generated within the mitochondria of cells. Essentially, ATP is generated from the breakdown of glucose through central metabolic pathways known as glycolysis. This small molecule is incredibly versatile and essential; it acts as the primary building block from which energy is extracted to power the countless activities that constitute life.

Glucose is the "currency" of energy in the biological world; glucose fuels biological systems just as money fuels economies. It can be stored, traded, and used across different life forms, providing the essential energy for growth, reproduction, movement, and even mental processes. In plants, glucose is produced through photosynthesis, a miraculous process of nature where sunlight, carbon dioxide, and water are converted into glucose and oxygen.

This not only sustains the plant itself but also forms the basis of the food chain. Animals and humans consume plants (or other animals that have eaten plants) and obtain glucose through their diet. Once consumed, the glucose is transported via the blood to cells throughout the body, where it makes its invaluable contribution to meeting the organism's energy needs.

Glucose has the uniqueness of universality and simplicity. Despite the vast differences among living organisms, glucose is a common denominator, another example of the interconnectedness of life and the unity in diversity. It powers vitality in life, such as developing muscles in animals, allowing them to move and migrate; fueling the brain, enabling complex thoughts and emotions; and sustaining plants, letting them grow and produce oxygen. The role extends beyond providing immediate energy. It is stored as a reserve and is available when the energy intake is low. Glucose is stored for future use in the form of glycogen in animals and starch in plants. This strategic energy management is crucial for survival across diverse living conditions and environments.

In humans and other mammals, glucose is regulated and utilized by cells, and this activity is finely tuned by insulin. This hormone plays a key role in managing blood sugar levels, allowing cells to absorb glucose and use it for energy, thereby ensuring that the body's functions operate smoothly. Insulin's role in regulation highlights the delicate balance within our bodies, ensuring that glucose serves its purpose as a vital source of energy without reaching harmful levels. Impaired glucose metabolism can lead to diabetes. The significance of glucose is felt daily. It is the reason we feel energized after eating a meal and why managing our blood sugar levels is essential for maintaining good health and preventing diseases. The brain, which is particularly reliant on glucose, consumes about half of the body's glucose energy, endorsing sugar's critical role in supporting mental functions. In essence, glucose is more than just a molecule; it's a fundamental life force that energizes, connects, and sustains the living world. Its omnipresence in the diet of organisms across the planet enhances its importance as the fuel of life, a cornerstone of biological energy that powers the vast diversity of life on Earth.

Water—The Necessity of Life

If glucose is the fuel of life, water is the elixir of life. About 60 per cent of our body weight is water, and it plays a critical role at the cellular level. Water is a transport system, carrying essential nutrients and oxygen to our cells and flushing out waste products. It's also the medium where vital chemical reactions take place, allowing our bodies to function properly. The unique properties of water make it a perfect fit for life. For instance, water's high heat capacity helps regulate our body temperature, preventing it from overheating or getting too cold. Because water is a superb solvent, it can dissolve a wide range of molecules, making them readily available for biological processes. This quality is crucial for everything from nutrient absorption in plants to the chemical reactions that power our muscles.

The importance of water extends far beyond our bodies. It's the foundation of healthy ecosystems, with plants relying on it for photosynthesis and animals depending on it for survival. Even the search for life beyond Earth focuses on the availability of liquid water as a key ingredient for survival. Yet, this vital resource isn't infinite. Losing too much water through dehydration is life-threatening. Our bodies can barely survive for a few days without water, as even mild dehydration disrupts essential functions.

Interestingly, the origin of water on Earth is still debated by scientists. Some theories suggest that water arrived on Earth via comets or asteroids early in our planet's formation. Recent research even explores the possibility that water was already present within the building blocks that formed Earth. Understanding water's origin can shed light on the potential for life elsewhere in the universe, further endorsing its importance as the lifeblood of our planet and potentially of others.

Proteins—The Building Blocks

Proteins are called the building blocks of life for a reason! These complex molecules are essential for almost everything our bodies do. Proteins are constructed from smaller units called amino acids. There are twenty different amino acids, which can be linked together in many combinations to create a vast array of proteins, each with a specific function. Our bodies can naturally build up or synthesize some amino acids, but other essential amino acids must be obtained through our diet. The remaining are considered non-essential, but they can still be helpful for various bodily functions. Regardless of origin, these amino acids are constantly broken down and rebuilt to create new proteins as needed, ensuring a continuous cycle of repair, growth, and function. Proteins play a critical role in our metabolism by acting as enzymes, which are like tiny molecular machines that speed up the countless chemical reactions within our cells to ensure that everything runs smoothly.

The Vital Principle—Life Force

The nature of the "life force" remains one of the great mysteries of existence. Could it be an undiscovered scientific phenomenon, a metaphysical entity, or a spiritual presence? Perhaps it's an intricate blend of all three, a bridge that connects the material to the immaterial, the known to the unknown, and the finite to the infinite. As our understanding of the universe continues to evolve, the concept of a life force challenges us to look beyond conventional frameworks. It invites us to consider a more holistic view of life, one that factors in not only the physical aspects of existence but also the philosophical and spiritual dimensions. The life force then becomes more than just a subject of inquiry; it becomes a symbol of our quest to understand life in its fullest, most profound sense.

While not measurable like physical processes, the life force can be seen as a metaphor for the dynamic, complex, and interconnected processes that sustain life. It adds depth to our understanding of what differentiates the living from the non-living, highlighting the less tangible aspects of existence. Life force is known by various names across cultures and disciplines, or simply as "energy". It has long been a subject of intrigue and speculation, as it spans across science, philosophy, and spirituality. Its enigmatic nature raises profound questions: Where does this life force come from? What could its intrinsic nature be?

In modern science, the concept of a life force has been replaced by concrete theories such as bioenergetics, the study of the transformation of energy in living organisms. This field acknowledges that while living organisms abide by the same physical laws as inanimate matter, they exhibit complex energy transformations that are unique to life.

From a scientific standpoint, the idea of a distinct life force is viewed rather sceptically because it eludes empirical methodology, which is the key to scientific inquiry. Modern biology explains life through a series of biochemical processes, from the cellular respiration that energizes organisms to the genetic codes that dictate their form and function. Yet, this scientific understanding, as comprehensive as it is, often leaves untouched the more profound question of what animates these processes—what turns a mere collection of molecules into a "living" being?

Philosophically, the concept of a life force is seen as relevant in metaphysics. Here, the life force is not just a physical entity; it is something that transcends it. In many Eastern philosophies, this force is closely linked with consciousness and is considered the very essence of being. The Upanishads, for instance, speak of a universal life force, *prana* in Sanskrit, which literally means "life force" or "life energy", and which connects all beings, suggesting an underlying unity in the diversity of life. It is the principle of vitality. It is often associated with breathing but is considered more than just the physical aspect of breathing. *Prana* is the subtle energy that permeates all forms of life. It is also referred to as *jiva*, *jivatma*, or self, which merges with universal consciousness to become "oneness" (Advaita).

Many cultures have conceptualized a life force or vital energy believed to be the essence of life. These concepts, although different in detail, share the common belief that this life force is what distinguishes the living from the non-living. In traditional Chinese culture, *Qi* (Chi) is the vital force that flows through all living things. It is considered to be the essence of existence. Qi is a central concept in Chinese medicine and martial arts, emphasizing the flow and balance of energy within the body because it unites the body, mind, and spirit. The French philosopher Henri Bergson introduced the concept of "élan vital" in his philosophy. It refers to the vital impetus, life force, or impulse that drives evolution and creativity. Wilhelm Reich was a psychoanalyst and a student of Sigmund Freud. He proposed the concept of "orgone energy". He believed that this life force energy exists within living organisms and even in the atmosphere.

In trying to understand the essence of life, it's essential to integrate ancient wisdom with modern scientific knowledge. While concepts like Qi and *prana* are often considered metaphysical and difficult to validate by current science, they offer a holistic understanding of life, emphasizing balance and harmony. Modern biology, on the other hand, provides a tangible and measurable understanding of life through genetics, cellular biology, and biochemistry. This scientific perspective offers insights into the physical processes that characterize living organisms, from metabolic activities to genetic inheritance.

Kundalini: According to various spiritual traditions, particularly Yoga-Vedanta, there is another concept of life force or energy referred to as the *Kundalini*. It is the primal energy located at the base of the spine. We can imagine this as a dormant but powerful force within us, visualized as a coiled serpent sleeping at the base of the spine. The practice of various yogic disciplines is to "awaken" this potent energy. When awakened through practices such as yoga or meditation, this energy rises through the spine in a channel called the *Sushumna*, activating centres of energy within us called *chakra*s.

Chakras and Nadi: An important concept of yoga is the *chakra*, which is a part of the subtle (*sukshma*) body. There are seven *chakra*s: *mooladhara* (at the base of the spine), *swadhisthana* (in the sacral bone), *manipura* (at the navel), *anahata* (in the chest or heart region), *vishuddha* (in the throat), *adnya* (the forehead), and *sahasrara* (top of the head). *Nadi* is the channel through which the life force or vital energy moves. Several *nadi*s connect the *chakra*s. The main three *nadi*s include *sushumna* (the central *nadi* that connects *mooladhara* to the *sahasrara chakra*). *Ida* and *pingala* (left and right of the spine) run parallel to the *sushumna nadi*, and they have solar and lunar dominance, respectively. *Kundalini*, the supreme energy, rises from the *mooladhara* through the *sushumna nadi*. When it reaches the *sahasrara*, then the yogi gets detached from the body and mind (see Fig. 4.3).

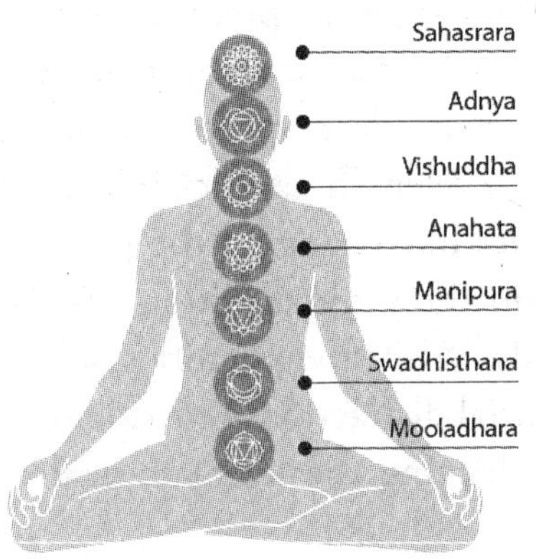

1. SAHASRARA *(Crown chakra)*
Located just above the crown of the head. It is the centre of pure consciousness, beyond the elements, connecting to divine energy and our highest self. This centre is also associated with Friday, the day of Venus, who embodies love, beauty, and spirituality. Connecting with the crown centre on Fridays is said to deepen spiritual practice. This is the last bastion of the Om Way as we surrender to the wisdom of pure consciousness.

2.AJNA or ADNYA *(Third-eye chakra)*
At this stage the practitioner is open, intuitive, and self-assured. This "centre of command" is located between the eyebrows. It is the seat of the controlled mind, of awareness, and 'beyond' the physical elements. It is the centre of knowledge, and intuition. This centre is aligned with Thursday, named after Jupiter, representing wisdom, expansion, and insight. Working with the third-eye centre on Thursdays is said to enhance its energies.

3. VISHUDDHA *(Throat or pharynx chakra)*
To speak or live the Truth. From the Anahata upwards, the human being is free of any physical or material desires or wants. It is the centre for self-expression and understanding. Aligned with Wednesday, it is associated with Mercury, symbolizing communication, expression, and intellect. Focusing on the throat centre on Wednesdays is said to help overcome desires and selfishness.

4. ANAHATA *(Heart chakra)*
The heart centre, spiritually, is the home of the inner, ubiquitous self. With emotional qualities such as peace, love, and openness, it is the core of selfless love. Corresponding to Tuesday, it is named after Mars and embodies courage, passion, and action. Engaging with the heart centre on Tuesdays is said to enhance its healing properties.

5. MANIPURA *(Navel, solar plexus chakra)*
Located at the navel, has a strong inner fire (Agni), which helps us to digest our food and life's experiences. Its function is to optimize our power to help us navigate our lives with strength and determination. Aligned with Monday, this centre is associated with the Moon and represents emotions, intuition, and nurturing. Balancing the solar plexus on Mondays is said to be beneficial for overall well-being.

6. SWADHISTHANA *(Sacral or pelvic chakra)*
Located in the pelvic region, this is the control centre of fluidity, adaptability, creativity, emotions, sexual energy, and the unconscious, and helps us regulate our emotions and desires. This centre is linked to Sunday, the day of the Sun, symbolizing creativity, vitality, and self-expression. Focusing on this centre on Sundays is said to amplify these effects.

7. MOOLADHARA *(Root or base chakra)*
This is located at the base of the spine, connected to the earth element. It governs the four basic or primal centres of energy, of the urges of a human being: food, sleep, self-preservation, and sex. This root centre is associated with Saturday, known as "Saturn's day", representing stability, security, and discipline. Working with the root centre on Saturdays is said to enhance its energy.

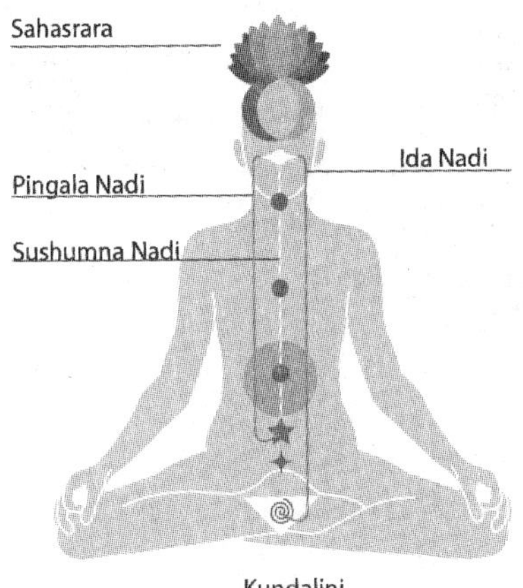

Nadi is the channel through which the life force or vital energy moves.

Fig. 4.3. Chakras and Nadis.

In modern science, there is no direct evidence to show the *Kundalini*, *sushumna*, and *chakras* as physical entities. These might seem exotic or complex, but at their core, they speak of a universal human experience: the quest for enhanced capabilities, deeper understanding, connection, and fulfilment. Whether it is a literal energy or a metaphor for human potential, *Kundalini* offers a captivating concept for those seeking deeper spiritual understanding and innovative scientific explorations.

Those who put in the austere and rigorous practice of yoga and meditation may develop *siddhi*s, which are extraordinary and extrasensory powers that can arise from deep spiritual practice (see Fig. 4.4). These can range from telepathy and clairvoyance to more "magical" profound abilities like changing the physical form or levitation. However, within the yogic tradition, *siddhi*s are regarded as double-edged swords. While they signify spiritual progress, they are also seen as potential distractions or roadblocks and will act as deterrents from achieving the ultimate goal of liberation.

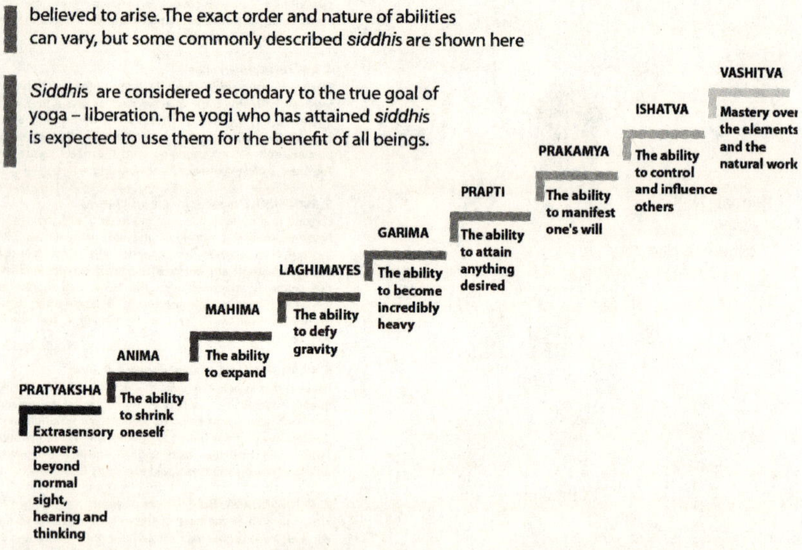

Fig. 4.4. The manifestation of *siddhis*: Great powers, greater responsibilities.

An interesting "take" on the concept of *siddhi*s, if we wish to see an analogy, is demonstrated in the sci-fiction 2014 film *Lucy*. The film portrays a dramatic approach. Lucy gains extraordinary abilities as a synthetic drug unlocks a greater percentage of her brain capacity. She transcends physical limitations by manipulating matter, accessing information from the environment, and even achieving telekinesis. However, this rapid power surge lacks the discipline and self-awareness fostered by yogic practices. Lucy's journey is marked by confusion and a struggle to control her newfound abilities. The film warns us by showing the dangers of possible misuse of powers like *siddhis*. These powers are acquired after following a very hard and disciplined spiritual path. They cannot be dissipated. This also gives us a message that the commodification of spirituality is not the way to evolve while transitioning from modern science to meta-science.

Creating Life?

Creating life from non-living materials or synthesizing life in the laboratory is one of the most ambitious scientific endeavours of this age. The attempt to create life in a test tube has been a focal point of scientific inquiry that touches on the diverse aspects of the origin of life, synthetic biology, and the fundamental principles that define living systems. Various efforts have been made to synthesize life or life-like systems in laboratory settings. These efforts range from attempts to recreate the conditions that might have led to the emergence of life on Earth to the construction of synthetic cells that mimic biological functions.

However, as research continues to push the boundaries of synthetic biology, biochemistry, and systems biology, the challenges of scientific understanding, technical capability, and ethical considerations remain. Modern science may gradually overcome these obstacles. However, the effort will always require careful reflection on the philosophical implications and responsibilities entailed in bringing new forms of life into existence.

Stanley Miller and Harold Urey conducted the landmark experiment in 1953 to simulate the conditions of the early Earth's atmosphere. They exposed a mixture of gases (methane, ammonia, hydrogen, and water vapour) to electrical sparks to mimic lightning. The experiment produced amino acids, the building blocks of proteins, demonstrating that organic compounds essential for life could form under prebiotic conditions. They successfully produced organic molecules from some of the inorganic components thought to have been present on prebiotic Earth. This experiment laid the foundation for prebiotic chemistry.

A team led by Craig Venter synthesized the first bacterial genome from scratch and inserted it into a recipient cell whose DNA had been removed. This achievement marked a significant step towards creating synthetic life, although the host cell itself was not synthesized. Creating a minimal cell provides insight into the essential components of life and opens avenues for building synthetic cells with specific functions.

Scientists have also been working on creating protocells, which are simplified models of living cells that exhibit some characteristics of life, such as compartmentalization, metabolism, and the ability to grow and divide. While protocells can demonstrate basic life-like properties, they do not constitute fully living systems. While significant progress has been made in constructing life-like systems and understanding the chemical origins of life, scientists have not yet created a fully synthetic life form from scratch. Each effort, however, contributes to the knowledge of life's fundamental principles and brings the researchers closer to understanding and potentially synthesizing living systems in their entirety.

The effort to create life from non-living materials, to replicate life in a laboratory in a test tube, raises very direct scientific, ethical, and philosophical questions. The complexity of living systems, the presence of consciousness, and the intricacies of evolutionary processes are hurdles that scientists have to face. These endeavours raise very profound and basic questions about the definition of life itself, the potential applications and risks of synthetic biology, and our responsibilities as "creators" of living systems. These obstacles come from the inherent complexity of living systems, our incomplete understanding of life's origins, and ethical considerations.

One fundamental issue is the absence of a universally accepted scientific definition of life. We have a vast range of processes and properties, from metabolic activities and reproduction to consciousness and adaptability. The ambiguity surrounding what precisely constitutes life complicates efforts to synthesize it. Life also exhibits emergent properties that are not fully understood—properties that arise from the complex interactions among a system's parts. Several scientists believe that consciousness emerges from the activity of neural networks, but how? That remains a mystery! How are systems integrated? Even the simplest living organisms are incredibly complex, with integrated systems for energy production, growth, waste disposal, and reproduction. Replicating this integrated complexity in a synthetic context has to be quite a task. Creating or replicating the molecular machinery of life is the next big challenge. This would include replicating DNA, protein synthesis, and metabolic pathways, which would involve intricate coordination and regulation. Creating or replicating these processes is not just a matter of synthesizing the molecular components but also ensuring that they function together harmoniously. Nature has managed this seamlessly; can humans replicate that?

Prebiotic chemistry is a branch of chemistry that studies the chemical processes and reactions that occurred on Earth before life emerged. It focuses on understanding how simple organic molecules could have formed under these conditions and how these molecules might have eventually given rise to life. Scientists have made significant progress in understanding prebiotic chemistry, but many questions about the origins of life and the conditions that might have led to the emergence of life on Earth remain unanswered. A defining characteristic of life is its ability to replicate and evolve. How can such synthetic systems be designed that can self-replicate and undergo natural selection? The exact pathways through which inanimate materials transitioned to living entities are subjects of speculation and research. There are legitimate concerns about the safety of creating life in the laboratory, including the risk of accidental release into the environment and the potential for misuse of synthetic biology in harmful ways. Could there be a recreation of science fiction in multiple Frankensteins? These concerns also question the consequences of creating new forms of life, including ecological impact, unintended or unknown side effects, and the moral status of synthetic organisms.

Finally, this endeavour to create life raises the key question about science's role in "playing God".

Mythological Perspectives on the Beginning of Life

Mythology, religion, and ancient philosophy have explained the beginning of life in various ways. Life is often depicted as a gift from divine entities or as a result of cosmic events. For instance, in Greek mythology, Prometheus created humans from clay and animated them with life. Similarly, many indigenous cultures view life as a cycle, deeply connected to the rhythms of nature. Another Greek myth is that in the beginning, there was chaos, from which emerged Gaia (earth), Tartarus (the underworld), and Eros (love). Gaia then gave birth to Uranus, the sky. Together, they produced the Titans and other primordial deities who later populated the earth. According to Hawaiian mythology, all living beings were the progeny of the goddess Papa and the sky father, Wakea. They created the first humans, and from them, the Hawaiian people were born. Nuwa, the Chinese goddess, formed humans from yellow clay and brought them to life, as goes the Chinese myth. The Sumerian god of wisdom, Enki, and Ninhursag, the earth goddess, created humans. Enki shaped the first humans from clay, and then Ninhursag breathed life into them. In some Egyptian myths, the god Atum began with the primordial mound that emerged from the waters of chaos, and from this mound, all life originated. In Norse mythology, the gods Odin, Vili, and Vé found two trees, Ask (an ash tree) and Embla (an elm tree), on the seashore, and they transformed them into the first man and woman.

These particular myths do vary in their details, and they speak of the creation of human beings more specifically. But essentially, the use of clay or trees is significant, implying the use of the natural elements and the breath of life coming from the creator. They all speak of a divine or supernatural being responsible for creating the first humans. In the Christian Bible, God created all living beings. The Book of Genesis gives an elaborate description of the creation of life. After God created the sun, moon, stars, and earth, he filled the seas with living creatures and the skies with birds. Next came the land animals, and finally, God created "man in his own image". God formed Adam, the first man, from the dust of the ground. Then He created Eve, from Adam's rib, to be his companion.

Philosophical Perspectives

As humanity progressed, these mythological views gradually gave way to philosophical inquiries on "What is life?" and these continue to evolve, drawing from interdisciplinary fields that include biology, physics, cognitive science, and ethics. Philosophers and thinkers have consistently expanded the dialogue on "What is life?" by integrating insights from the sciences, ethics, and technology. Their diverse perspectives highlight the complexity of defining life, the moral considerations of altering or creating life, and the deep philosophical questions about consciousness, identity, and our relationship with the natural world.

Aristotle attempted to define life in more empirical terms. His concept of the "soul" or "psyche" as the essence of life, responsible for growth, reproduction, and purposeful movement, laid the groundwork for later biological exploration.

In the Upanishads, life is explored through introspective questioning. The sages ponder deeply on the nature of existence and consciousness, asking, "What is it that, by knowing which, everything else becomes known?" (*Mundaka Upanishad*). In the *Chandogya Upanishad*, it is stated that by knowing Pure Consciousness, everything is known. Such profound aphorisms highlight the intrinsic curiosity about life's true essence. It is further explained that life is a cycle of birth and death perpetuated by ignorance of our true nature. Life is an opportunity for us to inquire into the nature of reality and seek self-realization, enlightenment, and liberation (*moksha*). In the Bhagavad Gita, it is said, "For the soul, there is neither birth nor death at any time. It has not come into being, does not come into being, and will not come into being. It is unborn, eternal, ever-existing, and primaeval. It is not slain when the body is slain." This verse explains the Vedic understanding of life as a manifestation of the eternal soul and suggests the continuity of life beyond the physical body, which is matter and all matter has to die or disintegrate.

Contemporary philosophers have expanded the dialogue to consider not just the biological basis of life but also the ethical, existential, and technological implications of living systems.

Daniel Dennett, a prominent philosopher and cognitive scientist, proposes that consciousness arises from physical processes in the brain and is a form of information processing. Dennett's views challenge traditional notions of a sharp distinction between life and non-life, suggesting a continuum of complexity and consciousness across different forms of life. Thomas Nagel, in his book *Mind and Cosmos*, argues that consciousness and subjective experience pose fundamental challenges to a purely physicalist view of evolution. Nagel calls for an expanded conceptual framework that can account for the mental and moral dimensions of life.

The American astronomer Carl Sagan said, "The nitrogen in our DNA, the calcium in our teeth, the iron in our blood, the carbon in our apple pies were made in the interiors of collapsing stars." "The cosmos is within us. We are made of star stuff. We are a way for the universe to know itself." Sagan highlighted the cosmic connection and the material continuity of life with the universe.

Consciousness Perspective

We cannot discuss life without reflecting on consciousness, which is one of the most profound and elusive topics. It has been a subject of fascination and inquiry across various fields, including biology, psychology, philosophy, and neuroscience. While we understand the biological processes that sustain life, the nature of consciousness—the subjective experience of being—continues to prompt intense debate and research. The simple fact is that consciousness is subjective, and one cannot describe or discuss the subjective without objectifying it! It's a phenomenon that has perplexed and fascinated scientists, philosophers, and thinkers for centuries as they try to understand and explain its relationship to life. The concept of a life force is closely linked to the study of consciousness. Understanding consciousness—a feature that is often seen as a definitive attribute of living entities—requires us to explore both the physical processes in the brain and the philosophical perspectives on the mind and self-awareness. These are dealt with in greater detail in Chapter 5.

Often considered a hallmark of higher life forms, particularly humans, consciousness adds another dimension to the definition of life. It is awareness—the capacity to experience or be aware of our existence, environment, and thoughts. This includes the ability to perceive, feel, and have subjective experiences. A higher level of consciousness, or self-awareness, involves being aware of oneself as a distinct entity in the world. This includes self-reflection and introspection.

Schrödinger's Vision of Life

Erwin Schrödinger's seminal work *What is Life?*, published in 1944, stands as a monumental inquiry into the physical aspects of living cells, particularly focusing on the principles of heredity and the nature of biological order. It is a foundational text at the intersection of physics and biology, posing questions that continue to inspire scientific inquiry into the nature of life. While his speculative answers have been refined and expanded with the advancements of molecular biology, genetics, and beyond, the core idea that life can be understood through the laws of physics persists. The Nobel Prize-winning physicist, best known for his contributions to quantum mechanics, ventured into the domain of biology to address fundamental questions about life from a physicist's perspective. He proposed an "aperiodic crystal" as the hereditary material (foreshadowing DNA) and highlighted how life, by constantly exchanging energy with its environment, maintains order (negative entropy) against the natural tendency towards disorder, influencing our understanding of the physical basis of life. The core ideas that Schrödinger presented in his work impacted the development of modern biology significantly. Recent research continues to explore and expand upon these concepts. His ideas inspired a generation of physicists and chemists, including Watson and Crick, to explore biological problems, leading to the rapid growth of molecular biology in the mid-20th century.

Schrödinger was inspired by different intellectual traditions to understand the nature of reality. He had a keen interest in Vedanta and Hindu philosophy. His understanding of Eastern thought influenced his philosophical views and his approach to the nature of reality, consciousness, and the universe. He was particularly interested in the concept of unity or nonduality, which emphasizes the interconnectedness and unity of all things, transcending the conventional distinctions between the self and the other, subject and object. The Vedantic concepts of the unity of all things and the one-ness of existence resonated with Schrödinger because the idea connected with the concept of the interconnectedness of particles through quantum entanglement and the wave-particle duality, where particles can exhibit both wave-like and particle-like properties simultaneously. He explored parallels between Eastern notions of consciousness as fundamental and universal and the role of consciousness in quantum mechanics, where the act of observing or measuring plays a central role in determining the behaviour of quantum systems.

The Quantum Perspective

Quantum physics, with its exploration of the subatomic world, has opened doors to understanding the profound and often non-intuitive aspects of the universe. At the quantum level, particles can exist in multiple states simultaneously, and the act of observation itself can influence the outcome—a concept known as the observer effect. Some theorists and philosophers have drawn parallels between this quantum behaviour and the concept of a life force or energy that permeates all things. They suggest that just as quantum particles are the fundamental building blocks of matter, a life force could be a fundamental component of life, operating beyond classical physics. This idea also touches upon consciousness playing a role in shaping reality, a point of view that resonates with many spiritual perspectives.

Quantum realities explore how quantum biology and quantum theories are reshaping our perception of biological processes and consciousness and the implications of this. Quantum biology means applying quantum mechanics to biological objects and problems. It studies how quantum phenomena, typically observed at the atomic and subatomic levels, play a role in biological processes.

The Debate and Challenges: Many quantum consciousness theories are speculative, and mainstream scientists view them with scepticism. Quantum biology could potentially complement traditional biology by providing explanations for phenomena that are inexplicable by classical physics alone. Understanding quantum phenomena in biological processes could lead to revolutionary advancements in areas like medicine, energy efficiency, and technology. In terms of consciousness, it opens up new dimensions for understanding the mind-brain relationship, bridging materialistic and dualistic views. The integration of quantum theories with biological understanding represents a unique, though controversial, field of study. While quantum biology offers insights into life's fundamental processes, its application to consciousness remains largely theoretical. As research progresses, these quantum perspectives could significantly alter our understanding of life and consciousness, challenging traditional biological paradigms and expanding our conceptual boundaries.

Ilya Prigogine on What Is Life?

Ilya Prigogine, a Belgian physical chemist and a 1977 Nobel laureate, made very valuable contributions with his research to our comprehension of life and the dynamics of living systems, offering a novel perspective on how order and complexity emerge from the interactions of a system with its environment.

Prigogine's concept of dissipative structures refers to systems that are far from equilibrium and yet can maintain a steady state through the continuous exchange of energy, matter, and information with their environment. These systems are characterized by their ability to self-organize. This self-organization is a hallmark of living systems, seen in phenomena ranging from the cellular level, such as the metabolic processes within a cell, to the ecological level, such as the organization of ecosystems.

Life as a Dissipative Structure: Viewing life itself as a dissipative structure helps us understand its dynamic and complex nature. Living organisms sustain themselves through the continuous flow of energy from their environment, for example, through photosynthesis in plants or food intake in animals, and the exchange of matter and information within the system, such as cellular respiration and neural communication. Prigogine's work highlights the interconnectedness of living organisms with their environment. He suggests that the evolution of life is driven by the constant interplay between the internal dynamics of organisms and the external forces of their surroundings. Prigogine also contributes to a broader philosophical and ecological discourse on the nature of life, emphasizing the importance of sustainability and resilience in living systems. It emphasizes the necessity of maintaining the delicate balance between stability and flexibility, allowing for the continuous adaptation and evolution of life in response to environmental changes.

Prigogine's work is a reminder of the importance of viewing life within the broader context of its environmental interactions, highlighting the ongoing flux between order and chaos. His research into the nature of complexity, self-organization, and the emergence of order resonates with certain philosophical ideas, including those found in Vedanta.

Vedanta emphasizes the interconnectedness and one-ness of existence, viewing the universe as a dynamic and evolving whole. Similarly, Prigogine's work highlights the interconnectedness and interdependence of systems, where order and complexity emerge from the interactions between components. While Prigogine's scientific work and Vedanta are distinct in their approaches, both offer insights into the nature of reality, its complexity, and the dynamics of change. Prigogine's exploration of dissipative structures and complexity theory provides a scientific framework for understanding the emergence of order and organization in the universe, and Vedanta offers philosophical perspectives on the interconnectedness and unity of all things.

The contributions of Schrödinger and Prigogine represent two pivotal approaches to understanding life. Both scientists challenge traditional views of life, proposing that life's essence cannot be fully explained by classical physics or simple biological models. Their work encourages a multidisciplinary approach, blending physics, chemistry, biology, and philosophy.

Other Scientists on What Is Life?

Many contemporary researchers and theorists have made significant contributions to exploring the essence and complexity of life. The consensus of several scientists is that essentially, life thrives on organization (building blocks from cells to organs), metabolism, response to stimuli, reproduction (flow of genetic information), homeostasis, harnessing energy, and finally, growth and development.

As mentioned earlier, Craig Venter, a prominent figure in genomics, is known for his role in sequencing the human genome and creating the first synthetic bacterial cell. Over thirteen years ago, he was said to have created a synthetic life form. His work raises fundamental questions about the definition of life and the potential for creating life in the laboratory, challenging our understanding of life's boundaries.

A theoretical biologist and complex systems researcher, Stuart Kauffman's work at the Santa Fe Institute has explored the origins of life and the concept of "autocatalytic sets", networks of molecules that are capable of self-replication. His ideas about self-organization and the emergence of complexity in biological systems contribute to ongoing debates about the nature of life. Exploring the origins of life with autocatalytic sets is ongoing research.

A chemist at the University of Glasgow, Lee Cronin's research focuses on the origins of life and the creation of inorganic living systems. He attempted to create "inorganic cells" that exhibit life-like behaviours. Cronin's work challenges traditional views on the chemical prerequisites for life and explores the possibility of a broader definition of life. George Church, a geneticist and molecular engineer, has made significant contributions to synthetic biology, genome sequencing, and gene editing technologies. His work on editing the genomes of living organisms and efforts to resurrect extinct species touches on fundamental questions about the manipulation of life and the ethical implications of such capabilities.

An astrobiologist and theoretical physicist, Sara Walker investigates the origins of life and the search for life on other planets. Her interdisciplinary approach seeks to understand life as a phenomenon of information processing and to develop a general theory of life that can be applied both on Earth and in extraterrestrial contexts.

Christof Koch is a neuroscientist and chief scientist at the Allen Institute for Brain Science. Koch studies consciousness and its relationship to the physical processes of the brain. His work on the neural correlates of consciousness contributes to the broader question of what distinguishes living beings endowed with subjective experiences from non-living matter. A biochemist and molecular biologist, Gerald Joyce's research on the evolution of RNA enzymes in the laboratory provides insights into the potential mechanisms through which life could have originated from simpler molecular systems. His work on "self-sustained replication" of RNA molecules addresses key aspects of the transition from non-life to life.

These scientists, among others, represent the modern efforts to understand "What is life?" through the lenses of genetics, synthetic biology, astrobiology, and consciousness studies. Their works continue to push the boundaries of our understanding, highlighting the complexity, diversity, and profundity of life.

Prospects for Living Robots?

Predicting the future of technology, especially in fields as rapidly evolving as artificial intelligence (AI) and machine learning (ML), is always speculative. However, based on current trends and advancements, can we make some educated guesses about the future capabilities of robots and AI systems? AI and robotics have made significant strides in recent years. From natural language processing to autonomous vehicles, the capabilities of these systems are expanding rapidly. It's reasonable to expect continued advancement in their complexity and functionality.

Machine learning, particularly deep learning, has enabled AI systems to learn and adapt in ways that were not possible a few decades ago. Future AI could exhibit even more sophisticated learning and decision-making capabilities. Despite advancements, the leap from sophisticated AI to consciousness is a massive one. Consciousness involves self-awareness, subjective experience, and qualia—the individual instances of subjective, conscious experience. These are not just technical challenges, but they also involve philosophical and biological questions that AI has not yet begun to address. It is theoretically conceivable that, with enough complexity, a form of "artificial" consciousness could emerge, but this remains in the realm of speculation and science fiction.

Self-replicating machines is also a concept that has been explored in science fiction and theoretical computer science. While simple forms of self-replicating machines could be developed, this is a far cry from biological reproduction. The self-replication of robots would be a controlled, engineered process, vastly different from the natural reproductive processes of living organisms.

Scientists at the University of Vermont, in collaboration with Tufts University, have created Xenobots, a new type of living robot. These millimetre-sized creatures are made from frog stem cells and can move, heal themselves, and even work together. They are designed using AI and show promise for tasks like environmental cleaning and delivering medicine (see Plate 19).

The development of highly advanced AI and robots will raise significant ethical questions. If robots and humans could successfully work together as cobots, they could add good value, but issues like autonomy, rights, and the moral status of AI systems will become increasingly important and contentious. The advancement of AI and robotics is likely to be accompanied by a robust regulatory framework to manage and guide their development, especially as it pertains to safety, ethics, and societal impact.

While advancements in AI and robotics will undoubtedly continue, leading to more sophisticated and capable machines, the emergence of truly conscious AI or entirely self-replicating robots remains speculative and faces enormous technical, philosophical, and ethical hurdles. The journey from current AI capabilities to the scenarios depicted in science fiction like *I, Robot*, a 2004 film projecting a 2035 scenario in Chicago, involves not just technological advancements but also fundamental breakthroughs in our understanding of consciousness and life itself. In the foreseeable future, AI and robots are likely to become more integrated into our lives, performing increasingly complex tasks and perhaps even exhibiting behaviours that "mimic" consciousness. They will be designed and structured to be super-efficient, enough to give us immense complexes with our physical and cognitive limitations! But will they be capable of subjective experience? Whether they will actually be "conscious" or "alive" in the biological and philosophical sense is a very speculative question. The future of AI and robotics is poised to be fascinating and transformative, but it is likely to be different from what is often portrayed in science fiction. What raises the alarm right here, right now, however, is a news flash, "Robot commits suicide in South Korea because it was made to do a lot of work" (Chakravarti, *India Today*, 5 July 2024). *So what next?*

The Individual Perspective

As we reflect on all aspects of this profound question of "What is life", we appreciate that it is entirely an individual and personal perspective of how we understand and see life. Taking the wide canvas of scientific inquiries of Schrödinger and Darwin, which delve into the biological and evolutionary aspects, to the philosophical musings of Aristotle, the depth of the Vedanta philosophy, and the views of various thinkers deepens our appreciation of life's complexity and its interconnection with the universe. Connected to all the questions that we posed at the outset, and after having tried to understand the beginning of the world and the journey through the centuries in this dynamic era, we have tried to understand life.

This brings us to the essence: as individuals, as we think and reflect, we interpret. Because once we have come to this stage of studying ourselves, we stumble when we try to get into deeper levels of our understanding or try to introspect. We are aware that these are all very complex and dynamic processes that we, as human beings, constantly function with. But the complex feature is our instrument of cognition, our mental capacity. The "mind", the much more vital unit of our psychophysical system, of the body–mind complex, is perhaps the most complex instrument that we have to deal with. Everything that we question, analyse, understand, and believe we know is inextricably linked to the mind. What is the mind, is our next question, and it is what we now need to understand as we continue with our journey, transcending modern science to meta-science.

CHAPTER 5

What Is the Mind? From Thinking to Knowing

The mind is everything. What you think, you become.

—Buddha

Once we understood how we evolved as human beings, we tried to get a feel for what life itself is. But as we look a little deeper than our biological and physical constructs, then we face one of the most complex and tantalizing questions that persist in the mind: "What *is* the mind?" As human beings, we are aware and conscious of our capacity to think, emote, understand, and question; then, at some point, we invariably wonder, how exactly do we think? This question has rankled in the minds of philosophers, scientists, medical professionals, especially neurologists, and thinkers, among others, through the centuries. Prima facie, *we know* that we are conscious and self-aware human beings. We also recognize that the mind is more than just the brain's ability to trigger cognitive functions or neurological processes. It is the core of our ability to think, reason, and understand. It also transcends all these functions and represents the amalgamation of all our thoughts, memories, emotions, and understandings that shape our perception of the world and ourselves. It is what makes us distinctly human.

As we try to unravel the mysteries and functions of the mind, we need to explore the relationship between, and the distinction between, the brain and the mind. Addressing the question, "What is the mind?" is difficult without grasping its relationship with the brain. This means understanding the mind as an abstract entity, distinct from the brain, the physical organ. This distinction is essential for us if we want to realize the concepts of consciousness and self-awareness. While often used interchangeably in casual conversation, the brain and mind are distinct yet interlinked concepts within neuroscience, psychology, and philosophy. We also need to trace the evolution of thought through Western and Eastern philosophical and psychological perspectives. This journey from modern science to meta-science will help us see the diverse points of view that have shaped our understanding of the mind and consciousness.

Fundamental Concepts

The brain and mind are thought to be synonymous, but they are not. The brain is a concrete physical organ, an object that can be seen and observed, while the mind is an abstract entity that lacks a quantifiable physical structure. The mind is part of the experience, related to consciousness, perceptions, thoughts, emotions, and memories. Most of us believe that the mind and the brain are inextricably linked because the brain and mind together form the cornerstone of human cognition, behaviour, and experience.

Brain

The brain is our biological control centre. This 1,300–1,400 gm weight physical entity is the foundation of our neurological processes. Made up of billions of neurons and glial cells, this complex organ is intricately wired to regulate everything from motor control and sensory perception to the more subtle nuances of emotional responses. It consists of approximately 86 billion neurons that form a vast network of neural connections, estimated to be in the range of 100 trillion synaptic connections. The intricate web allows for the most complex processes of thought, emotion, and, essentially, experience. Neuroscience has successfully mapped how specific regions of the brain correlate with various functions.

The brain's wrinkled outer layer, called the cerebrum, is divided into two hemispheres, left and right. Each hemisphere is further divided into four lobes—frontal, temporal, parietal, and occipital—with each lobe responsible for specific functions (see Fig. 5.1). The left-brain/right-brain dominance theory, although very popular, is a significant oversimplification of the structure and function of the brain. While there may be some specialization, with the left hemisphere typically associated with language, logic, and analytical thinking and the right with creativity, spatial reasoning, and emotions, both sides work extensively together for most tasks. Many times, this demarcation is also compared with the nature of Eastern and Western cultures as a divided globe.

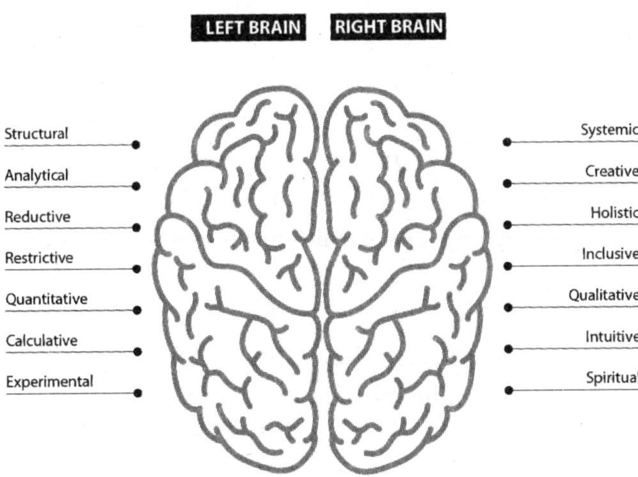

Fig. 5.1. Left and right brain.

This is an interesting imagination; the theory tends to categorize people as "left-brained" or "right-brained" based on the supposed dominance of some functions. However, our brains are far more complex, and most activities involve cooperation between both hemispheres. The enduring appeal of this concept likely stems from its easy application to understanding ourselves and others. Nonetheless, it doesn't reflect the intricate reality of how our brains function. Neurons are specialized cells, the basic working units of the brain, designed to transmit information through the body. These cells communicate with each other through synapses, where neurotransmitters are released to cross the gap between neurons, thus allowing the transmission of signals. The neural networks in the brain are formed by groups of interconnected neurons that process the information received from the external environment or other parts of the body. These networks are responsible for everything, from simple reflexes to complex cognitive processes like learning, memory, and decision-making.

The brain's neurons and neural networks are organized into different regions and structures, each with specialized functions. The cerebral cortex manages higher-order functions such as thought, reasoning, and memory. It's divided into four lobes, and each is responsible for different aspects of cognitive and sensory processing. The limbic system, including the hippocampus and amygdala, is critical to emotion and memory. The brainstem controls basic functions such as breathing, heart rate, and sleep. Interestingly, the neural networks are not static; they change in response to learning and experience, a phenomenon known as neuroplasticity. This adaptability is fundamental to the brain's ability to grow and change throughout a living being's life, influenced by experiences, learning, and even injury.

Recent advances in neuroscience show how neurons and neural networks function. In May 2024, researchers from Google and Harvard mapped the human brain in amazing detail. They created a 3D map using electron microscope data and AI tools that covers a neural network of roughly 57,000 cells and 150 million synapses (see Plate 20). This study can offer a deeper understanding of the inner workings of the human brain, which can help treat some psychiatric and neurodegenerative diseases.

Innovations in imaging technology, such as functional magnetic resonance imaging (fMRI) and positron emission tomography (PET), have helped scientists observe the brain in action and understand how its different parts communicate. Neuroplasticity continues to reveal how the brain can reorganize itself, offering hope for recovery from brain injury and effective learning strategies. Ongoing research also explains how neural networks are involved in specific disorders; therefore, it should be possible to develop targeted treatments for conditions such as Alzheimer's disease, dementia, epilepsy, Parkinson's disease, schizophrenia, and depression.

Mind

The mind is not a physical entity; it functions in the realm of consciousness, emotions, thoughts, beliefs, and imagination. It is often considered the manifestation of brain activity, although philosophical and spiritual interpretations extend beyond this framework. The attributes of the mind, including thinking, reasoning, and problem-solving, allow us to formulate ideas, make decisions, and process and experience emotions; they influence how we feel and respond to different situations. The mind allows for imagination, visualization, and creativity and helps us to "look" at possibilities that go beyond current realities. It interprets sensory information and perceptions and contributes to our understanding and interpretation of the world around us.

Modern neuroscience often views the mind as an emergent property of brain activity, with mental states correlating to specific neural processes. Philosophers and psychologists explore concepts like consciousness, free will, and the self, which are subjective experiences beyond mere physical processes. The brain's functions are grounded in biological and physiological activities, while the mind has a wide range of cognitive and emotional experiences arising from and interacting with these activities. Understanding the full extent of their relationship is one of the greatest challenges in both science and philosophy.

The Brain and Mind Are Different

The distinction between the brain and the mind is fundamental in neuroscience, psychology, and philosophy of mind. Although the brain and mind are closely related and interactive, they represent distinct aspects of the human experience. A thematic infographic makes it easier to understand (see Fig. 5.2). The brain is a physical organ composed of matter and tissue; the mind is non-physical or immaterial, representing mental phenomena that arise from brain activity. The brain controls bodily functions, processes sensory information, regulates emotions, and coordinates behaviour. It performs these functions through electrochemical signalling among neurons and complex neural networks. From a biological perspective, the brain is the seat of cognition and behaviour. The mind is involved with cognitive processes, subjective experiences, and mental states. It includes thoughts, perceptions, emotions, beliefs, memories, desires, intentions, and other aspects of inner experience. It is said to be the *seat of consciousness.* According to some philosophical and spiritual traditions, consciousness projects through the mind. Unlike the brain, which is tangible and observable, the mind is subjective and introspective, representing the inner world of thoughts, feelings, and experiences.

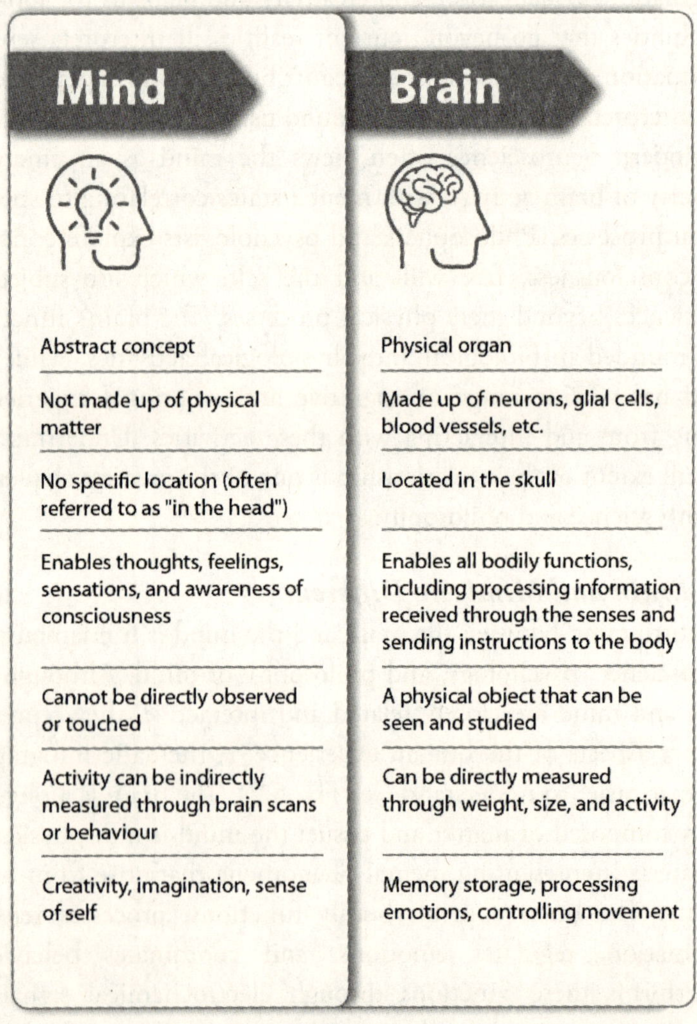

Fig. 5.2. Mind and brain.

The brain can be studied objectively through neuroscientific methods such as brain imaging, electrophysiology, and neuroanatomy. It processes objective information. The mind's activities are accessible only to the individual experiencing them because they represent the subjective experience of thoughts, feelings, and perceptions that arise from brain activity.

What we realize, therefore, is that the brain and the mind are distinct and yet intricately interconnected, and they influence each other in very complex ways. Their relationship is crucial for comprehending human cognition, behaviour, and consciousness.

Where Is the Mind?

This is a tricky question. From a scientific point of view, the mind is closely associated with the brain, and neuroscientists study how different parts of the brain correlate with different mental functions, suggesting a strong connection between the mind and the brain, but there is no location of the mind in the physical organ called the brain. While materialists believe that the mind is entirely a "product" of the physical brain, the dualist perspective suggests that the mind and body are separate entities, and the mind is not found in the physical realm. It is even suggested that the mind may not be confined to the brain alone but may manifest at every level of cellular life, as we discussed in Chapter 4.

Traditionally, we know that the mind is associated with the brain, and many of its functions are tied to specific areas within the brain. However, the exact nature and location of the mind are unknown. Recent scientific studies suggest that the mind is more like a virtual machine running on distributed computers, with brain resources flexibly allocated. David Rudrauf at the University of Iowa discovered that a patient who lacked three regions of the brain, which are thought to be essential for self-awareness, was still self-aware. This suggests that mental functions might not be tied to fixed brain regions and that self-awareness or consciousness is not a product of the brain.

Researchers at the Washington University School of Medicine have found a connection between the brain areas controlling movement and those involved in thinking, planning, and control of involuntary bodily functions such as blood pressure and heart rate. This suggests a literal linkage of body and mind in the structure of the brain. Our understanding of the mind and its connection to the body is constantly evolving. The mind–body connection, which is built into the structure of the brain, will continue to expose more about this complex relationship.

At this stage, we go from modern science to meta-science, as we go from the concrete explanations of biology and physiology into the abstract dimensions of consciousness and cognition. This re-emphasizes that what makes us truly human is not just the physical organ encased within our skulls but the intangible, mysterious mind. When we talk of the mind, we transcend the physical boundaries of neural tissue. The mind is an abstract concept, and this abstraction makes it a subject of great intrigue and debate across various disciplines, including psychology, philosophy, neuroscience, and artificial intelligence.

Our inner experiences—thoughts, memories, feelings, and the subjective experience of *being*—are monitored by the mind. This includes our capacity for reasoning, problem-solving, creativity, being self-aware, and the ability to reflect upon ourselves and our experiences. The interpretative process is a subjective mental capacity uniquely tailored to each of us individually, influenced by our memories, emotions, and cognitive processes. This subjective quality makes the mind very deeply personal and private, accessible only to us, individually, as we experience it.

The distinction is critical. The brain can be observed and studied physically, but the elusive mind can neither be physically localized nor seen even under a microscope. The brain provides the necessary biological substrate for mental functions. The mind has a much broader scope, and it manifests beyond mere neural activity. For instance, the process of decision-making or problem-solving illustrates mental capabilities, which are not solely reducible to electrical impulses and synaptic connections. The interactions between the mind and brain are central to human experience. They represent a fascinating and complex area of study that sits at the intersection of neuroscience, psychology, philosophy, and even the emerging field of neurophilosophy. The interactions span a range of phenomena, including how our thoughts influence our brain's physiology and how the brain's workings manifest as thoughts, emotions, and consciousness.

The Mind's Mechanisms

Emotions and Cognition

Emotional intelligence is our ability to recognize and understand emotions, both ours and those of others. It plays a vital role in personal development, managing relationships, and overall mental well-being. This is crucial to healthy relationships. Our ability to regulate emotions is the key to emotional intelligence.

If we are self-aware, then we recognize and understand our emotions better and also figure out the related causes and effects. Emotions can influence our thoughts and behaviour very significantly. Managing them is most challenging because biases and preconceived notions influence our ability to think and act rationally. We store facts and events according to their relevance and the extent of their impression. Emotional experiences leave deeper impressions, and we remember them more vividly and for longer periods. There is a strong link between emotions and memory. We are also more likely to store information that is meaningful or can be related to existing knowledge. Emotions play a major role in how we process information and make decisions. They colour our judgements and influence the decisions we make. They can significantly influence our thoughts and behaviour and lead to quicker, more intuitive decisions. Emotions also introduce biases. Fear might lead to our being overly cautious, while happiness might make us reckless! Biases and preconceived notions influence our ability to think and act rationally and sometimes result in irrational choices. Emotions help us navigate our interactions with others. Understanding the emotions of others requires empathy, social skills, and responding appropriately to others' emotional states. Regulating and processing emotions, therefore, is essential for healthy social interactions and relationships.

Emotions and cognition continually interact with and influence each other. The relationship is integral to how the mind functions; it is dynamic and bidirectional. This interplay influences our day-to-day experiences and has profound implications for our overall psychological well-being. Emotions significantly impact our various cognitive functions, such as attention, memory, and problem-solving, dictating what we pay attention to. Cognition is critical to how we manage and regulate our emotions.

Focused attention is crucial for assimilating information effectively. Positive emotions are conducive to learning by increasing motivation; negative emotions are not. The former can increase creativity and encourage broader and more flexible thinking. The balance between emotion and cognition is crucial for our mental health. An imbalance between the two can lead to various psychological disorders. Anxiety disorders, for example, can stem from overemphasis on emotional responses, while certain cognitive disorders might involve diminished emotional processing.

Emotions are crucial to how we make decisions and interact socially, and they greatly influence the choices we make. They provide a framework through which we interpret and respond to the world around us. If we understand the role of emotions, then, possibly, we can appreciate our behaviour and work towards effective personal and social functioning. Sometimes we make intuitive decisions, called "gut feelings". This is evident when we make quick judgements even when the information is incomplete.

Gut feeling and heart feeling: The experiences or emotional sensations of "gut feeling" and "heart feeling" suggest that our body communicates with us emotionally and intuitively, transcending physical sensation. Most often, these are spontaneous and instinctive responses.

"Gut feeling" is scientifically explained as an activity of the enteric nervous system (ENS), sometimes referred to as the "second brain". The ENS is a complex system of about 100 million neurons embedded in the lining of the gastrointestinal system. It's capable of functioning independently, and it plays a crucial role in digestion, but its influence extends to affecting the mood and well-being of an individual because it is connected with the central nervous system (CNS) through the vagus nerve. The gut microbiome can therefore influence emotional responses and cognitive functions, and that explains the physiological basis behind a "gut feeling". This complex communication between the gut and the brain highlights the importance of gastrointestinal health for overall emotional and psychological well-being. Studies on the gut microbiome suggest that our mental health, including conditions like anxiety and depression, can be significantly affected by our gut health.

Similarly, "heart feeling" describes emotions and decisions that are deeply felt but not entirely articulated through logic or reason. While the heart does not seem to have any known mechanism capable of cognition, it does have a complex network of neurons, neurotransmitters, proteins, and cells that communicate with the brain, influencing emotional and cognitive processes. Research into heart-rate variability and the heart's electromagnetic field is beginning to uncover how our emotional states can influence physiological responses and vice versa. The heart's electromagnetic field, the largest in the body, can affect our emotions and those of people around us.

The idea of the heart as a source of wisdom and insight has roots in various philosophical and spiritual traditions. For example, in Sufism, the heart is the spiritual centre of the human being, capable of directly perceiving the truth. This suggests that true knowledge and understanding transcend rational thought, residing instead in a deeper, intuitive awareness. Philosophically and spiritually, both "gut feeling" and "heart feeling" challenge the belief that we can understand the world only with the rational mind. Knowledge and truth can be accessed intuitively and as conscious experiences. This perspective is found in various spiritual traditions that emphasize the importance of listening to our body and emotions as a way of accessing deeper truths and guidance.

People often recount moments when a "gut feeling" saved them from a dangerous situation or guided them to make a decision that was not immediately logical but ultimately beneficial. At times, it is seen as refusing to go by a decision that seems entirely logical and rational but does not seem "right" intuitively or due to a "gut feeling". There are many instances of people who, "follow their heart", and find paths in life that are more fulfilling than if they had followed solely rational considerations.

"Gut feeling" and "heart feeling" challenge the boundaries between the physical and the metaphysical. They encourage a holistic view of human health and decision-making and emphasize the importance of integrating intuition and emotion with rational thought. Our eyes can also "feel". They are more than mirrors reflecting emotions; they're active windows to the soul. Sparkling depths convey joy, narrowed glares signal anger, and science hints they might even reveal the brain's immune system. This exciting frontier could unlock new treatments for neurological conditions. Remarkably, studies show we can accurately read a range of emotions, from joy to sorrow, simply by gazing into another person's eyes.

Memory

One of the fundamental tools of brain–mind functioning is memory—the storehouse. We acquire new knowledge, which is sorted and stored. Experiences are encoded and converted into memory, stored, and then retrieved or accessed on recall. We recall past experiences and use that information to function in the present and plan for the future. Assimilating information and storing or saving it in the memory "bank" is a vital mental activity. We store, retrieve, and utilize information that we gather from our experiences. This is central to how we understand the world, interact with it, and build our personal narratives. However, this function is not always purely rational. Emotions are very intricately interwoven with cognition and reasoning. The influences of emotions, biases, and preconceived notions, along with subjective experiences, play a crucial part.

From a biological perspective, memory is primarily an attribute of the brain. The formation of a memory involves various neurobiological processes, including the strengthening of synaptic connections between neurons. This process, known as synaptic plasticity, is fundamental to learning and memory.

Specific regions of the brain are said to be involved in different types of memory. For example, the hippocampus is crucial for the formation of new explicit memories (facts and information); the amygdala is involved in emotional memories; and the cerebellum plays a key role in procedural memory (skills and tasks).

We need attention and perception to store information in memory. Attention decides what is significant enough to be saved as a memory. Without focused attention, much of what we experience would simply pass by without being stored. Consciously or subconsciously, we choose which information is significant enough to be remembered. We have both short-term and long-term memories. The short-term, or working memory, stores and retrieves limited information for immediate use. In the long-term memory, we store information about facts and events, skills, and tasks more permanently. We store facts and events as per the relevance and extent of the impression of the experience. Emotional experiences, for example, leave deeper impressions and are thus more memorable. We are more likely to store information that is meaningful or can be related to our existing knowledge.

From a psychological and experiential perspective, memory is an attribute of the mind. Recalling memories, for example, is a conscious effort and a function of the mind. This includes both voluntary recall, such as remembering a name, and involuntary recall, a memory triggered by a smell or sound, for example. Interestingly, memories are not stored as exact replicas of experiences but are reconstructed as we recall them. The mind's interpretive processes can influence how we perceive and recall memories. These are often integrated with existing knowledge and emotional states.

Memory cannot be attributed solely to either the brain or the mind; it is a product of their interaction. The brain's structures and functions enable the physiological aspects of memory, while the mind contributes subjective experience, interpretation, and conscious recall. This interplay highlights the complexity of memory as a fundamental aspect of human cognition and experience. The relationship between the brain and the mind in the context of memory, therefore, is dynamic and interactive. The neural substrates in the brain provide the physical basis for memory, while the mind has the subjective experience of remembering, including the feelings, thoughts, and context associated with memories.

Forgetting is as natural as remembering. Sometimes forgetting is due to the information not being "encoded" or saved effectively, while at other times it is due to our inability to retrieve the information. Another deep phenomenon connected to memory is "haunting memories". These memories linger in the recesses of our consciousness, like ghostly apparitions, and refuse to fade away easily. They evoke emotions and sensations that persist over time. Depending on past experiences, they tend to remain embedded very deeply in our subconscious and unconscious minds. Nightmares and unexplained anger are our reactions as and when these memories surface. They are either connected to personal experiences or those of others towards whom we feel empathy.

Brain, Mind, and Sleep: A Powerful Trio

Sleep is the most essential and irreplaceable experience of a living being. Much more than physical rest for the body, sleep is fundamental to our existence and survival. From the point of view of the brain, it takes up vital tasks that are essential for our optimal functioning and mental well-being. As a meticulous curator, the brain consolidates memories by making strong connections between nerve cells. Much like "pigeon-holing", or compartmentalizing and categorizing, the brain organizes "files" for better storing and retrieving. It sifts through the experiences of the day and stores them, ready for recall! What does sleep do, then? It is the custodian, and at the same time, it clears away the accumulated junk that we might have gathered during our waking hours. This natural purging is part of the healing or correcting activities that ensure overall mental health and vitality. Sleep is also a catalyst for learning and assimilating creative ideas. The mind processes information gathered throughout the day and makes unique connections and in-depth insights. Adequate sleep is indispensable for our mental and emotional balance. We are aware of the consequences of sleep deprivation—stress, irritability, and fatigue, to say the least—although a concentrated and focused mind can manage extremely well with minimal sleep. The yogis of yore were known for the hours spent in deep meditation, and with that, they did not need the quantum of sleep advised for common people like us. For saints and realized people, the hours of deep meditation are far more rejuvenating than sleep.

Sleep cycles, or stages of sleep, have distinct patterns of brain activity and eye movements. The usual pattern is to begin with light sleep as we transition from wakefulness to sleep. Muscle activity decreases, and brain waves slow down. This takes us to the next stage, with a further decrease in muscle activity, till we can transcend to the stage of deep, restorative sleep. REM, or rapid eye movement, sleep is associated with dreaming and heightened brain activity. There is increased heart rate and blood pressure and temporary paralysis of the muscles—nature's way of ensuring we do not act out our dreams! The body, however, remains relaxed. We may rerun this cycle several times through the night. Non-REM sleep is when we consolidate our memories. Sleep cycles are crucial to our overall health, as each stage of sleep contributes to restoring our physical and emotional well-being. Disrupted sleep cycles lead to various sleep disorders and affect our cognitive functioning, moods, and overall health. In Vedic literature, especially in the *Mandukya Upanishad*, the states of dream and deep sleep are discussed very specifically in relation to the witness of all these states of consciousness (see Box 9.2 in Chapter 9).

Earlier, scientists had noted that wakefulness is essential for human consciousness. However, the brain connections underpinning wakefulness were not clear. Researchers have now mapped a neural network that can sustain human wakefulness. The fMRI study mentioned earlier in this chapter indicates that the two vital aspects of consciousness, wakefulness and awareness, might be linked.

Dreams

Dreams are a fascinating and complex aspect of human consciousness that has intrigued scientists and philosophers through the ages. They are related to both the brain and the mind and represent a unique state of consciousness. The study of dreams is a subject of interest in various disciplines, including psychology, neuroscience, philosophy, and even literature, as each discipline offers insights into the nature and significance of dreams. Neuroscientific research shows that dreams predominantly occur during the REM phase of sleep, when brain activity is high and resembles that of being awake. Various areas of the brain are involved in dreaming, including the amygdala, which is associated with emotions, and the hippocampus, which plays a role in memory. This suggests that dreams are a product of brain activity.

But dreams are also a key aspect of the mind; they are our personal, subjective experiences or projections that occur in the mind during sleep, and they present as narratives, images, ideas, emotions, and sensations. They can include elements of our daily experiences, thoughts, fears, desires, and much more—a complex interplay between our conscious and subconscious minds.

Dreams occur in the mind's eye, so to speak, within the realm of our subconscious. Modern research tends to view dreams as a by-product of brain activity during sleep, possibly playing roles in memory consolidation, emotional regulation, and the processing of daily experiences. While they might be products of brain activity, the experience of dreaming happens in our consciousness, in a space that is not physical but rather psychological and emotional. Despite being integral to our dreams, we are also witnesses to the roles we play in them. Waking, dreaming, and deep sleep are all states of consciousness.

There have been great endeavours and many efforts to interpret dreams, and this phenomenon has been a subject of debate among psychologists, neuroscientists, and philosophers. Some theories, like those proposed by Sigmund Freud, suggest that dreams are expressions of our deepest desires and fears, serving as a window to our unconscious mind. Carl Jung views dreams as a way for the psyche to communicate important messages to the individual, often using symbolic language. Dreams can have personal significance to us, possibly reflecting our thoughts, worries, and circumstances and events in life.

Some researchers suggest that dreams help in consolidating memories and processing and integrating new information with existing knowledge. They believe that dreams may help us process emotions and work through conflicts, worries, and desires in a safe, simulated environment. They might also serve a cognitive function, contributing to problem-solving and creativity by allowing the brain to explore different scenarios and ideas. Recent research continues to explore the function of dreams with studies using neuroimaging techniques to examine brain activity during dreaming.

There is growing interest in how dreams might support emotional and cognitive health, as well as in the potential for lucid dreaming, where the dreamer is aware of dreaming and can exert some control over the dream. The lucid dreamer can voluntarily direct attention and is aware of where he or she is. By conventional definitions, the dreamer is awake and asleep simultaneously. Lucid dreaming is the prerogative of religious and mystical schools because it points to another state of consciousness. According to Madhu Tandan, in her book *The Logic of Dreams*, "Perhaps the onset of lucid dreaming initiates a process in which an actual separation takes place between a part of the self that consciously reflects that it is dreaming and another part that participates or is involved in the dream itself. The dreamer shifts from being only a participant in the dream to also becoming its observer."

Daydreaming is a conscious but relaxed state of mind that is not focused on the world outside, even while we are in the waking state. We may imagine scenarios, replay events, or envision future possibilities. Daydreaming is a form of wakeful introspection or fantasy that can provide a break from routine, stimulate creativity, or help in problem-solving. The difference is that dreams that occur during sleep are more vivid and less under our conscious control than daydreams. Both types of dreaming, however, reflect the brain's and mind's capacities to generate rich, immersive experiences beyond immediate reality.

The mind monitors our inner experiences—our thoughts, memories, feelings, and the subjective experience of being. This includes our capacity for reasoning, problem-solving, creativity, and self-awareness, which is the ability to reflect upon ourselves and our experiences. The interpretative process is a subjective mental capacity uniquely tailored to each of us individually, influenced by our memories, emotions, and cognitive processes. This subjective quality makes the mind very deeply personal and private, accessible only to us individually as we experience it.

Brain–Mind Interplay

Nature and Functions of the Mind

The mind influences how we interpret information, how we make sense of the world, and how we interact with it. Many psychosomatic diseases originate in the mind. The mind draws its conclusions with the power of reasoning, which is based on the information sorted or available, and with the faculty of logic. Its true nature is linked to the ongoing integration of cognition and emotions. This enables the complex network of our behaviour, actions, and reactions, empathy, creativity, and the ability to judge.

But the functions are not a clear-cut division. Encoding, storing, and retrieving memory—these are the functions of the mind and brain working together. Under these heads, the mind manages a wide range of processes and activities. The physiological mechanisms of the brain and the cognitive functions of the mind work together. While the brain provides the physical substrate for memory processes, the mind influences and interacts with these processes through attention, perception, emotion, cognition, and conscious awareness. The mind perceives and processes sensory information. To create a coherent representation, it has to select specific stimuli and filter out others. We think, communicate, express feelings, make decisions, and consciously remain aware of all the functions of the mind, singularly and collectively.

The Realm of the Mind

The intricate interplay between human consciousness on the one hand and the biological functions of the brain–mind structure on the other is a fascinating subject of ongoing research. The core of this psychophysical system is the brain. Its capacity to process sensory information, orchestrate movements, and maintain vital functions without our conscious input plays a crucial role in our survival. Yet, there exists another dimension—the realm of the mind. This abstract concept manages our thoughts, emotions, and intentions of which we are conscious. The mind shapes and monitors decisions based on past experiences, beliefs, desires, and rational deliberations. The mind influences our actions, guided, as it is, by conscious thought while also being vulnerable enough to be swayed by the undercurrents of our subconscious. Our deepest motivations and habits reside here, in the interplay between conscious and subconscious processes, and sometimes this propels us in directions that we might resist rationally.

In certain scenarios, actions without the mind's conscious involvement are the brain's autonomy. Reflex actions, such as the jerk of a hand withdrawn from a hot surface, mediated by the spinal cord's reflex arcs, are the brain's ability to act independent of our conscious will. Similarly, the autonomic nervous system quietly regulates the heartbeat, digestion, and other life-sustaining processes without drawing our attention from the mind, unless there is any cause. This is how we can focus our mental energies elsewhere. Notwithstanding these apparent independent functions, the relationship between the brain and mind can be quite turbulent.

Habits and addictions lead to great tension when the brain's reward circuits undermine our conscious intentions. What does this mean? Habits are our behavioural patterns that are performed automatically, with little or no conscious thought. Habits can be efficient and productive, but they can lead to tension if they conflict with conscious intentions or goals. Typically, addictions are compulsive cravings despite negative consequences. These rise because of the brain's reward circuitry for reinforcing pleasurable experiences, particularly the release of dopamine in response to the addictive substance or activity. This behaviour disregards any conscious efforts to abstain or control. It is a reminder of the simplest of the advice we have heard through time: think before you act!

In her classic book *Molecules of Emotion*, Candace Pert challenges traditional views of the brain–mind–body connection. She proposes a radical idea: our emotions aren't just fleeting feelings but rather the product of neuropeptides, chemical messengers that course throughout our bodies. These "molecules of emotion", as Pert calls them, influence not only our brains but also our cells, creating a network that shapes our overall well-being. Pert also suggests that the body itself is a kind of unconscious mind. Memories and emotional experiences, she argues, may be stored not just in the brain but within the body's cells through receptor changes. This mind–body link has profound health implications. By understanding how emotions influence our entire body chemistry, Pert argues, we can develop new approaches to treating emotional and physical ailments. While some controversies arose over Pert's ideas, her work has helped pave the way for a more integrated understanding of mind, body, and emotions in health and wellness. Pert's experiences with meditation and yoga are likely to have influenced her theories. These practices emphasize the mind–body connection and the body's capacity for self-healing, which is similar to her idea of emotions as embodied experiences.

Mind and Behaviour

The mind and behaviour are directly connected since the mental processes drive our actions and responses. Our mind influences how we perceive and interpret situations; the consequent emotional responses affect our behaviour, which in turn reflects our thoughts, emotions, and intentions. Our actions are driven by the cognitive processing in the mind. The mind generates desires, needs, and goals that direct our behaviour. Motivation is often rooted in mental processes like the desire for reward, avoiding discomfort, or pursuing personal values. Understanding how the mind influences behaviour is fundamental in psychology. Every mind is unique, with its own personality traits, temperament, and cognitive abilities. Thus the vast differences in behaviour.

The most obvious feature of the mind, however, is its relentless activity. Unless we are gifted with the ability to go into deep sleep, the mind is constantly swirling. Aside from being the hub of all thoughts, emotions, and perceptions, it is also the source of unrest and turmoil, with incessant thoughts, worries, and stresses that can affect "normal behaviour". Behavioural inconsistencies are clear indicators of some imbalances, which could be genetic, social, neurological, physiological, or psychological. Treatment and management typically involve a multidisciplinary approach, including therapy, medication, lifestyle changes, and support networks. Two prominent disciplines explore the mind and its complexities: psychology and psychiatry. They share a common interest in understanding and treating the complexities of human thoughts, emotions, and behaviours. However, their approach, methodology, and focus are significantly different.

Food, Mind, and Thoughts

It may seem an unnecessary tangent to discuss the relationship between food and the mind. But it is important. Food plays a significant role in influencing the mind and thoughts due to its impact on brain function, mood regulation, and cognitive processes. The connection between nutrition and mental health is complex. Nutrient-rich foods can support cognitive function, mood regulation, and mental well-being, while a poor diet can cause cognitive deficiency and mental health issues.

The brain is highly energy dependent. A balanced diet that provides a steady supply of glucose (from carbohydrates) or sugar levels helps with concentration, memory, and mood. Research suggests a strong connection between the gut and the brain. The gut microbiome, influenced by diet, can impact mood and cognitive function. A healthy gut contributes to better mental health. Certain nutrients in food help with the production of neurotransmitters in the brain. For example, tryptophan, an amino acid found in protein-containing foods, helps the body make serotonin and maltonin, which help to regulate sleep cycles, moods, and stress. Essential vitamins and minerals, such as B-vitamins, vitamin D, magnesium, and omega-3 fatty acids, play crucial roles in brain health. Deficiencies can lead to cognitive deficits and mood disorders.

Chronic inflammation in the body is often triggered by processed foods, and it can negatively impact mood and cognition function.

Anti-inflammatory foods, such as fruits, vegetables, and omega-3-rich fish, may help reduce inflammation and support mental well-being. Dehydration can impair cognitive performance, leading to difficulties in concentration and memory. Adequate hydration is essential for optimal brain function. Antioxidant-rich foods, including fruits, vegetables, and nuts, protect against oxidative stress, which can damage brain cells. Antioxidants help preserve cognitive function as we age. Research suggests that dietary patterns can influence the tendency towards mental health disorders. For example, a Mediterranean-style diet, rich in whole grains, vegetables, fruits, and lean proteins, has been associated with a lower risk of depression and cognitive decline. Emotional states influence our food choices. The consumption of certain foods can influence emotions. Emotional eating, where individuals eat in response to stress, sadness, or other emotions, can impact both mental and physical health.

In the Indian philosophy of Yoga and Ayurveda, the traditional medical system practised through centuries, the identification of *sattvic*, *rajasic*, and *tamasic* foods is important. These classifications are based on the qualities they impart to the mind and body, just as the *gunas*, the three fundamental attributes of the human being are: *sattva* (pure and good, calm and in harmony), *rajas* (passionate, active, restless), and *tamas* (inert, ignorant, lazy) (see Fig. 5.3). Food is not only nourishment for the body but also an influence on our mental and spiritual state. The purpose is to consume *sattvic* foods and minimize or avoid *rajasic* and *tamasic* foods.

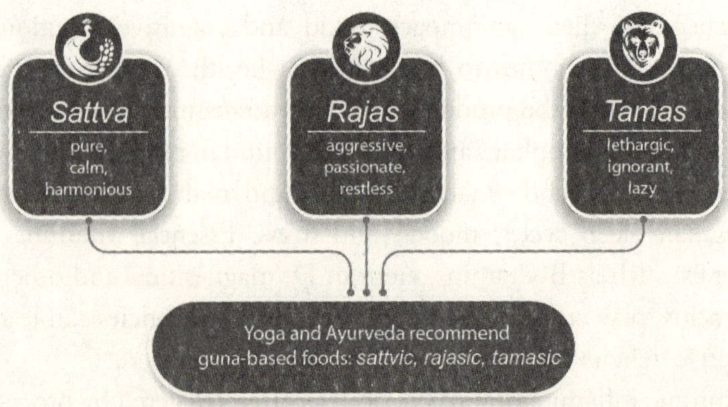

Fig. 5.3. The three gunas.

Sattvic foods are pure and clean and are believed to promote a calm and peaceful state of mind, enhance meditation, and support mental focus and clarity. They generally include what are referred to as anti-oxidant foods. *Rajasic* foods are said to be stimulating and passionate, often associated with restlessness and excess energy. Apparently, they increase mental agitation, distract from inner peace, and lead to a restless mind. These include spicy and pungent foods, excessive caffeine, heavily processed and fried foods, and foods high in sugar. *Tamasic* foods are heavy, often associated with ignorance and lethargy, and tend to induce a state of mental inertia, confusion, and lack of awareness. They include over-processed, stale, and spoiled foods and those with little nutritional value. These classifications are part of a holistic approach to health and well-being and have cultural and philosophical significance in Indian traditions, among others.

In many traditions, offering "grace" before a meal is the norm. It is in the form of a thanksgiving for the food we are about to receive, and it is meant to generate a 'sattvic' mind towards the food. Shloka 15.14 of the Bhagavad Gita is commonly chanted by several people before consuming food. This verse acknowledges the divine presence of the Cosmic Being within us who digests and assimilates food. Chanting this verse before meals is a way to express gratitude for the food we are about to receive and to remind ourselves of the sacred act of eating. It serves as a spiritual practice to cultivate mindfulness and appreciation for the nourishment provided by the food we consume. Shloka 4.24, which is also popularly chanted prior to taking food, is a very unifying thought in this "Genome to Om" journey as it essentially dedicates all actions to the divine, understanding that the entire universe and the actions within it are Its manifestations. It highlights the importance of performing actions selflessly and with devotion, like the oblations in the sacrificial fire, ultimately leading to the realization of one's true spiritual nature and union with the divine.

Studying and Treating the Mind

The Role of Psychology

Psychology, rooted in both philosophy and biology, emerged as a distinct scientific discipline in the late 19th century. Its focus is to understand the mental functions, behavioural processes, and underlying mechanisms of human interaction with the environment. It includes various theoretical approaches, including cognitive, behavioural, psychodynamic, and humanistic perspectives. It covers a vast range, from basic cognitive processes to the complexities of social interactions and development. Psychologists resort to experimental studies, testing, observation, and counselling. They focus on understanding both the normal and abnormal functioning of the mind to help individuals cope with stress, life challenges, and psychological disorders through counselling and psychotherapy. In psychology, consciousness is studied in relation to perception, cognition, and mental health, exploring how our conscious experiences shape our behaviour and interactions with the world.

Beliefs and attitudes are products of the mind. Our beliefs about ourselves, others, and the world influence our choices, actions, and interactions with others. Not all mental processes are conscious. The unconscious mind influences our behaviour through hidden biases, automatic responses, and underlying psychological mechanisms. Understanding these processes is essential in psychology. Mental health conditions, such as anxiety, depression, or psychosis, influence behaviour. They lead to changes in moods and perceptions. The mind and behaviour are influenced by both genetic factors (nature) and environmental factors (nurture). Both genetics and experiences shape our mental processes and actions. Society and culture have a profound impact on the mind and behaviour. Social norms, expectations, and cultural values shape our thoughts, attitudes, and actions.

Contributions of Psychiatry

Psychiatry is a branch of medicine that focuses on the diagnosis, treatment, and prevention of mental health disorders. Psychiatrists are medical doctors with specialized training in mental health. They have the authority to conduct physical examinations, order and interpret laboratory tests and brain imaging studies, and prescribe medications. They are trained to understand the complex interplay between the mind and body and to prescribe medication as and when needed. The approach is mostly clinical, focusing on the biological and neurological aspects of mental disorders. Psychiatrists treat major depressive, anxiety, and bipolar disorders, as well as schizophrenia, among other such ailments of the mind. Modern psychiatry treats mental conditions by manipulating brain activity, using pharmaceutical medicines mainly as stimulants or suppressants, and managing secretions, neurotransmitters, hormones, and other biochemicals. However, treating the brain is not the same as treating the mind. Treating the mind and mental conditions also needs psychotherapy, cognitive-behavioural therapy, possibly brain-stimulation therapies, and most importantly, changes in lifestyle through counselling and meditation. Psychologists and psychiatrists often collaborate to address issues with both the mind and the body to ensure a holistic view of human well-being.

Modern Psychology and Neuroscience

Wilhelm Wundt, often credited as the father of experimental psychology in the 19th century, established the first psychology laboratory. Sigmund Freud, in the early 20th century, introduced psychoanalysis, focusing on the unconscious mind, and worked on understanding the complexity of human behaviour and mental disorders. Also in the 20th century, John B. Watson and B.F. Skinner introduced the idea of behaviourism. They focused on observable behaviour and rejected introspection. In the mid-20th century, with the "cognitive revolution", there was a shift back towards internal mental processes. Eminent thinkers like psychologist Jean Piaget and linguist Noam Chomsky emphasized the roles of mental processes and structures in understanding behaviour and language.

Today, Western psychology and philosophy are highly interdisciplinary, integrating insights from neuroscience, computer science, artificial intelligence, and other fields. Current research focuses on understanding the biological bases of behaviour, the nature of consciousness, cognitive processes, and the interplay between genetic, environmental, and cognitive factors in shaping behaviour and mental health. Contemporary philosophy of mind continues to deal with questions about consciousness, identity, and the mind–body problem, often intersecting with studies from neuroscience and cognitive science. Ethical considerations, particularly about artificial intelligence, neuroscience, and mental health, are increasingly prominent.

The approach of modern psychology and neuroscience comes from empirical and scientific perspectives. Thoughts are understood as the outcomes of the activity of the brain, where neural processes and biochemical reactions give rise to consciousness, cognition, and the mental phenomena that we experience as thoughts.

Each of these perspectives views the origin of thoughts differently, reflecting the diverse ways we seek to understand the complexities of the mind and consciousness. This journey shows a gradual but significant evolution in our understanding of the human mind and behaviour. From the abstract concepts of the soul and rationality in ancient times to the empirical and integrative approaches of today, Western philosophy and psychology have evolved continuously, reflecting on and contributing to our broader understanding of what it means to be human.

The Impact of Freud and Jung

Sigmund Freud and Carl Jung were two of the most influential figures in the history of psychology, particularly known for their theories regarding the unconscious mind. Their explorations opened up new dimensions in understanding human behaviour, thought processes, and emotional life. Freud introduced the idea of the unconscious mind as a repository of feelings, thoughts, urges, and memories outside of conscious awareness. He believed that the unconscious mind influences much of our behaviour and mental processes. According to him, the unconscious is primarily composed of repressed experiences and underlying drives, including sexual and aggressive impulses (the id). Conflict between these primal drives, societal norms (the superego), and the mediating ego forms the basis of much of the psychological tension and neurosis.

Techniques like dream analysis, free association, and slips of the tongue (Freudian slips) are central to Freud's method of exploring the unconscious.

Carl Jung was initially a follower of Freud, but he developed his independent theories and diverged significantly from Freud, particularly in his conceptualization of the unconscious. Jung introduced the idea of the collective unconscious, a level of unconscious shared among all humans, comprising innate, universal archetypes. These archetypes are symbolic elements and patterns that manifest in dreams, myths, and behaviours across cultures. He also elaborated on the concept of individuation, a process of psychological integration wherein a person reconciles the personal unconscious with the collective unconscious, leading to a more harmonized and balanced self.

While Freud focused on repressed memories, rooted in childhood experiences and sexuality, Jung was more expansive and spiritual. He explored a wide range of human experiences, including religion, mythology, and art, trying to understand the collective aspects of the human psyche. Jung talked of synchronicity (meaningful coincidences) and emphasized the role of spiritual and existential factors in an individual's psyche.

Both Freud and Jung had a deep impact on psychology and psychotherapy. Freud's theories initiated various forms of psychotherapy, including psychodynamic and psychoanalytic therapy. Jung's ideas contributed to the development of analytical psychology and influenced fields beyond psychology, including literature, art, and religious studies. While some aspects of Freud and Jung's theories have been criticized in contemporary psychology, their contributions to understanding the complexity of the human mind and the role of the unconscious remain influential. Many modern theories and therapeutic practices incorporate or build upon their foundational ideas.

Western Perspectives: Aristotle, Plato, and Others

Thinkers from various cultures, civilizations, and philosophies, over the centuries, have offered unique perspectives, reflecting diverse understandings of the human mind, consciousness, and the nature of reality. Tracing these perspectives is a shift from philosophical speculation to empirical science and a move towards the growing integration of both approaches in contemporary thought.

Early Greek philosophers Plato and Aristotle laid the foundational ideas about the soul, mind, and rationality in the West. Aristotle's concept of the mind, or "nous" in Greek, was an essential component of his philosophical system, offering a comprehensive view of human cognition, psychology, and the nature of knowledge. He saw the mind as the faculty responsible for rational thought and knowledge. He distinguished between the passive mind, which has the potential to receive knowledge, and the active mind, which actualizes this knowledge. For Aristotle, the mind's primary function was to understand and contemplate. This included not only reasoning about practical matters (practical wisdom) but also understanding theoretical principles (theoretical wisdom). Plato proposed a dualistic universe where the "forms" (perfect and immutable ideas) exist apart from the physical world. Thoughts arise as the soul contemplates these forms, suggesting that thought originates in a non-physical realm of perfection and is a reflection of eternal truths. Aristotle disagreed with his teacher, Plato, and argued for a greater empirical understanding that thoughts are the result of sensory experiences processed by the mind. He believed the mind had the potential to form thoughts through interaction with the physical world, emphasizing a more material and experiential origin of thought.

In his works, particularly *De Anima* (On the Soul), Aristotle outlined a hierarchy of souls: the nutritive, the sensitive, and the rational soul. This rational soul, including the mind, is unique to humans, capable of rational thought, and distinguishes humans from other animals. He made a distinction between the active intellect ("nous poietikos") and the passive intellect ("nous pathetikos"). The active is a formative force that actualizes the potential for thought, while the passive can be shaped by knowledge. This distinction has been subjected to various interpretations and debates by scholars, particularly regarding the nature and immortality of the active intellect.

Aristotle also connected the mind's functions to ethics and happiness. He believed that the highest form of happiness is achieved through the rational part of the soul, which pursues intellectual virtues and contemplates truth. Aristotle's concept of the mind was integral to his broader philosophical system and connected with his theories of knowledge, ethics, and biology. His emphasis on empirical observation and the rational processing of sensory experiences led to later developments in philosophy, psychology, and the natural sciences.

The mediaeval Christian world, with thinkers like St Augustine and St Thomas Aquinas, integrated religious doctrine with philosophical ideas about the mind and soul. Influenced by Aristotle, Aquinas argued for the unity of the soul and body, seeing the soul as the form of the body. This period saw a fusion of faith-based perspectives with philosophical reasoning, exploring issues like free will, moral responsibility, and the nature of divine and human understanding. During the mediaeval period, alchemy, a precursor to modern chemistry, also explored the nature of the human psyche. Alchemists like Paracelsus explored the transformation of matter and spirit, contributing metaphorically to our understanding of psychological transformation.

The Renaissance revived interest in individualism and humanistic perspectives. René Descartes, the 17th-century French philosopher, mathematician, and scientist, was a foundational figure in Western modern philosophy. Known for his dualistic theory of mind and body, he famously proposed "*Cogito, ergo sum*" ("I think, therefore I am"). He suggested that the capacity for thought is the most undeniable proof of our existence, and he emphasized doubt and reasoning as the roots of knowledge. For Descartes, thoughts originate from the interaction between the immaterial mind and mechanical body. John Locke and David Hume focused on experiences and sensations as the basis of knowledge, moving further into empirical territory. The 17th-century English philosopher Thomas Hobbes had a materialistic and mechanistic view of the mind and consciousness. His masterwork *Leviathan* was published in 1651. He worked with political philosophy, which influenced his understanding of human nature and the relationship between the mind and the body.

During the 17th and 20th centuries, many thinkers made significant contributions to the understanding of psychology and philosophy and the complex relationship between mind and consciousness. Immanuel Kant, a German philosopher, introduced critical philosophy, investigating the limitations and structure of human knowledge and experience. He talked of the nature of perception, consciousness, and the relationship between the mind and reality. George Berkeley, an Irish philosopher, presented two metaphysical theses: idealism (everything that exists is either a mind or depends on a mind for its existence) and immaterialism (matter does not exist). For him, all physical objects are composed of ideas; this is found in his motto, *esse est percipi* (to be is to be perceived). He thus suggested that the external world exists only through perception and consciousness.

The American philosopher and psychologist William James contributed significantly to the study of consciousness in the 19th century. He helped to set up psychology as a formal discipline and pragmatism in philosophy, placing great importance on empirical experience and practical consequences. His book *The Principles of Psychology* explored the functional aspects of consciousness and the nature of subjective experience. Also in the 19th century was German philosopher Friedrich Nietzsche, who, with his dramatic statement, "God is dead, and we have killed him…." challenged traditional views of reality and morality. He explored human consciousness, authenticity, and creating meaning in a seemingly meaningless universe.

Edmund Husserl was a 20th-century Austrian philosopher who founded the philosophical movement of phenomenology. He focused on the first-person perspective of conscious experience, the structures of consciousness, and how we perceive and interpret the world. The 20th century was the era of existentialism, or exploring the issues of human existence. The French existentialist philosopher Jean-Paul Sartre explored themes related to human freedom, consciousness, and authenticity. He emphasized the subjective experience of existence and, therefore, the responsibility of individuals to create their own meaning.

Many indigenous cultures often see thought as inseparable from the body, spirit, and the natural world. Thoughts are not just individual mental activities but are interconnected with the community, ancestors, and the environment. For instance, the Australian Aboriginal concept of "Dreamtime" relates to the origin of the universe, where the thoughts and actions of the ancestral beings shape the landscape, laws, and living beings.

These insights of some of the ancient to contemporary thinkers show the diverse ways in which philosophical traditions have grappled with the nature of the mind and consciousness. They laid the intellectual groundwork for modern psychology, philosophy, and neuroscience and provided a rich historical context for contemporary discussions about the human mind. To integrate the diverse perspectives, we need a balanced acknowledgement of the strengths and limitations of both objective inquiry and subjective experience.

Eastern Perspectives: From Kapila and Patanjali to Confucius

The ancient Sankhya philosophy, one of the six classical schools of Indian philosophy, was founded by Sage Kapila around the 8th–7th century BCE. References to this ancient dualistic orthodox school can be found in the Rig Veda and the Upanishads. It provides a comprehensive framework for understanding the nature of consciousness, existence, and the cosmos. It explains the structure of the mind or mind stuff, which includes the intellect and the ego. This philosophy views reality as composed of two independent principles: Purusha (Self) and Prakriti (Nature). Purusha is the unchanging, pure consciousness, unaffected by the modifications of Nature. Prakriti is dynamic and the material cause of the universe. It is the primal substance that evolves as the "First Cause" and manifests as the physical world. An overview of different schools of thought in Indic philosophy is given in Fig. 5.4.

According to Sankhya, the universe evolves through the interaction of Purusha and Prakriti. Prakriti evolves from its unmanifest state to the manifest. The ultimate goal of life is liberation from the cycle of

Various schools of philosophy developed in the Indian subcontinent. The key philosophical thoughts include six theist *darshana*s based on Vedic thought and three atheist philosophies based on Charvaka, Buddhist and Jain thought.

THEIST DARSHANAS

NYAYA
Investigates knowledge & logic.

NYAYA: Founded by sage Gautama. Emphasizes logic and knowledge acquisition as the path to liberation (*moksha*). The basic premise is that suffering is caused by ignorance.

VAISHESHIKA: Founded by Sanskrit philosopher Kashyapa. Proposed an atomic theory of the universe guided by a divine force. Merged with the Nyaya school became known as the Nyaya-Vaisheshika school of Indian philosophy.

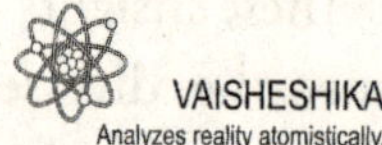

VAISHESHIKA
Analyzes reality atomistically.

SANKHYA
Explores cosmic unity through reason.

SANKHYA: Founded by sage Kapila. Dualistic system, distinguishing between eternal and pure consciousness (Purusha) and the first cause, the matter of the physical universe and nature, which is within time, space and causation (Prakriti).

YOGA: Founded by sage Patanjali, expounded in the *Yoga Sutras*. Aligns with Sankhya philosophy, adding the concept of a personal God and emphasizing practical methods for *moksha*.

YOGA
Physical, mental & spiritual practices.

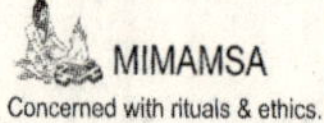

MIMAMSA
Concerned with rituals & ethics.

MIMAMSA: The earliest work, *Mimamsa-sutra* founded by sage Jaimini. Focuses on interpreting the Vedas, particularly rituals (Karma Kanda).

VEDANTA: This is the monistic school of Indian philosophy, deals with the later portion of Vedic literature called the Upanishads, the knowledge section or the Jnana Kanda. It has three main sub-branches :

- Dvaita (dualistic) by Madhavacharya
- Vishishtadvaita (qualified monism) by Ramanujacharya
- Advaita (monism) by Shankaracharya

VEDANTA
Explores ultimate reality& liberation.

ATHEIST SCHOOLS OF PHILOSOPHY

CHARVAKA

CHARVAKA: The ancient school also referred to as the *nastika* or atheist, rejects the authority of the sacred scriptures. Founded by sage Brihaspati, this materialist school recognizes only what can be directly perceived.

BUDDHISM: The philosophic thought based on the teachings of Gautam Buddha. It posits the transcendental nature of life, of reincarnation, and denies the existence of soul or God.Emphasizes understanding suffering and its cessation to attain liberation (*nirvana*).

BUDDHISM

JAINISM

JAINISM: Founded by Mahavira, emphasizes non-violence (*ahimsa*) and the liberation of the soul (*jiva*) from the material world (*ajiva*).

Fig. 5.4. Major schools of Indian philosophy (Darshanas).

birth and death (*samsara*). Consciousness also means transcending the limitations of the ego and the mind and recognizing the underlying unity of all existence to achieve liberation, or *moksha*. We can attain this liberation by understanding that the Self is distinct from the modifications of nature, and thus withdrawing or distancing ourselves from the material world takes us towards liberation. The Advaitic or non-dualistic perspective proposes only one universal consciousness, and that is the individual being in its intrinsic reality. This is exemplified by the Great Statement, "That Thou Art", That (Brahman) is You (Atman).

Patanjali's Yoga Sutras offer deep insights into the nature of the mind, consciousness, and the path to spiritual liberation. The mind (*chitta*) is composed of thoughts and emotions, and it's in a constant state of flux, which can be a source of suffering and distraction. Patanjali's great statement is "Yoga for *chitta vritti nirodhah*", which means Yoga restrains the mind (*chitta*) from taking various forms (*vritti*) and thus results in the cessation of all fluctuations, making it stable and conscious. This is the key principle to the practice of yoga towards self-realization. These teachings have shaped spiritual practices and influenced psychological and philosophical thought worldwide.

Another important perspective from the East comes from Buddhism, founded by Siddhartha Gautama (the Buddha) in the 5th century BCE. Buddhism explores the mind extensively, studying the nature of the mind and suffering. The central concept is *shunya*, which means nothingness, emptiness, or a void. This is a fundamental concept in Buddhism and Zen Buddhism. According to this, all phenomena lack inherent, independent existence. Everything is dependently originated, interconnected, and devoid of any fixed, permanent essence. The teachings of the Buddha include the Four Noble Truths and the Eightfold Path, which outline the origins of suffering and the way to overcome it. Mindfulness, meditation, and living with awareness, compassion, and wisdom are the recommended ways of living life if we want to avoid suffering and sorrow. These practices are practical even in contemporary Western psychology, particularly in treating stress, anxiety, and depression.

Zen Buddhism became prominent in China and Japan. It emphasizes direct experience and insight (*satori*) through meditation. Zen teachings focus on the nature of the mind, encouraging everyone to experience reality directly. Taoism originated from the teachings of Laozi and Zhuangzi in ancient China. Taoist philosophy proposes simplicity, spontaneity, and harmony with nature, focusing on the flow of life and the interdependence of all things.

Confucianism, founded by Confucius, places significant emphasis on ethics, social harmony, and the proper conduct of relationships, whether familial, social, or governmental. Mediaeval Islamic philosophy, proposed by Islamic scholars like Al-Farabi, Avicenna (Ibn Sina), and Averroes (Ibn Rushd), played a crucial role in preserving and expanding upon Greek philosophical works. It essentially posed various philosophical perspectives, such as the nature of reality, the relationship between faith and reason, free will, ethics, the existence of God, and the nature of the soul. In Sufism, the mystical arm of Islam, the state of mind that transcends the ego is reached through meditation, invocation, chanting, and dancing, aiming to get a direct experience of divine love and one-ness. Christian mystics have described states of ecstatic union with the divine, emphasizing contemplative practices that transcend ordinary consciousness and bring direct, experiential knowledge of God.

Transcending from Thinking to Knowing

Origin of Thoughts

A very pertinent aspect of this discussion on the mind is the subtle but vital difference between thinking and knowing. The key questions here are: How and from where do thoughts arise? How do we perceive them in our minds? What is their nature and their potential to be "transferred"? There are no definitive or universally accepted answers to these questions, which are being actively researched from a scientific and biological perspective in neuroscience, cognitive science, psychology, and now in technology and AI. Neuroscientists assert that thoughts first emerge from the complex interactions within the brain's neural networks. The brain receives sensory inputs from the external world, processes them, combines them with stored memories, interacts with centres of emotion, and generates outputs in the form of thoughts, emotions, and actions. This process is influenced by various factors, including genetics, environment, and individual experiences. The brain can also create internal representations of reality, such as memories, imaginations, and dreams.

From a psychological perspective, thoughts are the expression of mental states, such as beliefs, attitudes, intentions, and motivations. Cognitive processes, such as perception, attention, memory, reasoning, and problem-solving, as well as mood, emotion, and personality, affect our thoughts. Social and cultural factors also influence them. From a philosophical perspective, thoughts are the manifestation of consciousness, which is the subjective experience of being aware and having a sense of self. Thoughts are the vehicles of rationality, logic, knowledge, creativity, intuition, and wisdom. They are also the basis of free will, morality, and ethics. From a spiritual perspective, thoughts are the reflection of the soul, which is the essence or core of our being. Thoughts are the connection between the physical and the metaphysical, the material and the immaterial, the temporal and the eternal. They are the connection between Genome and Om. Thoughts are the means of communication with the divine, the source, or the higher power, whatever we may call it or believe in. Spiritual traditions have various teachings and practices to cultivate positive, peaceful, and compassionate thoughts and to transcend the illusion of separation and duality.

How We Think

Thinking is so fundamental to our mental activity that it is difficult to even question the process or its origin. We manipulate and organize the information that we gather through our sensory experiences, or what we can recall from the storehouse of memory, which is only experiences stashed away for use as and when needed. But from where and how do thoughts arise? This question consistently perplexes us. Plato and Aristotle grappled with the nature of reason and consciousness. Ancient India's exploration of thought goes beyond just its origin. The Vedas and the Sankhya, introduce *manas* as the mind stuff or the thinking faculty, while the Upanishads go deeper with *chitta*, the underlying source of thoughts shaped by past experiences. These texts see thoughts as products of the individual self, but true liberation comes from realizing its one-ness with the universal Self. Stanford psychologist Barbara Tversky suggests that to understand how people think, one must understand how people act and come to understand the world through their spatial reasoning. She challenges the long-standing notion that language is the true catalyst for thought—that thinking is impossible without language. She argues that it is not verbal communication at the root of thought. Instead, spatial thinking gave rise to the myriad systems of written and oral communication we employ today. Spatial thinking is the foundation of thought, and it evolved long before language. This "internalization of movement" may have eventually led to the development of language and other communication systems.

Colombian neuroscientist Rodolfo Llinás believes that the brain's core function is prediction based on thinking. Our ability to anticipate danger, find resources, and navigate the world may have helped to evolve emotions, thoughts, and even consciousness. The question of who controls our thoughts is equally intriguing. Some argue that our thoughts arise unconsciously, influenced by emotions, past experiences, and environmental stimuli. We may then become aware of these thoughts as if they appear "out of thin air". Others believe that we do have some control over our thoughts. Through meditation and self-reflection, we can consciously choose the thoughts we wish to focus on and which to let go. The likely answer lies somewhere in between. Our thoughts are a complex interplay of unconscious and conscious processes.

While the origin and control of thoughts are fascinating, how exactly we think can be broken down into distinct categories or types of thinking (see Fig.5.5). We have multifaceted thinking capabilities. Critical thinking helps us sift through information and form sound judgements, while analytical thinking allows us to dissect problems and find their root causes. This analytical breakdown then becomes the springboard for divergent thinking, where we can brainstorm a wide range of possible solutions. However, not all situations call for an explosion of ideas. Convergent thinking helps us narrow down the possibilities generated by divergent thinking, allowing us to choose the most effective solution. Finally, concrete thinking is crucial for putting our plans into action, ensuring we can follow through and achieve the desired outcome.

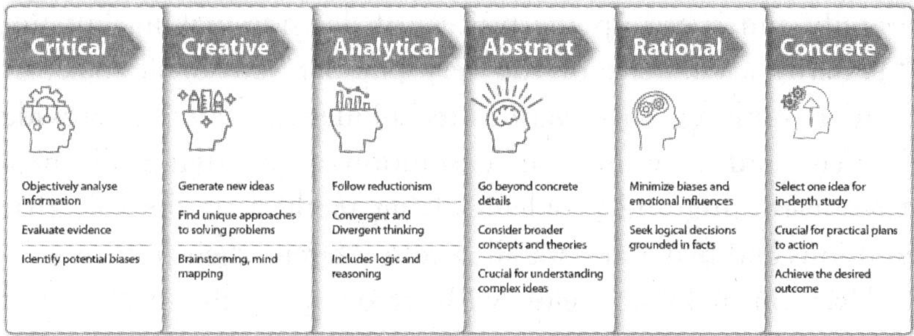

Fig. 5.5. Ways of thinking.

Cognitive processing is crucial to how we think. We can perceive and interpret the data that our senses bring to us from the world outside. Our brain filters and processes this information, and our mind interprets it, influenced as it is by our past experiences, our beliefs, and our expectations. Our ability to focus our attention determines what we process actively, what we register but do not act upon instantly, and what we ignore and discard completely. This ability to focus selective attention is important because the senses bring in all the experiences that they can assimilate, without selection or discrimination. The ability to focus attention is not entirely the activity of the brain. This is also not entirely a mental activity; it is not only the outcome of the mind. This is a conscious process, and the connection to consciousness and self-awareness is crucial here, which we shall discuss in detail. But prima facie, metacognition, or thinking about thinking, is how we approach information, process it, and solve problems. It involves self-awareness about our cognitive processes and the ability to regulate and adapt them according to the task at hand.

From a simple scientific perspective, we can look at the strategies that the mind adopts for processing information and solving problems. From sensory input, the data goes to short-term memory and then to long-term memory. Each stage plays a specific role in handling and storing information. In solving problems, our mind either uses algorithmic thinking, the step-by-step logical approach that guarantees an accurate solution; the mental short-cut or "rules of thumb or heuristic methods" that simplify and speed up the process of decision-making but not necessarily accurately; or the very popular "out-of-the-box" or creative thinking. These various techniques can certainly provide different and even innovative solutions. Underlying all these decision-making and problem-solving methods is the rule that emotions and biases are not allowed to exercise their influence.

Decision-making begins with recognizing the need for a choice, followed by a careful consideration of various factors. The ability to make decisions, reason, and process rational thoughts highlights the sophistication of the entire mental process. This ability draws upon deeper, subconscious understanding, balancing these elements to navigate the myriad choices we face in life. However, decision-making is not solely a product of rational thought. Intuition plays a vital role. This is the subconscious aspect of reasoning, the intuitive process based on our experiences and internalized knowledge.

Knowing or Having Knowledge

Knowing is the quest for truth. It is the culmination of the process of thinking—a state of understanding that emerges after either validating or falsifying our thoughts through empirical evidence, logical reasoning, or realization. Knowing is the experiential state of conscious certainty, of having specific information or understanding. Knowing is the state of confident affirmation, of being aware of the information, of having the facts. It expresses confidence in our understanding or beliefs. It is the endpoint of the cognitive process, which carries a sense of conviction or certainty about the information that we have garnered.

Knowing is only possible with evidence, experience, or justification for our beliefs or assertions. Knowing means a level of understanding and certainty gained through learning, experiencing, or reasoning. We can say that it is more static than thinking, which is a dynamic process. We can *know* something as a fact (declarative); we can *know* how to do or say something (procedural); or we can *know* through personal experience (experiential). Thinking can lead to knowing as we acquire knowledge. Whereas, knowing can influence thinking. What we know or believe shapes how we think or process new information. Scientific knowledge, while often regarded as a closer approximation to "truth", is inherently provisional. It evolves with new discoveries, technologies, and methodologies that challenge and refine existing paradigms. Humility and honest openness are essential for us to accept the evolutionary nature of scientific truth. We have to acknowledge that what we know today may be revised or overturned tomorrow.

Knowing can be explicit, such as consciously recalling facts or concepts, or implicit, such as intuitively understanding a situation or recognizing patterns. *Knowing is the assertion of knowledge from the position of the knower.* There is no speculation or guesswork involved in knowing. The Chinese philosopher Confucius rightly said: *Knowing the known is knowledge and knowing the unknown is wisdom.*

Difference between Thinking and Knowing

Thinking and knowing are two distinct processes, interrelated but distinct aspects of cognition. Thinking is the foundation of inquiry, a dynamic process of manipulating information and forming ideas, while knowing is understanding or awareness that often results from the process of thinking. The relationship between the two is fundamental to how we process information, form beliefs, and acquire knowledge.

In philosophy, these concepts go into epistemology or the study of knowledge. Philosophers have long debated what constitutes knowing (epistemic justification) and how it is different from mere belief or opinion. Knowledge is traditionally seen as a justified true belief. Thinking could be questioning our beliefs, and knowing is holding on to what we believe. From a psychological perspective, thinking involves cognitive processes like perception, memory, language, and reasoning. Knowing is often connected to memory and learning, where information is stored and retrieved.

Thinking is based on existing knowledge and on creativity and intelligence, the two attributes inherent in us. Thinking is an acquired or cultivated character. The idea is the product of creativity and critical thinking. The basis for generating novel ideas arises from in-depth knowledge related to the subject, creativity, and thinking capacity. And in the entire process, intuition plays a vital role.

To graduate from thinking to knowing, we have to validate, internalize, and transform information or ideas generated through thought into knowledge. This transformation is central to learning and understanding. We acquire information, which is raw, unprocessed data, through reading, observing, listening, or experiencing. We actively engage with this information through critical, analytical, or logical thinking. We evaluate the validity, relevance, and implications of the information. As we think through the information, we experience moments of insight or understanding. We make connections and grasp or recall concepts and deeper meanings. We shift from merely processing information to internalizing it. This is where the information begins to "make sense". When we can internalize and integrate the information into our existing cognitive framework, it becomes knowledge and is stored in a way that we can readily access and use it in the future. What does it take to *know* something? It's not enough just to believe it—we don't know the things we're wrong about. "Knowledge seems to be more like a way of getting at the truth. The analysis of knowledge concerns the attempt to articulate in what exactly this kind of 'getting at the truth' consists." In ancient epistemology, true knowledge was considered as seen. Therefore, those who could see knowledge are called "seers".

Knowledge provides the foundation, and experience enriches our understanding. In Vedic philosophy, experience has two subtle connotations: *anubhava* and *anubhuti*. *Anubhava* is experiencing through the senses and the mind. What we perceive through our sense organs and the experience of the perceptions is *anubhava*. It also refers to knowledge or skill gained through doing a job or activity. *Anubhuti* is a physical or emotional feeling. It is a general insight caused by a particular experience, a realization, or a mental experience. For example, the direct experience of seeing the sunset.

The mind is constantly fluctuating and wavering and is super active, and these fluctuations are obstacles to deeper understanding and inner peace. This is seen as a distraction that can cloud true understanding and knowledge. The yogic approach to thinking and knowing is rooted in ancient philosophy and spiritual practice. It is a means of acquiring a deep, intuitive, and holistic form of knowing that transcends intellectual understanding and leads to spiritual enlightenment and self-realization. In yoga philosophy, the mind and the process of thinking and knowing are deeply linked with the broader goals of wisdom, self-realization, and spiritual growth. Yoga emphasizes the importance of understanding the nature of the mind and its inclination to achieve higher states of consciousness and spiritual liberation. This involves practices such as concentration, meditation, and introspection. By cultivating awareness and mastery over the mind, we can transcend the fluctuations of thought and emotion and reach a state of inner peace, clarity, and union with the divine or higher self. The path to self-realization is in three simple stages: first, listen to the truth, that is, listen to (or read about) the truth as conveyed in the scriptures; next, reflect on it or contemplate with focused attention; and finally, practise it and abide by it. A crucial goal in yoga is to overcome ignorance, which is the misunderstanding of the nature of reality and the Self. Through dedicated practice, we move from superficial knowledge to true, profound understanding.

These methods of knowing, which transcend empirical verification, are based entirely on internal experiences and meditative insights. While such claims challenge scientific validation, they allude to exploring consciousness and the potential capacities of the human mind.

Quantum Thinking

The approach of quantum science to how we think is a relatively new and speculative field. It tries to understand cognitive processes through the principles of quantum mechanics rather than classical physics. It recognizes that, at the quantum level, the fundamental behaviours of particles—superposition, entanglement, and uncertainty—might give us insights into the complexities of the human mind, including the process of thinking. Superposition in quantum mechanics suggests that particles can exist simultaneously, in multiple states, until they are observed. This implies that the mind is capable of holding multiple thoughts or several divergent ideas in a state of potentiality because all of them cannot be interpreted or processed and implemented at once. With our focused attention, we consciously select one thought at a time, making it the definitive state of mind at that moment. That is the convergence point. This could provide a quantum explanation for the ability to process and generate complex and diverse thoughts. As F. Scott Fitzgerald stated, "The test of a first-rate intelligence is the ability to hold two opposed ideas in mind at the same time and still retain the ability to function."

Quantum entanglement, the subject of the Nobel Prize 2022 in physics, is where particles become interconnected in such a manner that the state of one particle, regardless of its distance, instantaneously influences the state of another particle. Some theorists speculate that quantum entanglement can explain telepathy or the deep, instantaneous "connections" that we feel sometimes, suggesting a quantum basis for empathy or shared understanding. Quantum tunnelling is when a particle passes through a barrier, which, according to classical physics, it shouldn't be able to. In the process of thinking, this concept could explain how insights or solutions to problems appear to "tunnel" through mental blocks or consciously appear as moments of sudden inspiration or creativity. Applying quantum mechanics to cognitive processes is still very theoretical and continues to be debated within the scientific community.

The Mind-Matter Unification Project, led by Nobel laureate physicist Brian Josephson from Cambridge University, explores unconventional connections between consciousness and the quantum world. It challenges materialistic views by proposing that intelligence exists in nature and may be associated either with brain function or other natural processes. The project investigates whether consciousness and quantum phenomena are intertwined beyond classical physics. It draws inspiration from ancient Eastern wisdom traditions, seeking common ground between rigorous scientific methods and holistic approaches to deepen our understanding of consciousness and reality.

Layers of Awareness

Conscious, Subconscious, and Unconscious Mind

These terms describe the mental processes that help us to perceive the layers of mental functioning and human cognition and the interplay between the conscious and unconscious processes. However, the exact nature and functioning of these mental aspects are still debated in psychology, neuroscience, and philosophy. In lay terms, the conscious mind takes in everything that we are currently aware of and are actively thinking about. When we are awake and alert, our conscious mind is thus engaged, and it includes the perceptions, thoughts, sensations, and feelings that we are experiencing. It is responsible for our rational thinking, decision-making, problem-solving, and day-to-day awareness. The ability to pay focused attention and to be aware of the external world in real time, as well as of internal experiences and activities such as reading a text, solving a math problem, or having a conversation, are all part of our conscious mental processes.

The subconscious mind lies beneath the surface of conscious awareness and has a storehouse of information that is not in our immediate awareness but can be brought to the conscious level with effort. This includes memories, beliefs, habits, emotions, and automatic behaviours that influence our thoughts, feelings, and actions. The subconscious mind can shape our behaviour and responses, often without our being aware of it. How do we uncover this level of the mind? We can access it through hypnosis, meditation, and psychoanalysis. In psychology and psychotherapy, accessing and understanding the contents of the subconscious mind is essential to addressing traumas, phobias, and behavioural patterns. Dreams are often a window into the subconscious mind. The subconscious mind may process emotions, memories, and unresolved thoughts, which can manifest as dreams.

The unconscious mind is another reservoir of thoughts, feelings, desires, and memories at the deepest level and is not accessible to our conscious awareness. It operates outside our conscious control and is often considered the repository of repressed or suppressed thoughts and emotions. According to psychoanalysts, the unconscious mind influences behaviour through defence mechanisms, dreams, and slips of the tongue (Freudian slips).

Neural Correlates

What are the different emotions that correlate with activity in specific regions of the brain? Three brain structures appear most closely linked with emotions: the amygdala, the insula or insular cortex, and a structure in the midbrain called the periaqueductal gray. Fear, anxiety, and fight or flight response are operated in the amygdala; disgust, pain, and social emotions are operated in the insular cortex; emotional response, adapt behaviour, decision-making, and cognitive control are in the prefrontal cortex; fear, laughter, and pain modulation are from the periaqueductal gray.

Neuroscientists study the neural correlates of consciousness to understand how specific brain processes are connected with conscious experience. The objective is to uncover the areas and networks within the brain that are active during different types of mental activities, such as dreaming, meditating, or problem-solving. Stress, beliefs, and emotions influence physical health because of the powerful interplay between the mind and the brain. Emotions are closely tied to brain activity. Neuroimaging has revealed that different emotions correlate with activity in specific regions of the brain.

A prominent theory linking quantum mechanics to consciousness is the Orchestrated Objective Reduction (Orch-OR) Theory, proposed by Roger Penrose and Stuart Hameroff. This theory suggests that quantum computations in the microtubules within brain neurons could account for the emergence of consciousness and, thus, the process of thinking. It explains that these quantum computations are "orchestrated" by the brain's neural structure and "reduced" to definite outcomes, contributing to conscious experience and decision-making. Many researchers are sceptical of the direct relevance of quantum mechanics to the macroscopic processes of the brain, given the brain's warm, wet, and noisy environment, which is generally considered hostile to sustained quantum phenomena. However, exploring these ideas is part of the concerted effort to understand the mysteries of consciousness and cognition, pushing the boundaries of interdisciplinary research. Quantum approaches to consciousness continue to be widely researched.

Several scientists see consciousness as a product of complex neural networks and brain processes, while others see it as an emergent property that cannot be fully explained by neuroscience alone. The Integrated Information Theory proposes that consciousness arises from the integration of information across a network of neurons. The Global Workspace Theory suggests that consciousness results from the interplay of various brain networks that integrate and process information. Researchers are exploring the brain areas and neural processes associated with consciousness. The prefrontal cortex, for instance, is crucial for self-awareness and executive functions, while the thalamus is thought to play a key role in controlling awareness.

Consciousness and the Body–Mind Connection

The fundamental questions are about the nature of consciousness and the mind–body problem—how can a non-physical mind arise from a physical brain? How can subjective experiences and self-awareness arise from neural processes? These questions highlight the complexity of human consciousness—a phenomenon that appears to arise from and yet operates on a plane different from the tangible workings of the brain. The mind–brain interaction challenges our understanding of the self and free will.

Is There Free Will?

We feel we have free will. We are free to decide. Between options A and B, we have the freedom to choose either. But this is an oversimplification. It is not necessarily as obvious as that. Human experience is said to hinge between awareness and choice. We see the world around us, interpret it through our worldview or our tinted lens, and then make decisions that shape our lives and the lives of others. But these choices are not necessarily truly free; they could possibly be predetermined by the very act of our being aware.

On the one hand, awareness seems fundamental to free will. Neuroscientist Daniel Wegner gives us the "ironic process theory", which states that if we suppress thoughts, we actually make them more prominent. If we are aware that we wish to avoid something, then the very act of focusing on avoiding that something takes a hold on our minds. This obviously limits "free" will because our awareness itself can influence our choices. Philosopher Sam Harris argues that our brains make decisions unconsciously before we can even become aware of them. Through experiments measuring reaction times, Harris suggests that choices may be predetermined by neural activity, and that limits our free will. There is a view that our choices are influenced by genetics and experiences, but we still have the capacity to weigh options and make decisions based on our values.

The Theistic argument is that we are not aware of what action is coming, in which case exercising our will only happens when we become aware of the action. God is omniscient, so God already knows what action we will be taking; thus, our will is no longer free. There is another aspect as well. It is stated that God gave us the gift of free will, but we misused it; we freely committed sins, and thus we suffer. The "original sin" was Adam exercising his free will to eat the apple in the Garden of Eden, specifically disobeying the instructions.

As far as Eastern philosophies or religions, like Hinduism, Buddhism, and Jainism are concerned, as human beings we are governed by the laws of *karma*, the cause and effect laws, and within that there is no free will. If we believe in the laws of *karma* and reincarnation, then every action is an event in time. Every event in time is preceded by a cause. If the cause is there, the effect must come. Causal determinism indicates that we may feel we have free will, but we do not have it. If our decisions or actions are part of cause and effect, then there can be no free will. Swami Vivekananda says, "The will is not free—it is a phenomenon bound by cause and effect—but there is something behind the will which is free." One analogy is of the cow tethered with a rope to a pole. The length of the rope is the extent of its free will.

"Choiceless awareness" in Zen Buddhism emphasizes watching our thoughts and emotions without judgement. This detachment allows for a more authentic response to situations, thus free will, as we are not driven by preconceived desires or the burden of past *karma*.

The debate continues. Perhaps the answer lies in a spectrum where awareness both informs and is informed by our choices, and that shapes who we are. We know that human experience hinges on the interplay between awareness and choice. But are our choices free? The question of free will is very caught up in diverse points of view. On the face of it, in today's day and age, we would say a very definite yes. As individuals, we are free to make choices and take decisions, and that is also independent of external constraints. But is that really so? The moral, ethical, social, and religious constraints hold us back, and no human being can operate entirely on his or her free will.

Consciousness Is

Philosophical perspectives have dealt with these matters in both dualistic and non-dualistic ways. Various theories have been proposed: physicalism argues that mental states are ultimately physical states; dualism places the mind and body as distinct entities, there is a clear demarcation between the mind and the physical brain, suggesting that while they interact, they are fundamentally different; and non-duality declares that *consciousness is*; it is the unchanging existence, while the body with the brain and the mind, or the body–mind complex, the psychophysical system, are transitory and function within the limitations of time, space, and causation.

Consciousness operates on a spectrum of three definite states. These include: first, the waking state, or the state of full alertness and focused attention; second, dreaming during sleep, or even altered states induced by substances or meditation; and third, deep sleep. Neuroscience studies these levels through brain wave patterns, neural activity, and how different brain regions interact. For us, as conscious human beings, the three states are tangible experiences. The *Mandukya Upanishad* explains these three states of consciousness and beyond, the fourth, which is intrinsic to this deeper understanding (see Box 9.2 in Chapter 9).

Non-duality, as rooted in Eastern philosophical traditions, positions an essential unity that underlies all phenomena. It challenges the Cartesian dualism of mind and body, suggesting that consciousness is not a by-product of neural processes; it is the fundamental aspect of reality. The non-dual approach to the brain and mind does not see them as separate entities. They are different expressions of the same underlying reality. This aligns with emerging scientific theories, such as the quantum and holonomic models of consciousness, which suggest that the brain functions more as an interface or mediator of consciousness than its source. These theories propose that consciousness pervades the universe, with the brain acting to localize and process this universal consciousness into the subjective experience of being.

Thinking and knowing are complementary in non-duality. Thinking can initially help us to remove false ideas of self and reality, and knowing ultimately leads us to the direct realization of non-duality, beyond conceptual understanding. Thinking is the function of the mind that can perpetuate the illusion of separation and duality if we do not understand our experiences with detachment or as observers. Knowing implies the direct experiential realization of non-duality. It means recognizing the underlying unity of the self and the universal.

Philosophically, consciousness raises questions about the mind–body problem, free will, and the nature of subjective experience.

Self-awareness is not only being aware but also knowing ourselves as individuals, separate from the environment and other individuals. This includes recognizing our emotions, thoughts, beliefs, and behaviours and understanding how they are perceived by others. It's also linked to self-reflection and introspection.

Thomas Nagel, in his article "What Is It Like to Be a Bat" in *Philosophical Reviews* (1974), has rightly stated that consciousness makes the body–mind problem intractable. When I am in a conscious mental state, there is "something it is like" for me to be in that state from the subjective or first-person point of view. When I am, for example, smelling a rose or having a conscious visual experience, there is something it "seems" or "feels" like from my perspective. An organism, such as a bat, is conscious if it is able to experience the outer world through its (echo-locatory) senses. There is also "something it is like to be a conscious creature" whereas there is nothing it is like to be, for example, a table or tree. David Chalmers talks of the "hard problem" of consciousness, referring to the difficulty of explaining why and how we have qualia, or conscious experiences. Subjective experiences, or qualia, are the first-person aspects of consciousness, such as the redness of red, the taste of chocolate, or the pain of a headache. These experiences seem to go beyond mere physical processes and cannot be fully explained by understanding the brain's mechanisms or neural correlates alone. As David Chalmers also says, "The hard problem of consciousness is the problem of experience." How can physical processes in the brain give rise to subjective experience? We all have subjective experiences. Conscious states include states of perceptual experience, bodily sensation, mental imagery, emotional experience, and more. There is "something it is like" to see a vivid green, to feel a sharp pain, to visualize the Eiffel Tower, to feel a deep regret, and to think that one is late. Each of these states has a phenomenal character, with phenomenal properties (or qualia) characterizing what it is like to be in the state.

In his talks on "Consciousness the Ultimate Reality", Swami Sarvapriyananda explains that understanding consciousness is not about blind faith or belief but rather about realizing and seeing something beyond ordinary experiences. Our true self is the *witness* of the body and mind, separate from them, existing and aware. Consciousness is not an object but rather the awareness with which objects are experienced. Consciousness cannot be reduced to the mind, body, behaviour, or language, as shown by various thought experiments. "The ups and downs of the mind, the pleasure and the pain, the frustrations and the anxieties and fear...they are objects of my awareness. I am not them."

Why is it so difficult to explain consciousness? Because consciousness is not objectifiable, we cannot say consciousness is "this or that". It can be pointed out phenomenologically, but it cannot be explained biologically. We have often referred to the concepts of direct experience and first-person experience in our discussions; here we are stating that consciousness cannot really be explained. We are all aware that we are conscious. The questions asked are: How does it happen? How does the brain produce consciousness? The key question here is: Is consciousness explicable as matter? There are no final answers; the questions remain. While various thinkers and experts continue to debate its nature, there are some approaches and perspectives that stand out in trying to understand consciousness.

Certain schools of thought try to establish a relationship between objects and consciousness. The Charvakas, or the materialists categorically, state that the activities of the brain generate consciousness. The Charvaka school was a philosophical movement in ancient India, and they held strong consumeristic, materialistic, and atheistic views. The Charvaka philosophy sees life as finite, with no afterlife, and prioritizes maximizing pleasure in the present life, even if it involves unconventional means. It is primarily concerned with empirical observation, as several modern-day scientists do, and they reject any beliefs that are not empirically verifiable.

Theistic belief has consistently emphasized that consciousness generates the objective world. Every theistic religion emphasizes the divine origin of consciousness and finds its significance and connection with the concepts of creation and salvation. The atheist and agnostic reject these beliefs, as they find no proof of the existence of God. A collection of religious ideas emerged in the late 1st century AD among Jewish and early Christian sects, called Gnosticism. These emphasized personal spiritual knowledge (gnosis) above the proto-orthodox teachings, traditions, and authority of religious institutions. Pleroma, in Gnosticism, refers to the fullness of divine existence or the totality of divine power, while *shunya* in Buddhism represents the concept of emptiness or void.

The dualistic approach is at the core of ancient Sankhya Philosophy. It posits two realities, essentially God and me, or consciousness and matter. This is panpsychism, where consciousness and matter co-exist, as do Purusha and Prakriti. These energies interact within all of us. This accepts that consciousness is a fundamental reality and cannot be reduced to brain activity. Time, space, matter, energy, and consciousness are all fundamental realities.

An unsolved bet! David Chalmers and Christof Koch had a bet in 1998, as Koch asserted that in twenty-five years, consciousness would be explained as an activity of the brain. David Chalmers, the one who has posed "The hard problem of consciousness", said this was not possible. Because consciousness cannot be reduced to the activity of the brain. When the question was revisited twenty-five years later, in June 2023, Christof Koch admitted that there was no explanation yet of how the brain can give the subjective experience of consciousness. The bet ended on 23 June 2023, in New York City, with Chalmers declared the winner. Both researchers agreed that consciousness is still an ongoing quest and that, despite all efforts, researchers still do not understand how our brains can produce consciousness. With all the innovations in S&T and the rapid developments in AI, the one thing that is not explained and thus has not been created is consciousness. AI machines are not conscious. The programmes of AI, the robots, and the most sophisticated gadgets that we use to resolve most of our physical and material issues, are not conscious.

The body eventually perishes, but consciousness does not. Non-duality offers a unique framework for understanding the continuity of consciousness and identity despite radical changes in physical embodiment. If consciousness is indeed fundamental and not strictly bound to the physical structures of the brain, then the transplantation of the brain into a new body does not inherently alter the core self or consciousness. Instead, it asserts the adaptability and resilience of consciousness, capable of maintaining continuity of self-awareness, identity, and personal history across dramatically altered physiological conditions. This is the core of the reincarnation theory: the body dies, perishes, and ends, and consciousness is and remains.

Adopting a non-dual perspective re-evaluates cognitive science and psychology, particularly in understanding perception, memory, and the nature of the self. This point of view asserts that mental processes and conscious experience cannot be fully explained by examining the brain in isolation but must consider the broader context of consciousness as a non-localized, fundamental aspect of reality. It supports altered states of consciousness, meditation, and other practices that reveal the fluidity and interconnectedness of the subjective experience. The non-dual approach proposes meditation, self-inquiry, and contemplation to directly experience the non-dual nature of reality and transcend the limitations of dualistic thinking. This perspective reimagines the relationship between the brain, mind, and consciousness. It enriches our understanding of human consciousness and identity, and we have to then see the interconnectedness of all life and the universe itself. Through the non-dual perspective, we can investigate the deepest questions about what it means to be conscious, alive, and fundamentally connected to the world around us.

Scientific Research and Practical Applications

Mind and Meditation

Meditation, originally rooted in ancient traditions, has transcended its origins and become very significant, particularly for mental health. This practice, once primarily associated with the spiritual journey towards enlightenment as detailed by Patanjali through the stages of concentration, meditation, and absorption, now enjoys widespread acclaim for its health benefits. Meditation and mindfulness mean focused attention on the present moment or guided imagery to create a sense of mental calm and relaxation. They enhance our sensitivity to the thoughts and emotions of others, generating empathy and understanding. Some meditation techniques lead to altered states of consciousness. The relationship between the mind and meditation is reciprocal. While meditation greatly influences the machinations of the mind, the mind's capacity to engage with meditation practices determines the depth and effectiveness of the practice.

Research confirms that regular meditation can significantly reduce stress and anxiety. If pharmaceutical medicine is for the body, meditation is the right medicine for the mind. A study published in "Psychosomatic Medicine" found that meditation reduces sympathetic nervous activity and lowers the levels of stress hormones in the body. Changes in the electroencephalogram (EEG) correspond to a state of relaxation. Meditation helps with anxiety disorders and depression. A systematic review and meta-analysis of randomized controlled trials suggests that mindfulness meditation can bring about moderate reductions in psychological stress, including anxiety and depression (Michaelsen et al. 2023).

Beyond stress and mood disorders, meditation enhances cognitive functions. Research in the *Journal of Cognitive Enhancement* (29 February 2024) states that blended mindfulness interventions can improve attention, memory, and academic achievements in students, making them an invaluable tool for enhancing mental clarity and focus. Meditation improves sleep quality. Mindfulness meditation can help those struggling with sleep disturbances by facilitating better sleep patterns and reducing symptoms of insomnia. Meditation is a bridge between traditional practices and modern health needs. Meditation keeps us connected to our inner reality; it can enhance cognitive function. Several studies have observed positive associations between meditation practice and increased grey matter density in brain regions linked to attention, memory, and emotional regulation.

Neurotransmitters: Meditation practices might affect the production and activity of certain brain chemicals. Studies suggest it can boost dopamine, which is linked to feeling good and motivated, and serotonin, which contributes to mood and well-being. Meditation might even help regulate glutamate, important for learning and memory, and gamma-aminobutyric acid (GABA), a neurotransmitter, a chemical messenger in the brain, a brain chemical known for its calming effects. Interestingly, the levels of neurotransmitters we already have can also influence our meditation experience. For example, having higher baseline dopamine might make it easier to focus during meditation. This two-way street between meditation and neurotransmitters is intriguing, and scientists are actively figuring out the specific mechanisms at play to understand how meditation can impact our minds. However, these studies are ongoing, and more research is needed to fully understand the cause-and-effect relationship between meditation and cognitive benefits.

States of Mind

The human mind is not static; it shifts through various states from moment to moment and across our lifetimes. Understanding these different states is very important for us to remain aware of our states of consciousness and how we live our lives. The most obvious state of being awake means being alert, aware of the external environment, and engaged with the surroundings. We spend most of our active hours in this state, working, interacting with others, and absorbing information. Similarly, sleep, another natural state, including its various stages from light to deep and REM (rapid eye movement), is vital for our mental and physical health.

We can experience altered states of consciousness through various means. Besides meditation, exercise is a simple and healthy way to alter the state of mind. Some human beings resort to substance use, which sadly turns into substance abuse. Hypnosis is a trance-like state where we can experience increased focus, concentration, and suggestibility. Hypnosis is used therapeutically to modify behaviours, enhance motivation, and decrease psychological distress.

Flow states are often described as being "in the zone". When we are fully immersed and engaged in any activity, we function with a sense of effortless concentration, great enjoyment, and an absence of temporal awareness.

It is natural for the human mind to respond to and handle stress and pressure. But prolonged states of anxiety can adversely affect cognitive function, the ability to handle emotions, and overall health. Much more than just feeling sad, depression is a complex mood disorder that can devastate how we feel, think, and handle daily activities. It can significantly alter the state of mind and generate a debilitating and persistent feeling of sadness and loss of interest, among a variety of emotional and physical problems. Euphoria is the direct opposite of depression. It is an elated state of happiness and self-confidence. It is prompted by various stimuli, including success, love, exercise, and certain drugs. While generally positive, artificially induced euphoria through substance use can be addictive and harmful.

We can enjoy a mental state of being focused and concentrated, which brings higher productivity and efficiency. We can also have a wandering mind, with thoughts shifting away from a task towards unrelated concerns or fantasies. It is not a mere lapse in attention; mind-wandering can, actually, have positive outcomes! The spontaneous shifting of attention away from the task at hand and towards unrelated thoughts or associations can encourage creativity and problem-solving by connecting disparate ideas.

Understanding and recognizing these states of mind can help us to manage our mental health, enhance our ability to focus, and increase our capacity for creativity and joy. By cultivating awareness of our mental states, we can make informed choices that lead to healthier and more fulfilling lives.

Power of the Mind

The power of the mind is the mental processes and conscious thoughts that influence our perceptions, emotions, behaviours, and even our physical well-being. The power of the mind includes what is currently accepted by science, the predictive capabilities attributed to the mind, and the yogic powers described in various traditions. The mind can exert a degree of control over various aspects of our lives, including our mental and physical states. However, the extent to which the mind can control the brain's activity and its mechanisms is still open to discussion and debate.

In medical science, for example, the power of the mind is understood easily through the "placebo effect". When we experience improvement in our health condition after receiving a treatment that has no real therapeutic quality as such. This improvement is driven by our belief in the treatment's efficacy or in the person prescribing the treatment. The power of the mind influences the physiological response, such as pain relief. The power of the mind facilitates neuroplastic changes in activities such as learning new skills or recovering from a brain "stroke", through rehabilitation. Neuroplasticity is the brain's ability to reorganize and adapt by forming new neural connections. Learning, memory, and recovery from brain injuries are examples of neuroplasticity in action.

Cognitive-behavioural theories, including cognitive-behavioural therapy (CBT), emphasize the role of thoughts, beliefs, and behaviours in shaping our emotional and psychological well-being. According to these theories, if we change our negative thoughts, it can alter our emotions and behaviours. Athletes and performers often visualize and mentally rehearse techniques to improve their performance. By vividly imagining successful outcomes, we can boost our confidence and focus to ensure better results. The popular "mind over matter" theory that many of us apply in our day-to-day lives proves that the mind can exert control over physical matter to a great extent. These hypotheses may be technically categorized as parapsychology or psychic phenomena, but they are practical and proven for many of us. The mind can and does play a significant role in our health, well-being, and cognitive performance, but the precise nature of this role is not empirically definable.

Scientists acknowledge the mind's incredible capabilities, especially in its capacity for learning, memory, regulating emotions, and healing. Neuroplasticity, the brain's ability to reorganize itself by forming new neural connections throughout life, shows the mind's adaptability. This concept, that learning and experience can physically change the structure of the brain, which was once a radical idea, is now well-supported by evidence. Experiments with meditation and mindfulness are also well-accepted, as their positive effects on mental and physical health are discernible. Regular meditation decreases stress and improves attention and memory. The mind can anticipate future events, referred to as intuition or precognition. Mainstream science remains sceptical of these phenomena because there is no empirical evidence, although there have been intriguing studies that talk about the power of predictions. For example, the field of behavioural economics has given insights into how people make predictions about future events and how those predictions influence their decisions, often using patterns and past experiences to inform these foresights.

As mentioned in Chapter 4, yogic traditions speak of *siddhi*s, or supersensuous abilities, that get developed through advanced meditation and yoga practices, essentially the power of the mind. While these claims are largely outside the purview of scientific validation, there are numerous anecdotes and reports of yogis demonstrating extraordinary capabilities.

Swami Rama, a yogi tested at the Menninger Clinic in the 1970s, reportedly demonstrated the ability to control his heart rate and body temperature—abilities that hint at the profound control the mind can exert over the body. Sri Ramakrishna, a 19th-century Indian saint, would repeatedly go into states of *samadhi.* During his trances, Ramakrishna reportedly became unconscious, sat in a fixed position either for a short time or for several hours, and would then slowly return to normal consciousness. During one of his deep states of *samadhi*, Sri Ramakrishna's body became rigid and insensitive to external stimuli. Concerned about his health, his disciples sought the advice of contemporary medical professionals to examine him. Dr Mahendralal Sarkar, a renowned physician of Kolkata but sceptical about spiritual matters, was invited to examine Sri Ramakrishna during his *samadhi* state. As a direct disciple present at the time wrote, "Mahendralal Sarkar and other doctors examined him with the help of instruments and found no sign of the functioning of his heart. Not satisfied with that, his friend, another doctor, went further and touched with his finger the Master's eyeball, and found it insensitive to touch like that of a dead man." Dr Sarkar was a follower of modern science. He accepted a thing only after judging all its aspects. He was intrigued by the outcome of the medical examination of Sri Ramakrishna during *samadhi*. He had to acknowledge that the phenomenon was beyond the realm of ordinary medical understanding at that time.

Many thinkers often explore the potential of the human mind in speculative scenarios about its untapped capabilities. Telepathy, telekinesis, and enhanced cognitive abilities are common themes, suggesting a future where the boundaries of the mind are expanded far beyond current limitations. These resonate with the yogic view of the mind's potential to transcend ordinary physical and temporal boundaries, suggesting a shared curiosity and aspiration towards understanding and harnessing the mind's full power.

Exploring the mind's power bridges the gap between science and spirituality, between what is known and what is yet to be discovered. Scientific research continues to unravel the brain–mind complexities, offering glimpses into its potential to influence our health, perceptions, and reality itself. Yogic practices and philosophical insights view the mind's capabilities from a different perspective, suggesting that beyond the physical and cognitive lies a realm of extraordinary potential waiting to be unlocked. Whether through the meticulous evidence of science or the experiential wisdom of yoga, the ongoing endeavour to understand the mind's power pushes the boundaries of what we believe is possible if we continue to explore the depths of our consciousness.

The Healing Power of the Mind

The power of the mind to heal the body is recognized in modern medicine. There are several approaches, from psychotherapy and mindfulness to hypnotherapy and somatic practices. Scientific perspectives highlight the significant impact of mental states on physical health. Mindfulness-based psychotherapy emphasizes the importance of awareness and acceptance of feelings, thoughts, and physical sensations. It helps to deal with depression, anxiety, and pain; immunity and brain function also seem to improve.

Hypnotherapy relaxes the mind to make the subconscious more receptive to positive suggestions. It helps to manage pain effectively and improve conditions such as anxiety and depression. Gut-directed hypnotherapy can decrease symptoms of irritable bowel syndrome, emphasizing the mind's influence over the gut. Somatic experiencing and tension and trauma-releasing exercises are therapies focusing on the body's physical responses to stress and trauma. By bringing awareness to the physical sensations associated with traumatic experiences, these therapies manage to integrate and heal traumatic memories. Studies have shown that mindfulness meditation can lead to measurable changes in the brain, which are associated with reduced inflammation and improved health outcomes.

Yoga, as a holistic discipline, combines physical postures, breathing techniques, and meditation to improve various health conditions, further emphasizing the mind–body connection. In all these aspects of the mind, if we explore from both the scientific and spiritual perspectives, we can relate to the fascinating connection between empirical evidence and spiritual wisdom. We can gain a deeper understanding of the various states of mind, their significance, and their impact on our lives.

From a scientific viewpoint, psychology, neuroscience, and cognitive science help to explore the states of mind. These disciplines examine the physiological, neurological, and psychological aspects of our conscious experiences, categorizing states of mind into various forms such as wakefulness, sleep (including REM and non-REM sleep), and altered states induced by meditation, psychoactive substances, or mental health conditions. Cognitive science investigates how perception, memory, thought, and problem-solving occur. Research in cognition and consciousness is trying to understand how the brain integrates information, generates awareness, and transitions between different levels of consciousness. Neuroscience studies the activity of the brain and the patterns and chemical balances that underlie different mental states. Techniques such as fMRI and EEG have shown the correlation between brain wave patterns and states of consciousness, including the distinct states of wakefulness, sleep, and other altered states. Psychology addresses how emotions, thoughts, and behaviours influence and are influenced by various states of mind. Neurotheology is an emerging field that investigates the neurological implications of spiritual and religious experiences and seeks to understand how these states affect the brain and consciousness.

Spiritual traditions across the world offer meaningful insights into the states of mind, not only from psychological or physiological perspectives but as gateways to deeper understanding, enlightenment, or unity with the divine. The mind is a tool for achieving higher states of consciousness. The integration of science and spirituality enriches our understanding of the mind, proving that they are complementary. While science provides tools and methodologies to observe and measure the mind's workings, spirituality offers pathways to transcend ordinary consciousness and explore deeper dimensions of our existence.

Mind travel: One very fascinating aspect of the mind is the implication of its "travelling". This refers to imagination, daydreaming, or deep contemplation, not physical travel! The mind can engage in various mental activities, but it does not possess the capability to physically transport itself to different locations at will. This concept is either in the realm of science fiction or metaphysical speculation. The mind has the remarkable ability to imagine and create mental scenarios, memories, and experiences. Through such imagination, we can mentally "travel" to places and situations where we may have never been physically. This form of mental travel is a creative and cognitive process. We often mentally rehearse or "travel" through scenarios in the mind to enhance our performance or achieve specific goals. Daydreaming is a common mental activity where we allow our thoughts to wander freely. We may mentally travel to different places, scenarios, or even fantasies. This is a spontaneous and unstructured mental exploration. Linked in a strange way to mind travel is the concept of out-of-body experience (OBE). Scientists tend to disregard OBE because there is no empirical evidence. It is taken as a hallucination and the possible outcome of a wandering mind. According to recent findings by neuroscientists, OBEs can be induced by delivering a mild electric current to specific spots in the brain region called the angular gyrus. This resulted in the sensation for a woman that she was hanging from the ceiling, looking down at her body. In another woman, an electrical current delivered to the angular gyrus produced the uneasy feeling that someone was behind her, intent on interfering with her actions. We have discussed this more in Chapter 8.

Many people find these experiences debatable or doubtful, suggesting that they may be the outcome of altered states of consciousness. Metaphysical theories explore the idea of consciousness, or the mind, having the ability to leave the physical body and travel to other realms or dimensions. However, mainstream scientific understanding does not support these ideas. From a neuroscientist's perspective, perhaps one way to approach the idea of mind travel is to consider the speed of thought, which is related to the speed of the nerve impulses that carry information through the nervous system. The speed of thought is measured by reaction time, which is the time from the onset of a stimulus to the initiation of an action. Reaction time depends on the type, intensity, and complexity of the stimulus, the type and difficulty of the task, the attention and motivation of the person, and the individual differences in cognitive abilities. The average reaction time for humans is around 200 to 300 milliseconds (ms), or 0.2 to 0.3 seconds, for simple tasks such as pressing a button when a light flashes. For more complex tasks, such as making a decision or solving a problem, the reaction time can be much longer, ranging from seconds to minutes. The speed of thought, and by extension, the speed of mind travel, is a variable and relative value that depends on both internal and external factors. The essential fact to understand is that the power of the mind has a very vast and diverse range. Our mind has the ability and power to both heal and ail. It can "travel"; it can give us unique experiences that perhaps the mind itself cannot explain. Technology is getting deeper into trying to explain, scientifically, how these powers function.

Advances in Technology

Brain–machine interfaces (BMIs) and brain–computer interfaces (BCIs) are technologies that facilitate communication between the human brain and external devices. While both terms involve brain-device interaction, BMIs encompass a broader range of applications beyond computers, including robotic arms and prosthetics. BCIs, on the other hand, specifically focus on cognitive communication with computers, allowing users to send commands directly from their brain. Whether it's controlling a cursor or assisting individuals with motor disabilities, these interfaces continue to advance our understanding of brain function and enhance human–machine interaction. Neuralink's latest implantable brain chip is an example of BMI. This could help paralysed or disabled individuals control devices through thought. But several serious ethical issues are linked to both BMI and BCI.

Research in BCI is an exciting exploration of the potential of the human mind. On the verge of a revolution, offering a future where our thoughts can directly control the world around us. Led by talented researchers and institutes worldwide, BCIs are already blurring the lines between humans and machines. They are paving the way for a future where our thoughts hold the power to shape our reality in ways we can only begin to imagine.

Dr Miguel Nicolelis, a neuroprosthetics pioneer at Duke University, and his team are making significant strides in deciphering the complex language of our brains. Using ML, they are translating brain activity into actions. This technology will be immensely promising for individuals with paralysis, offering them the ability to control prosthetic limbs or communicate through BCIs. Dr Andrea Jackson at the University of California, San Francisco, is also focusing on using BCIs to restore sensory and motor function in paralysed individuals.

Dr Laura Frey at the Wyss Institute at Harvard University is developing non-invasive interfaces with user-friendly dry electrodes, making BCIs more accessible. Beyond medical uses, BCIs could revolutionize gaming experiences by providing intuitive control in virtual reality. Dr Benjamin B. Pesaran of the Nathan Kline Institute is investigating the neural basis of decision-making to develop more advanced BCIs for these applications.

Despite the challenges, the future of BCIs is brimming with possibilities. As research progresses at institutes like the Wyss Institute and BrainGate (Stanford University), these interfaces have the potential to become seamless extensions of ourselves, allowing us to interact with the world through the sheer power of thought. BCIs are thus also pushing us towards the realm of science fiction, with the possibility of full-blown thought transfer. Imagine directly transmitting our thoughts and eliminating the need for using words or gestures! Current BCIs can decode basic commands and rudimentary speech patterns, but capturing the complexity of a complete thought to transmit remains a significant hurdle. The Institut du Cerveau – Brain Institute (Paris) and RIKEN Brain Science Institute (Japan) are just a few examples of research centres around the world grappling with this challenge.

A multi-institutional collaborative research project, BrainGate, is focused on developing and testing medical devices that would restore communication, mobility, and independence in individuals with neurologic disease, paralysis, or limb loss. They use electrodes implanted in the brain to translate neural activity into commands for devices. They intend to use holography and ultrasound to read and write to the brain and to understand and interface with the brain. Some projects aim to decode speech or visualize mental imagery from brain activity to create new ways of communication for those who are unable to speak or type. Efforts are in progress to restore sight, hearing, and movement using BCIs that interpret the user's intentions and translate them into action. Direct brain stimulation could enhance learning, memory, or other cognitive functions. As technologies advance, there are ethical, privacy, and security concerns, such as the potential for unauthorized access to anyone's thoughts or intentions.

As we discussed earlier regarding the perils of misuse, these technologies can be disastrous if put in the wrong hands for the wrong purposes. The potential for thought transfer, for example, comes with its own set of ethical concerns. The breach of privacy is imminent, and the very definition of self in a world where thoughts can be shared freely needs careful consideration. Dr Rajesh P.N. Rao at the University of Washington and like-minded researchers are actively addressing ethical considerations around privacy and data security that are inevitably linked with the diverse potential uses of BCIs.

Another company, Neurable, has developed BCI-enhanced headphones, which allow users to control external software or hardware using their brain signals. How can BMI and BCI impact individual thoughts? How can the chip influence behaviour? Or, can the chip remotely take control of the brain? These may no longer be the ethical issues of science fiction only.

The Openwater project, still behind the wall of secrecy, is pioneering revolutionary non-invasive brain imaging technology. This MRI-like system offers unprecedented detail, reaching the level of individual neurons within the brain. The potential to decode dreams, thoughts, and even speech from brain activity makes Openwater a game-changer for medicine, mental health, and potentially, communication itself. Although the exact origins and inventors remain unclear, Openwater's open-source approach encourages collaboration and gives the promise of a future where the secrets of the mind are unlocked.

All the innovative advances in technology do bring us back to recalling how science fiction becomes reality in due course! There is no dearth of science-fiction stories and films that talk about the powers of the mind and brain signals. These stories cover various fascinating aspects such as mind travel, exploration of dreams and alternate realities, and the consequences of manipulating consciousness. These engaging narratives challenge our understanding of reality and the power of the mind. The film *Inception*, directed by Christopher Nolan, is a mind-bending thriller where a group of thieves use a technology that allows them to enter the dreams of others to steal their secrets. The film dives into the layers of the subconscious mind and the consequences of manipulating dreams. *The Matrix* series is actually a yogic sci-fi film series that explores the concept of a simulated reality created by machines to control humanity. In the first film, the protagonist, Neo, discovers the truth about the artificial world and embarks on a journey of self-discovery and resistance against machines. The key here is that the script and screenplay writers and directors must necessarily "travel" into the unlimited dimensions of fantasy to create and actualize these ideas, which they make "real" for the viewers, who then "travel" with them. We have a perspective here that we need to take cognizance of: the fact that when modern science spirals into the inner world of the human mind, it is already in the world of meta-science, and the synergy of Genome and Om is a reality.

Mind reading: The idea of reading others' minds, or telepathy, has fascinated thinkers for centuries, and they have explored this from various perspectives, including science and spirituality. Science has not provided conclusive evidence for the reality of telepathy or the ability to read others' minds in a paranormal sense. As human beings, we do have the natural capacities for empathy and for understanding the thoughts and emotions of others based on observable cues. These psychological and neurological processes give us the ability to connect with and relate to one another at a deep level. Belief in telepathy often stems from spiritual and metaphysical traditions, but it is not scientifically validated.

In mainstream science, there is no empirical evidence to support telepathy or the ability to read others' minds. Parapsychologists have conducted experiments on extrasensory perception (ESP), including telepathy, but the results have been inconclusive and often subject to criticism. The scientific community generally maintains a sceptical stance on telepathy due to the lack of replicable and robust evidence. Some New Age and metaphysical teachings suggest that individuals can develop psychic abilities, including telepathy, through training and heightened awareness. Individuals can develop an understanding of others' thoughts and emotions through empathy and the theory of mind. This is the ability to attribute mental states, such as beliefs and intentions, to others. It allows individuals to make educated guesses about what others might be thinking or feeling based on observable cues. Empathy involves sharing and understanding the emotions of others without necessarily "reading" their minds.

Some spiritual and metaphysical traditions propose the existence of telepathy or similar abilities as part of a broader concept of interconnected consciousness. From the pure and universal consciousness perspective, the ability to reach out to another living being, both mentally and physically, is a "no-brainer". In the oneness of universal consciousness, this empathy and bond are a given; they are there; only we may be unaware of them due to the veiled self-perception of a psychophysical system encased within time, space, and consciousness. Often, in our effort to reach out, seeking that one-ness, we listen to an inner voice.

The inner voice, also known as self-talk or inner monologue, is the internal dialogue or stream of thoughts that we experience within our minds. It is the ongoing narrative, commentary, or conversation that we have within ourselves, often without vocalizing the words. The inner voice can take various forms, including spoken words, phrases, or mental images. It represents the thoughts and reflections that occur within our consciousness. It can be considered a means of self-reflection, self-assessment, problem-solving, planning, decision-making, or the processing of emotions.

Sometimes, the inner voice is conscious and deliberate, like when we intentionally engage in self-talk to make decisions or analyse situations. This is when we are completely aware of this inner voice. At other times, the inner voice may be subconscious, with automatic thoughts and reactions that occur without our conscious effort or awareness. The most common fallback on the inner voice is when performing cognitive tasks such as reading, writing, or mathematical calculations, when we often use inner speech to guide our thinking.

The inner voice can significantly impact our mental well-being. Positive and self-compassionate self-talk can ensure emotional resilience and self-esteem. Negative and self-critical inner dialogue can be associated with or trigger anxiety, depression, and low self-esteem. Mindfulness practices involve observing and becoming aware of the inner voice without any judgement. We can then distance ourselves from the subconscious, automatic, or distressing thoughts that the inner voice triggers. Conscience and inner voice are often confused, but they are different in meaning and connotation. We are very aware of our deeply held moral principles, which we may have inherited from our parents, from society, or cultivated ourselves in the course of our lives. We are motivated to act upon them, and our character, our behaviour, and ultimately ourselves are assessed against those principles. Different philosophical, religious, and common sense approaches to conscience have emphasized different aspects of this characterization.

Change of mind and change of mood: A change of mind typically refers to a shift in our beliefs, opinions, or decisions. It could mean a deliberate effort to reassess and re-evaluate situations, decisions, or courses of action. A change of mood is a shift in our emotional state or attitude. Internal factors like hormones, neurotransmitters, and thought patterns can change our moods; external factors like stress, environmental conditions, and social interactions can affect our moods. Unlike a change of mind, a change of mood may not always involve conscious awareness. We might experience a sudden shift from happiness to sadness due to an unexpected event or trigger of memory. While both "change of mind" and "change of mood" impact our behaviour, they operate on different levels of consciousness.

Given the immense capacity of mind–brain activities and the unfathomable capacity of the mind in the role it plays in our functioning as human beings, the idea of "change" of mind in a different way fascinates us. This has given impetus, through the ages, to the use of herbs or drugs that can cause altered states of consciousness characterized by major changes in thought, mood, and perception. In modern terminology, these are hallucinogens. Such plants have been used since ancient times. They were popular for their healing effects and also for their religious significance. Mind-altering drugs were used very commonly, as were mind-altering materials and psychedelic drugs in ritual and spiritual practices, especially in the shamanic traditions of the East and the

West. These practices are apparently still common in some tantric traditions of Hinduism and Buddhism.

Cannabidiol (CBD), a cannabis compound distinct from the psychoactive tetrahydrocannabinol (THC), is being studied for its potential mood benefits. Early research suggests that CBD might lessen anxiety and depression by interacting with the body's mood-regulation system. However, these effects aren't guaranteed, and they require further investigation.

Psychedelic drugs, a "sub-class of hallucinogenic drugs", are of keen interest and research because of their strong effects on the mind and consciousness. These substances can significantly alter perception, cognition, and emotions and lead to "unique" mental experiences. With their ability to alter states of consciousness, the users of the drugs experience vivid hallucinations, changes in perception of time and space, and a heightened sense of interconnectedness with the world. Colours appear more vibrant, sounds more vivid, and textures more pronounced. Some psychedelic experiences include the experience of the loss of the sense of self or ego and therefore lead to a unity with the universe where all boundaries between the self and others are dissolved. Many individuals have overwhelming spiritual or mystical experiences during psychedelic trips. These experiences involve a sense of one-ness with the universe, encounters with divine entities, or a feeling of transcending time and space. Psychedelics are known to bring emotions and traumas to the surface, which can perhaps be therapeutic if guided by trained professionals. Many artists, writers, and scientists experience enhanced creativity and tremendous insights.

Research suggests that psychedelics primarily affect the serotonin system in the brain. They disrupt the brain's default mode network, which is associated with the sense of self and ego. This disruption contributes to the altered states of consciousness. Psychedelic-assisted therapy has shown promising results for conditions such as depression, post-traumatic stress disorder (PTSD), and addiction. The use of psychedelics is not without risks. Inappropriate or unsupervised use can result in adverse reactions like anxiety or panic attacks due to unexpected and overwhelming experiences.

An easy alternative to seeking relaxation and joy through substances like CBD, hemp, and opioids might seem appealing, but it's a risky shortcut. These substances can easily become addictive, impairing our judgement and increasing health risks. The feel-good effect is temporary and doesn't address the root of stress or unhappiness. The attempt to artificially change the state of mind or mood is not advisable.

Exploring Beyond

The Future of Understanding the Mind

The future of understanding the mind is an exciting frontier. The advances in various fields promise deeper insights and unique perspectives. This evolving understanding is likely to be shaped by interdisciplinary collaboration, technological innovation, and a continuous re-evaluation of existing theories.

Continued advances in neuroscience will undoubtedly provide more detailed and comprehensive maps of neural networks and how they relate to cognitive functions and behaviours. Optogenetics, high-resolution brain imaging, and neurogenetics have the potential to reveal the workings of the brain in unprecedented detail. Developments in AI and ML are not only providing tools for analysing vast amounts of neurological data but are also offering new models for understanding cognitive processes. The intersection of AI and neuroscience might lead to sophisticated models that simulate aspects of human cognition, offering insights into the mechanisms of learning, memory, and decision-making.

In psychology, there's likely to be a continued trend towards integrating different therapeutic approaches, informed by research in clinical psychology, neuroscience, and social psychology. Personalized and precision medicine approaches to mental health, based on individual genetic, environmental, and psychological factors, are expected to become more prevalent. Philosophical and ethical considerations are bound to evolve. As our understanding of the mind deepens, philosophical and ethical questions will become increasingly important. Issues surrounding consciousness and the moral implications of neuroscientific discoveries will require thoughtful consideration and debate. Although still speculative and controversial, some researchers are exploring quantum theories of consciousness. These theories might offer novel explanations for some of the more enigmatic aspects of consciousness and cognition.

Knowing with the Mind's Eye: Intuition

"Exploring Beyond" encourages an inquiry into the unknown territories of human consciousness and the universe. It suggests moving past the conventional boundaries of scientific understanding to explore areas like metaphysics, spirituality, and paranormal phenomena. This exploration seeks to understand the deeper layers of reality and consciousness, challenging our perceptions and beliefs about the nature of existence.

"Knowing with the mind's eye" refers to intuition—a form of knowledge that comes without the use of rational processes. Intuition is often described as a "gut feeling" or an "inner knowing" that guides decisions without conscious reasoning. Beyond intuition lies the concept of extrasensory perception (ESP), which includes telepathy, clairvoyance, and precognition. These phenomena represent an understanding or awareness that transcends conventional sensory experiences and rational thought processes.

The Vedic concept of visualization or perception in Yoga philosophy tells of the extraordinary claim of the abilities of yogis, the serious practitioners of yoga, to perceive truths beyond the sensory boundaries. This takes our search into the interconnections between mind, consciousness, and reality. Experiencing a divine presence or receiving spiritual insight through visual contact or perception is a sacred and transformative experience in Hinduism, Jainism, and Buddhism. It is believed that devotees can receive blessings, guidance, and inspiration, leading to spiritual growth and enlightenment. In the context of yoga practice, it also refers to the inward vision or perception attained with meditation or self-reflection. It means gaining insight into our true nature, connecting with inner wisdom, and experiencing a sense of unity with the divine or the universe.

Peace of Mind

In the final analysis, we human beings have a single-point focus and agenda, and that is to be happy. As behavioural scientist Steve Maraboli has said, "Happiness is a state of mind, a choice, a way of living; it is not something to be achieved, it is something to be experienced." If it is the state of mind that will ensure that we attain the ultimate goal of happiness, then we would automatically seek "peace of mind", because peace and joy are meant to be synonymous. Peace of mind refers to a state of mental and emotional calmness and tranquillity. It is a subjective experience characterized by a sense of inner harmony, contentment, and a lack of mental turmoil. Peace of mind is highly valued and sought after because it contributes to overall well-being and mental health. It entails a state where we are free from excessive anxiety, stress, and the mental burden of unresolved problems. It involves a sense of relaxation and ease. A peaceful mind is associated with clarity of thought and emotional stability. It allows us to think rationally and make decisions without being clouded by strong negative emotions.

Peace of mind does not imply the absence of challenges or difficulties in life. Instead, it involves the ability to accept life's ups and downs with equanimity and resilience. It's about adapting to change and coping effectively with adversity. If we are mentally peaceful and thus happy, we will automatically focus on the present moment rather than dwelling on the past or worrying excessively about the future. Mindfulness and being fully engaged in the present contribute to a sense of peace. We would cherish contentment and gratitude for what we have in life, appreciating the positives, and not constantly yearning for more.

On the ground, healthy and supportive relationships with others can contribute to peace of mind. Positive social connections provide emotional support and a sense of belonging. Taking care of our physical and mental well-being through practices such as exercise, relaxation, meditation, and self-reflection can foster peace of mind. Feeling at peace is often associated with living in alignment with our values and principles. When our actions and choices align with our core beliefs, we feel a sense of inner harmony. A peaceful mind is conducive to making mindful decisions that are in sync with our long-term well-being and happiness.

Peace, or "*shanti*" is the common thread, the chant in all ancient scriptures. All aphorisms begin with "Om" and conclude with "*shanti*", as that is the universal eternal prayer. Ultimately, peace of mind is a subjective experience, and what brings it about can vary from person to person. Different individuals may find peace through various means, such as meditation, spiritual practices, spending time in nature, engaging in hobbies, seeking therapy, or simply finding solace in quiet moments of reflection. *If peace of mind and happiness are states of inner tranquillity and equilibrium, then they would explain who we actually are.* In this entire process of trying to understand the mind, we have honed in on the core question of our existence: "Who Am I?"

Swami Vivekananda on the Mind:
How hard it is to control the mind! Well has it been compared to the maddened monkey. There was a monkey, restless by his own nature, as all monkeys are. As if that were not enough, someone made him drink freely of wine, so that he became more restless. Then a scorpion stung him. When a man is stung by a scorpion, he jumps about for a whole day; so the poor monkey found his condition worse than ever. To complete his misery, a demon entered into him. What language can describe the uncontrollable restlessness of that monkey? The human mind is like that monkey, incessantly active by its own nature; then it becomes drunk with the wine of desire, thus increasing its turbulence. After desire takes possession comes the sting of the scorpion of jealousy at the success of others, and last of all, the demon of pride enters the mind, making it think itself of all importance. How hard to control such a mind!

CHAPTER 6

Who Am I? The Body or the Soul

Knowing others is wisdom, knowing yourself is Enlightenment.
—Lao Tzu

With the creation of the planet, we, the *Homo sapiens*, came into being either through the process of evolution, as the special creations of a creator God, or as manifestations of the all-pervading one-ness. We have then reflected on life and on the mind and consciousness aspects of our existence. Now we face one of the most fundamental and profound questions: "Who am I? or "What am I?" This question has been as extensively researched by scientists and mathematicians as it has been by philosophers, psychologists, theosophists, theists, spiritualists, and the likes of you and me because we wish to know. We have different perspectives and insights into how we identify ourselves. Caught in our complex ego-personality, we fail to capture the essence of who we are because we are constantly evolving and changing with our environment, experiences, and the choices we make. Going back to the origin of life and what it is, we have to understand this "self", which is more than the body and its physiological structure. To find the answer to "Who am I", we must go beyond superficial classification and categorization and dive deeper to know who we really are.

The Perfect Machine

The most obvious and tangible layer of our identity is our body, comprising various organs, tissues, cells, and molecules that enable us to function and interact with the physical world. A long evolutionary process has shaped our anatomy, physiology, and genetics. Our body is also influenced by the environment, the climate we live in, the food we eat, the air we breathe, and the water we drink. Physically, we have a natural process of ageing, disease, injury, and death. This process challenges our sense of identity. Genetically speaking, we are a fascinating amalgamation of the biochemical functions that operate as per the information blueprint the DNA of the cells provides, with instructions that regulate both the structural and functional aspects of the organism! The human body, the near-perfect piece of sophisticated machinery, is connected by a tangled but extremely organized neural network, sending electrical signals at the rate of a hundred metres per second to the other cells in the body. It performs intricate functions super-efficiently, mostly without our conscious attention; we never have to tell the stomach, "Come on, start digesting that burger I just ate," or tell the heart, "Please pump more blood to my stomach, it needs to digest food." or tell the pancreas to secrete insulin! With unfathomable accuracy and speed, the three-pound brain, the crowning glory of this miraculous construct, communicates with every cell in the body within microseconds. Comprising billions of neurons and glia, the brain performs several voluntary and involuntary functions with absolute precision. It is the "mission control centre", directing the entire show, gathering information through small portals located there, and instructing participants (bodily cells and systems) to stage an appropriate response. Such is the intricacy and interconnectivity of this neural network that David Eagleman imaginatively quotes in his book *Incognito*: "If you represented each of these trillions and trillions of pulses in your brain by a single photon of light, the sum total would be blinding. The cells are connected to one another in a network of such staggering complexity that it bankrupts human language and necessitates new kinds of mathematics."

A new method of cell-to-cell communication within the brain was shared by Dr Hollis Cline, professor of neuroscience at Scripps Research, and her team. This study identifies hundreds of proteins travelling throughout the brain in tiny, membrane-bound sacs. This suggests that there is a new communication pathway beyond neurotransmitters and hormones; it throws more light on how brain cells "talk" to each other.

This complicated chemical–electrical network of trillions of neurons perched between our ears commands the entire nervous system, controls all the functions vital for survival, and is the base of the innumerable facets of our behaviour, thoughts, and experiences. And most of us, despite the advances in science and technology, are nowhere near deciphering its code, much less understanding its intricacies. If we reflect on the history of evolution and where we are today in this bipedal humanoid form, with the level of intelligence and understanding that apparently surpasses all the other expressions of life, we know that we have not sprung up instantly as a highly successful race; we owe our roots to almost all species that precede us.

Human beings and chimps are almost 98 per cent the same as regards the DNA structures, and the visible differences are very few. It is, however, amazing that even the chimp might not consider us to be congenial fellow beings, much less distant relatives! Although science is certain that the differences between humans and chimpanzees are nothing but genetic, it is mind-boggling to even imagine how growth and form are affected to such great lengths by these tiny molecules, which turn us into two distinctly identifiable species. What makes our human species so surprisingly unique? What is that distinctive characteristic that makes us what we are today—an identity that any and all of the other species on the planet would indeed envy if only their understanding could reach that level? What has helped us to develop this complex system, which can dwell deep into deciphering its own programming language? Searching for this identity over the ages, we have arrived at several conclusions, the outcome of progressive advances in the understanding of a range of subjects such as biology, chemistry, physics, psychology, philosophy, and medical science, among others. Our endeavour may just about touch the tip of the iceberg, but the search continues.

Who Am I, What Am I, or What Am I Not?

Diverting to more personal musings in the attempt to understand who "I" might be brings "me" to different conclusions. Addressing the individual in us, our hand inadvertently goes to the chest and points out "me" or "I" as our first identity; it never reaches the head, the abdomen, the hand, or the leg. Therefore, is this "I" located in the chest, at the heart of all physical existence? Or is it the thinking unit of the brain that governs or dictates our identity? Every conscious activity of the mind or body, such as, "I think", "I remember", "I feel", "I am acting", revolves around the tacit assumption that there is an individual "I" who is doing these things.

But intuitively, we know that it is not the body or any part of it that we talk about when we say, "I desire, I feel, or I aspire", for it is not only the head that desires, or the hand that feels the love that we feel when we hold our beloved, or it is not only the heart that aspires to fulfil ambitions.

It is some entity, beyond the confines of the physical body that experiences these feelings, desires, emotions, and aspirations. It is this "me" who is referred to when "I" respond to an emergency call or when "I" take the initiative and responsibility for ensuring that a particular task gets done. It is this "me" who can act out of vision, with courage and determination, caring and compassion, and perhaps the same "me" who can act out of impulse, out of negative feelings of anger, lust, and greed, and proceed to destroy either myself or those connected with me. Do both of these contrasting attributes apply to the same "I" present in me? Or are there two different personalities, or are they mere expressions of the emotions arising? How can we assign meaning to the words "motivation", taking "initiative" or "responsibility", or even the negative shades of "impetuousness" or "recklessness"? It is such human qualities as being able (or unable) to visualize the fruits of an action, or series of actions, and being able to set in motion a sequence of events with correctly persuasive arguments that separate the human from the mechanical. Correctly relating to another person's sensitive points, to other people's feelings, and utilizing due "emotional intelligence"—all these are most integral to us, and all of them ultimately depend on our functioning from "the heart", with the overpowering emotion of love. How can these qualities compare with the mechanical picture of our underlying reality? Can such a picture be trusted and expected to work hard on its initiative in times of stress?

Once we have reflected on "life", sought answers to "What is life?", and once we have discussed the complexities of the mind and the underlying consciousness, we have to simultaneously answer the inextricably connected lead question, "Who am I?" or "What am I?" We have to agree that "life" and therefore, we as living beings, cannot be simply compartmentalized between the animate and inanimate, nor can we be comprehensively defined by the multiple sciences, both singularly and collectively. Our lives cannot also be explained merely as linear chronologies in the process of evolution. The very question prompts responses that extend to the meaning, relevance, purpose, and goal of human life. The question "What am I?" therefore seeks an answer much beyond the limitations of physiology, biology, and the pure sciences. We must delve into the understanding and concept, or meaning, of being a human being as explained in different religions, philosophies, psychologies, or belief systems in the world. Human life, in particular, has an all-encompassing, very wide-angled perspective.

The human side of a healthy personality is one thing, and all explanations, whether scientific or philosophical, are only indications on which we can comprehend that the "I" is something much more than a mere physical entity. Behind the covering of the physical body and the subtle body, where sensory perceptions exist, is the real "I", the centre point of our existence. The age-old questions have motivated inquiry about ourselves and the universe around us. The answers to this incredibly complex riddle are not easy, with a dearth of even suggestive pointers and much less ascribed pathways to reach the knowledge of "I". We find theories galore and patchy knowledge scattered in bits and pieces, and then the interpretations of some of our intelligently evolved counterparts, which happen to stare back when we stand in front of the mirror to question, "Who am I?"

The empirical nature of modern science does not necessarily lend a helping hand in understanding the nature of this complex "creation", which is a medley of all actions, perceptions, feelings, emotions, and almost every aspect of its very existence. The crux of the enquiry lies in integrating all the feelings we have, the actions we perform, and the emotions we display. The entire investigation revolves around whether we are finite beings with a definitive beginning and end, or whether we existed before this physical birth and will continue to exist after the death of this physical body.

The question begs a clearer answer. Because, as we can see ourselves and each other as human beings, we appear to be both true and untrue at the same time! It is not true that we are these limited embodied beings, with name and form. What we see of each other is what we appear and try as we may, we cannot stop at that because that is a very superficial and transitory identification. Almost instinctively, we protest that we are not the persona that we are taken to be. Therefore, layer after layer, going inwards, we seek the true identity of this very same embodied form encased within name and form, limited by time, space, and causation.

This world that we live in, this entire universe, this body–mind complex, all living beings, all objects have an existence, or an identification through name and form, and everything or everyone exists within time and space. Simultaneously, as living beings, we experience everything within our awareness, our consciousness. Therefore, while trying to figure out who am I or what am I, the question that invariably arises is: What is existence itself? We know that our existence is never doubted. In fact, the greatest fear or unacceptable concept is to believe that we will cease to exist one day. Let us therefore understand what exactly we are and how we are perceived, both in terms of modern science and traditional belief systems.

Existence in the Psychophysical System

As the personality, as the "I" of the unique individual as we see ourselves, we are undoubtedly and almost unconditionally connected to the psychophysical system. And this "I" is born, goes from being an infant to a vibrant youth, matures into middle age, ages further to old age, and eventually dies. This process, with its variations, is unchanging. The variations and limitations are within the workings of time, space, and causation. At what age we will die, where, how, and why are the questions to which we do not find ready answers.

Science explains the construct of this body–mind complex in detail, and as cognitive thinking human beings, we can perceive, understand, and relate to this. We know that the body is not just a material and mechanical layer of identity but a vital and dynamic one. Thus, it is important to respect, appreciate, and care for the body, as it is the "laboratory" of our existence in this world and in this life. The body–mind, which is the source of pleasure, pain, emotion, sensation, and expression, enriches our experience and understanding of the world and ourselves. It is the medium of communication for conveying our thoughts, feelings, and intentions. As the vehicle of action, this body–mind complex enables us to achieve our goals, fulfil our needs, and help us realize our dreams. But we also realize that the body is not who we are; it is an integral part of our larger and more complex identity. As we go deeper to understand the core of our being, of our spirituality, and as we go deeper into the search for "who am I", the body–mind is our instrument to help us connect to a higher power, a deeper purpose, and a greater reality.

Identity and Its Evolution

The term "identity" is very significant in psychology, sociology, anthropology, and cultural studies, among others. But the mention of "identity" *per se* pops open the "Pandora's box" yet again, and we are compelled to dive deep into the nuances of identity that envelop us. The term derives its roots from the Latin word "*idem*", which implies "sameness". "Identity" refers to our understanding of who we are, with various aspects of our individuality and self-image. It is how we perceive and make sense of ourselves in relation to others and the world.

The notion of identity has evolved through history, influenced by shifts in societal norms, cultural dynamics, and our understanding of human nature. While the concept is as old as humanity itself and the very basis of human existence, it became a subject of increasing importance in the 18th–19th century when numerous philosophers and psychologists began engaging in discussions and presenting their points of view. A criterion of identity was introduced into philosophical terminology by Frege (1884) and was strongly emphasized by Wittgenstein (1958). Exactly how it is to be interpreted and the extent of its applicability are still debated (Stanford Encyclopedia of Philosophy).

In ancient civilizations, identity frequently revolved around tribal or communal associations, with the sense of self heavily influenced by the position within the group and by the adherence to customs, traditions, and beliefs that predominate. The individualistic concept of personal identity as we know it today has a less prominent role.

During the classical era, the Greek philosopher Plato formulated the concept of "self", although it has been at the core of the Upanishads. In his numerous dialogues, Plato argues that the genuine essence of a human being is the "rational soul", which signifies the intellect or reason forming an individual's soul and is detachable from the physical body. Later, Aristotle put forth a "practical" viewpoint of distinguishing identity in species, genus, and numerals. Species denote things that do not exhibit differences within each other, such as two men or two horses, for similar reasons. Things are categorized within the same genus when they belong to that genus; for instance, a horse and a man both fall within the "animal" or "living being" genus. Aristotle defined numeral identity as the situation in which multiple names or descriptions could be applied to a single entity. These early philosophical ideas were more concerned with the metaphysical nature of the self, whereas, in the post-Christian era, individual identity was beginning to be contemplated within a social framework.

In the mediaeval period, identity was often defined by an individual's social role, such as being a serf (slave), knight (soldier), or clergy member (priest). The Renaissance period marked a shift towards individualism and humanism, with an emphasis on personal achievement and self-expression. The 17th and 18th centuries enlightened the concept of identity with a significant shift.

John Locke, Jean-Jacques Rousseau, David Hume, and Immanuel Kant contributed significantly to the understanding of personal identity and the self. Locke and Rousseau introduced ideas about individual rights, personal autonomy, and the importance of self-identity. According to John Locke, the mind is a *tabula rasa*, or "blank slate". Thereafter, the sensations and reflections that the infant's mind experiences and gathers contribute to the knowledge system as it develops. From assimilating the simple ideas, most of our more complex library of information is constructed. David Hume utilizes introspective awareness to demonstrate that the self is a non-substantial "collection" of perceptions. Immanuel Kant rejects the common approach taken by Locke and Hume, as he disagrees that self-awareness provides objective insights into personal identity. He does, however, resonate with Hume that we never directly apprehend the self, which he terms "the systematic elusiveness of the 'I'". This implies that we naturally think of ourselves as enduring and unified individuals as per our cognitive processes, which do not directly conclude about the intrinsic nature of the self. The Upanishads, on the other hand, put forth this single-pointed focus, the goal of Self-realization.

The development of psychology and psychoanalysis in the 19th and 20th centuries deepened our understanding of identity. Sigmund Freud stated that the mind is divided into three parts: the id, ego, and superego. The id is the instinctual part that drives sexual and aggressive behaviour. The superego operates as a moral conscience. The ego is the realistic part that mediates between the desires of the id and the superego. The ego makes decisions and allows for the expression of natural impulses in socially acceptable ways. The interactions and conflicts among the id, ego, and superego create personality or identity. Freud revealed the role of the unconscious mind, or dreams, in shaping identity.

In the 20th century, identity became a central topic of discussion in psychology, sociology, and cultural studies. Erving Goffman, the sociologist, discussed how individuals behave in social situations, and this gave an idea of the sociological understanding of identity.

Stuart Hall, a cultural theorist, ventured into the complex nature of cultural and social identities. The cultural and ethnic identity of an individual gained prominence with greater recognition of the diverse cultural backgrounds and, therefore, the impact of globalization on cultural identity. Erik Erikson, a German psychoanalyst and one of the earliest psychologists in the 20th century, was particularly interested in the idea. He talked of the ego identity, which gives us a sense of continuity. He developed the theory of psychosocial development and explored the concept of identity and its developmental stages. He emphasized the importance of the formation of identity during adolescence. We believe that this sense of continuity emerges from the deep, unchanging sense of "self", which we shall discuss later.

The digital age of the 21st century has brought new challenges to the concept of identity. Online personas, social media, and digital footprints have added layers to how we tend to construct and present our identities. This is now referred to as "digital identity", which has raised questions about the authenticity, sustainability, and impermanence of online identities. Identity has also become a central theme in discussions of social justice and equality. Movements advocating the rights of marginalized groups, such as LGBTQ+ rights, civil rights, and feminism, have highlighted the importance of recognizing and respecting diverse identities. An evolving understanding of gender and sexual identity has challenged traditional binary notions, leading to more inclusive and diverse ways of defining identity.

Identity is not a static, fixed entity. Individual identity is shaped by multiple factors, including race, gender, class, and sexuality. This intersectional perspective has become central to discussions of identity and social justice. It continues to be a complex and evolving concept, influenced by ongoing societal, cultural, and technological changes. It encompasses personal, social, and cultural dimensions and is a subject of ongoing exploration and debate in various fields, including psychology, sociology, philosophy, and cultural studies.

We may question here why we have touched upon this spectrum of overviews. In trying to know "Who am I?", we have to sift through these perspectives because knowing "what or who I am not" becomes vital. Since ego, personality, and identity are the tools whereby we think we know who we are, we need to go deeper and understand them.

Various levels of identity are explained in the knowledge systems of today. However, we will first go to the most ancient knowledge system and see how it explains the concept of identity and "Who am I?"

Levels of Identity in the Upanishads

There is constant reference to "Who am I?" in the Upanishads. In the context of the human body–mind complex, knowing *who or what I am not* is scientifically and graphically explained in Vedic literature in the study of the human being, comprising the "five sheaths". In the *Taittiriya Upanishad*, the five "*koshas*" or sheaths are explained (see Fig. 6.1).

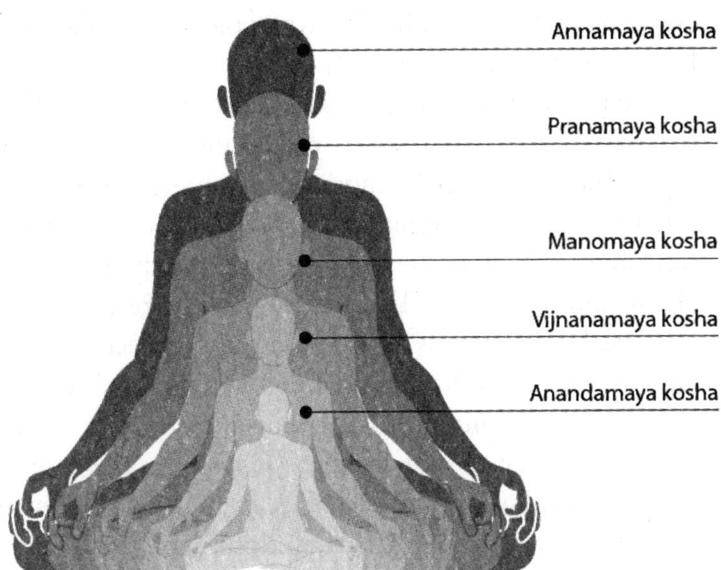

Fig. 6.1. Panch kosha.

1. *Annamaya kosha*: The first and outermost layer is the physical sheath, this physical body. Formed out of matter such as food and water, it is perishable like all matter. So it cannot be who "I" am. I do not identify with the degenerating matter.
2. *Pranamaya kosha*: The second sheath refers to the inner vital life force or energy—the inner sheath of *prana*. "Life" or living depends on this vital force. We cannot see the biological body, or the vital force and energy, as our intrinsic reality.
3. *Manomaya kosha*: The third level is the mental or psychological sheath, the one constituted by the mind, and thus is the mental "body". The sense of "self" develops here, but it is not the Self because the mind is unstable and ever-changing. "I" cannot be the fickle, ever-changing mind or the mind that "shuts down".
4. *Vijnanamaya kosha*: It refers to the intellect. This is the seat of intuition, the level of inner wisdom, and it takes us to the deeper states of consciousness. It is responsible for our inner growth. The intellect is a reflection of Pure Consciousness, and it pervades the whole body–mind in the waking state but disappears in deep sleep. It is known as the intellect sheath. It is also not the Self, because it is also changeable.

   We need to reflect on the third and fourth sheaths a little more because we can be confused between the mental and intellectual levels. The mind-stuff, comprising the inner mental organ, functions both as the agent and the instrument. Therefore, the mind and intellect are treated as separate, as the mind sheath and the intellect sheath. They have separate fields of operation, so to speak. The mind is focused "outward" collecting all sense data from experience and presenting it to the intellect, which is "inward". Cognition and intelligence are two concepts related to human mental activities. Cognition, the term used for understanding mental activities, is the overall term that includes all mental actions such as acquiring information and knowledge, thinking on and comprehending such information, the level of rationalizing or judging, and then problem-solving, as and when required. "Intelligence can be defined as a general mental ability for reasoning, problem-solving, and learning. Because of its general nature, intelligence integrates cognitive functions such as perception, attention, memory, language, or planning."

5. *Anandamaya kosha*: The fifth or innermost sheath is the one of "bliss". At this level, we can transcend the awareness of the physical body, the mind, and the intellect and experience being united with the universal one-ness. The intellect, when it goes further inward and catches a glimpse of bliss, is at this level. This state, known as the bliss sheath, is still different from the Self. To relate to this concept of "inner bliss" we have to be in a very steady and comfortable place in our spiritual identity, from where we can and do seek that level of inner calm and joy. The ultimate one-ness is beyond this state, where there is only Pure Consciousness, Self-awareness.

At this juncture, we take a few steps back and reflect on how we understand identities in this day and age, as they are widely discussed and explained by leading psychologists. We need to know how the physical, biological, social, psychological, religious, and spiritual identities are explained in contemporary terms before we can reach some definitive conclusions on "Who am I?".

Physical Identity

Our perception of physical identity is shaped by a complex interplay of all five senses. Vision plays a primary role, allowing us to distinguish individuals based on facial features, body shape, and posture. Touch provides equally important information through sensations like texture, temperature, and the presence of unique markings like scars or birthmarks. The remaining senses contribute more subtly. Body odour, a combination of individual biochemistry and external fragrances, can become a recognizable signature for those close to us. Taste preferences, while not directly linked to physical identity, can become associated with an individual—someone who consistently avoids spicy food, for example. Finally, hearing plays a role in recognizing voices, with variations in pitch, tone, and even minor imperfections contributing to how we identify others through sound. Interestingly, these senses interact. Visual cues like a familiar face might trigger memories associated with a person's scent, enhancing recognition and potentially evoking shared experiences. In conclusion, all five senses work together to create a unique sensory fingerprint for each individual we encounter, enriching our perception of the physical world.

Perception of Sight: The senses give us the experience of the world around us. What we see is a key sense perception for physical identity. Race and skin colour are significant components of physical identity—the attributes that distinguish us from others. These aspects have profound societal, historical, and biological implications, and they indicate and influence how we perceive others and ourselves and how our perceptions influence our interactions and experiences.

Race is a social construct that puts us into distinct groups based on certain shared physical and cultural characteristics, often including skin colour, facial features, hair texture, and more. Interestingly, we know that race is neither a physical nor a biological concept but a social one. There is no inherent genetic or biological basis for racial distinctions. Skin colour is one of the most visible and prominent physical characteristics used to slot us into racial categories. People with similar skin tones are often grouped as members of the same race. While race is a social construct, skin colour is a consequence of biological factors arising from the interplay of genetics, melanin production, and environmental influences. Skin colour is primarily determined by the amount and type of melanin produced in the skin. Melanin is a pigment that protects the skin from the harmful effects of ultraviolet (UV) radiation. People with more melanin tend to have darker skin, and those with less melanin have lighter skin. Due to its link with UV radiation, individuals from regions with intense sunlight, such as the equatorial areas, often have darker skin to protect against UV radiation. In comparison, people from regions with less sunlight typically have lighter skin, which facilitates vitamin D synthesis.

Historically, in too many societies, the colour of the skin has been used for discrimination, segregation, and grossly unfair and unequal treatment. It has given rise to racial stereotypes and biases. These stereotypes have led to prejudicial attitudes, discriminatory behaviour, and systemic inequalities. There is no single "correct" or "typical" skin colour because all of us, even within a racial group, can have a wide range of skin tones.

Race, the colour of the skin, the eyes, and hair are essential to physical identity. Acknowledging the biological basis while recognizing the social construct of race is vital for ensuring inclusivity, promoting equality, and challenging prejudice and discrimination based on physical attributes.

In the matter of identity as related to physical appearance, the extreme cases of narcissism, solipsism, or self-obsession cannot be overlooked, and this is very graphically demonstrated in the famous Greek myth of Narcissus, the son of the river god Cephissus and the nymph Liriope. He was obsessed with his good looks, and the fact that several women were enchanted by his looks, prompted Narcissus to rebuff all the advances. One of the female admirers Echo was so devastated by his arrogant rebuttal that she withdrew from the world and wasted away. The goddess Nemesis, in an endeavour to teach Narcissus a lesson, made him fall in love with his own reflection in a pool of water. He sat and stared at it till he withered and died.

Perception of Touch: The sense of touch holds a special place in shaping our physical identity. It is a complex sense that includes various aspects of physical interaction. It involves the sensation of pressure, temperature, texture, and pain, besides the emotional reactions that touch can generate. Through touch, we can gather very valuable information about the people we encounter. We interact with our environment, connect with others, and develop an understanding of ourselves through touch.

Babies learn about their own bodies and the world by reaching out and exploring through touch. The sensation of their own skin and the tactile feedback they receive when touching objects contribute to their growing awareness of self and external environment. Our perception of touch extends to how we perceive our bodies. The sensations of our skin, the contours of our bodies, and how we physically interact with the world shape our self-image and self-identity. The role of touch in shaping physical identity is also influenced by cultural norms and individual preferences. Different cultures have unique practices and customs related to touch, which can impact how we perceive our bodies and those of others. Additionally, individuals may have personal preferences for certain types of touch, which can affect their comfort level and sense of identity in various social and intimate contexts.

For physically challenged individuals, such as the blind or the deaf, touch takes on an even greater significance. It is a primary means of perceiving and navigating the world. Braille, sign language, and tactile communication methods are vital tools for shaping physical identities and facilitating connections with others.

Touch is a powerful means of social bonding and communication. Physical contacts, such as hugging, holding hands, and cuddling, foster emotional connections between individuals; they contribute to our sense of identity within the context of our relationships. For example, the warmth of a parent's embrace is incomparable in instilling a sense of security and belonging in a child. For some, the experience of touch may lead to a heightened awareness of physical sensations and body image concerns, while for others, it can be a source of empowerment and self-acceptance. In the context of touch and physical identity, the importance of consent and boundaries is crucial. While touch can be a powerful and positive force in shaping our physical identities, it must always be respectful and consensual. Respecting personal boundaries and consent is fundamental to basic ethics and to creating safe and empowering interactions.

Our physical identity, influenced as it is by the senses, indicates how we see ourselves and our place in the world. It guides our early development, fosters social connections, shapes our body image, and transcends cultural and individual variations. Understanding the profound impact of touch on physical identity emphasizes the significance of respectful and mindful interactions that honour personal boundaries and promote positive self-identity.

Perception of Sound: Sound is considered the most immersive and emotionally resonant of the five senses, and it plays a key role in shaping our physical identity. While primarily associated with hearing, the perception of sound extends beyond our ears, affecting our bodies, emotions, and even our sense of self.

Through hearing, we gain information about the environment, from the rustling of leaves in a forest to the disharmony of a bustling city. This auditory landscape is integral to our perception of space and time, contributing to our physical sense of place and presence.

Infants, even before they can articulate speech or have focused and pinpointed sight, learn to recognize and respond to familiar voices and sounds. This initial engagement with sound brings self-awareness, social connections, and the formation of an identity. Sound, like touch, is intricately tied to our emotions, often evoking strong feelings and memories. Musical compositions, for instance, can elicit joy, sadness, or nostalgia. These emotional responses to sound influence our physical state, affecting our heart rate, breathing, and posture. Sound is instrumental in the shaping of our emotional identity.

Our voices, or the sounds we produce, are closely linked to our physical identity. The pitch, timbre, and resonance of our voices not only convey our thoughts and emotions but also our gender, age, and cultural background. Voice is a powerful tool for self-expression. It also carries immense cultural and linguistic significance. The languages we speak and the accents and tones with which we communicate shape our physical identity in the eyes of others. Our ability to comprehend and produce language profoundly affects our social interactions and sense of belonging within a cultural context.

Individuals with hearing impairments navigate a unique relationship between sound and physical identity. Because their perception of sound is altered or absent, they improvise and adapt to alternative sensory pathways. Sign language, tactile communication, and assistive technologies play vital roles in shaping their physical identities and facilitating interaction.

Given the importance of sound in shaping our physical identity, we must engage with our auditory world mindfully. Paying attention to the soundscape around us, the nuances of our own voices, and the emotional responses triggered by music and speech can deepen our connection to our physical selves and the world. Sound encourages us to listen empathetically to others, recognizing the diverse ways in which it influences our environment and communication.

Biological Identity

Biological identity refers to the unique characteristics that define us, shaped by the interplay of our genetic and epigenetic makeup. The genetic makeup is encoded in its DNA (deoxyribonucleic acid), often referred to as the molecule of life, as it contains instructions of how an organism develops, survives, and reproduces. Epigenetic changes control gene activity in response to changing behaviour or environment, without changing the DNA sequence. This again relates to the Ayurvedic concept of individual nature or prakriti, which we discussed earlier.

We also discussed how our genetic makeup determines our physical traits, such as eye colour, height, and susceptibility to certain diseases. Additionally, it helps to identify us as members of a particular species, given the high degree of genetic similarity we share. The epigenetic changes affect our phenotypic or observable characteristics. These characteristics include body shape, colouration, behaviour, and physiological functions. Interestingly, phenotypic variation can occur even among individuals with the same genetic identity due to environmental influences. Environmental factors such as diet, climate, and habitat can significantly impact our biological identity.

Integrative physiology takes into account internal and external stimuli such as exercise, stress, environmental conditions, and diseases. What then defines us as individuals? Is it our genes, our physical structure, or the intricate interplay of all the biochemical factors?

Extensive research and development efforts have opened up for us both the structural and functional aspects of biological identity. When we study our biochemistry, we can gauge and perceive structural signposts. Genes and fundamental biomolecules constitute the genetic blueprint, further nesting within the complex milieu of cell organelles. These organelles collectively orchestrate the functions of a cell, and cells, in turn, join forces to form tissues. These tissues, like pieces of a puzzle, come together to construct organs. Organs, in their cooperative synergy, create organ systems that harmoniously function as a cohesive unit within the body.

Research has revealed the fascinating fact that our bodies are a complex ecosystem teeming with trillions of microbes that outnumber human cells by a staggering margin. Studies suggest that the human body harbours around 30 trillion human cells and a whopping 39 trillion microbial cells. These microbes include bacteria, archaea, fungi, and viruses. The gut microbiome, residing in the intestines, is the most populated, with other microbial communities thriving on the skin, mouth, and even the lungs. We have seen earlier how the microbiome plays a crucial role in various bodily functions. It aids digestion, regulates the immune system, and even influences mood and behaviour. The Human Microbiome Project (HMP) was launched in 2008 by the National Institutes of Health (NIH) to characterize the microbial makeup of healthy individuals. Dr Joshua Lederberg (Nobel laureate, 1958), a visionary microbiologist, is considered a pioneer in the field. Thus, the microbiome plays an important role in our biological identity.

What truly defines us as individuals, then? Is it solely the genetic code or the physical structure, or is it the intricate and dynamic interplay of all these factors combined? The answer lies in the fusion of genetics, structure, and physiology, which collectively weave together the intricate biological identity, making each one of us a unique and complex entity. This explanation or definition is as simple as it can be stated, keeping the physiology and biology of a human being at the forefront. But this is just the tip of the iceberg when it comes to knowing "Who am I?"

Social Identity

Social identity goes deeper in our effort to identify ourselves in the eyes of the world. Social psychology refers to our self-concept, which is derived from our membership in various social groups. These social groups can include categories such as race, ethnicity, gender, religion, nationality, profession, and more. The social identity theory, as developed by British psychologist Henri Tajfel in the 1970s, is one of the foundational theories that focuses on how we tend to define ourselves based on group membership as per the criteria of categorization, identification, comparison, and social context.

We are social animals. Therefore, we naturally categorize ourselves and others into various social groups based on shared characteristics or attributes. These categorizations help us to define our social identity and understand our place in society, which plays a very important role. Once we identify with various social groups, we emotionally invest in the groups and take on the norms, values, and behaviours as part of our self-concept. For example, we may identify strongly as members of a particular political party. Then there are odious comparisons, as it is human nature to compare! This comparison leads to in-group favouritism, where we view our own group more positively and may even discriminate against other groups. Such behaviour contributes to social biases, stereotypes, and prejudices. Social identity is highly context-dependent. Its prominence can vary significantly based on the circumstances. For example, identifying as a parent might hold greater significance within a parenting group but diminish in relevance when the same person is attending a professional conference!

In the digital world of today, social identity is intertwined with gender and how we present ourselves on online platforms. A spectrum of identities has now evolved, and these fall under the broader category of social identity:

1. **Sex and Gender Identity:** Sex and gender are two concepts that are often used interchangeably in everyday language, but they have distinct meanings in scientific, sociological, and cultural discussions. Sex refers to the biological and physiological characteristics that define humans as male or female. This includes chromosomes, hormones, internal and external reproductive organs, and secondary sexual characteristics that develop at puberty. Biological sex is determined at conception by the genetic contribution of the sperm cell to the egg and is usually assigned at birth based on visible genitalia. The differences in physical strength between men and women are primarily attributed to differences in hormones (testosterone), which influence muscle mass and fat distribution. Of course, language reflects culture too. "Woman" embraces "man" and "female" incorporates the "male", pointing to the inherent connection between the sexes. But this is all tongue-in-cheek! Genetics might not dictate emotions, and true strength comes in many forms, regardless of chromosomes.

   Gender is a complex social and psychological construct that includes the sense of self and identity. It is how we, as human beings, perceive ourselves in terms of being masculine, feminine, both, neither, or somewhere along a spectrum of gender identities. Gender identity may or may not align with the assigned sex at birth. Some people identify with the gender they were assigned at birth (cisgender), while others do not (transgender or non-binary). While gender identity can be and often is influenced by both biological and societal factors, it is a deeply personal and individual experience.

   As of 2023, there has been increasing recognition of the diversity of human gender identities. Not just LGBTQ+, but as many as 107 distinct gender identities are now on record. This development is a significant departure from the traditional binary understandings of gender (male and female) and highlights the complexity of human gender experiences. These identities include transgender, non-binary, genderqueer, genderfluid, agender, bigender, and two-spirit. Each of these identities reflects a unique way in which we perceive and experience our gender, with elements of identity, expression, and personal understanding. The number of gender identities can continue to evolve as society becomes more inclusive and as individuals gain the language and understanding to articulate their unique gender experiences. This ongoing evolution highlights the importance of respecting and affirming each other's self-identified gender, and our society must be ready to accept all gender identities.

It is important to recognize and acknowledge that gender identity is distinct from biological sex. The former is based on our internal understanding of our gender, and the latter is based on our physical characteristics. The uniqueness of gender identity is that it can be fluid and may evolve. In the context of social identity, gender refers to the roles, behaviours, expectations, and attributes that a society considers appropriate for individuals based on their perceived or self-identified sex.

2. **Digital Identity:** Social media platforms have become integral to our lives, and they play a significant role in shaping our digital identities and how we present and portray ourselves in the digital realm. We may or may not include various aspects of our social identity, such as gender, race, religion, and affiliations, even though the medium allows us to express our social beliefs and values. We are drawn in to connect with like-minded individuals and engage in discussions related to our social identity. Some individuals use digital spaces to explore or experiment with aspects of their social identity, such as gender expression, without the same societal constraints they might face in the offline or face-to-face world. Managing a digital identity in the context of social identity gives choices about what to share, how to project oneself, and when and how to engage in discussions related to social issues.

All such activities, as popular as they are today, have a downside. It is important, for example, to protect the digital identity in the context of social identity; we need to safeguard personal information and avoid potential harm or discrimination based on social characteristics. The diverse range of social identities in digital spaces has to be recognized. Online discrimination and harassment can get troublesome, even intimidating. Ironically, it is often noted that some people successfully camouflage or even monitor their digital personalities, which are literally made into a "persona" or "mask" that has been superimposed. This kind of situation can be extremely misleading and, at times, risky. Biometrics are used very widely in a world where connectivity is now almost entirely "digital". It includes measuring or recording physical characteristics for assimilating statistical analysis, which is used for identification and authentication. Biometric data includes:

- Fingerprints: The unique and permanent patterns of a human being's fingerprint are exclusive. No two human beings can have or do have identical ridges and valleys on the skin of their fingers! This is nature's miracle and uniqueness. This feature is used in the modern world to authenticate a person's physical identity.
- Facial recognition: The key facial features and landmarks, including the distance between the eyes, the shape of the nose, the contours of the lips, and the position of all these features, are integral to making us unmistakably identifiable.
- Iris scans: This process analyses and extracts specific features from the iris, including all the very technical and minute details such as the arrangement of crypts, furrows, and the unique pattern of the iris's stroma, which is the fibrous tissue that gives the iris its colour. Again, the eyes tell no lies, so this method is used for the most accurate form of physical identification.
- Voiceprints: We all seem to have unique voice prints as well. This methodology analyses specific vocal characteristics such as pitch, tone, speech rhythm, vocal tract length, and the unique patterns in our vocal cords and throat.
- Keystroke patterns: The rhythm and timing of our typing on a keyboard are also a means of detecting our physical identity. This kind of behavioural biometric technology

identifies us based on such features as dwell time (the time taken to press a key), flight time (the time taken between key presses), and the time intervals between keystrokes.

Today, this world is one big global village. Nothing and no one is out of reach. And given the varied dimensions that are available to us just within the realm of physical identity, it gets overwhelming. Over the centuries, and with the more recent events of the 20th and 21st centuries in particular, the wars, battles, and widespread skirmishes have left indelible scars of suspicion and distrust, to say the least. Therefore, it has obviously become imperative that physical identification be authentic and reliable. For worldwide travel, for example, the use of biometrics for fingerprints, facial recognition, and iris scans has become essential. The process of online transactions has made some of these physical identity techniques indispensable. However, there are always two sides to the coin. Even keystroke patterns can be accessed remotely to perpetrate scams!

3. **Virtual Identity:** "Virtual" or "unreal" is the "reality" of life today! We are as involved in an "online" life as we are in a face-to-face "real" life. Online identity is an offshoot of digital identity. It refers to the persona or representation we choose to project in the digital or online realm. How we present ourselves and interact with others in computer-mediated communication and in virtual communities is not necessarily how we are in our day-to-day "real" world. It is a digital persona that we create to engage with others on the Internet. Most often, usernames, images, profiles, and the information shared online are possibly caricatured or camouflaged. The virtual identity mirrors aspects of our "real" self, including name, interests, and characteristics. It serves as a bridge between our offline and online lives, apparently allowing us to maintain continuity.

People use virtual identities for various forms of communication, including social media interactions, email exchanges, participation in online forums, online shopping, and e-commerce. The Sanskrit word *avatara* refers to a deity taking a physical identity on Earth, often to guide humanity. Today, transcending religious roots, the contemporary popularity of avatars stems from their application to online representations of ourselves, reflecting the increasing importance of digital identities in the modern world. These virtual identities facilitate online conversations and relationships. An example is the online multimedia platform Second Life, where individuals can craft personalized avatars and engage with fellow users and user-generated content within a virtual, multiplayer online environment. It was founded in 1999 and came into existence in late 2002. Almost twenty years later, in 2021, approximately 64.7 million users were active. In the year 2023, there was an average daily user count of 2,00,000 individuals representing 200 different countries.

There are multiple gaming sites like Nintendo, Games Spot, Addicting Games, Epic Games, and tools such as Unreal Engine, Fortnite, or the very popular Pokemon, where people spend hours gaming, creating games, and interacting, all behind virtual personas. Millions of people, both young and old, are focused on either generating games or playing virtually. For many, it is their only way of life. Most individuals choose to maintain a level of anonymity or pseudonymity in their virtual identities, using nicknames or other avatars instead of their real names and images. This can offer a degree of privacy and protection, and several people find comfort in the freedom to interact anonymously using their unique avatars.

Understanding social identity is essential to addressing issues related to discrimination, prejudice, and social inequality, as it sheds light on the dynamics of group-based biases and conflicts. Researchers and practitioners in fields such as psychology, sociology, and intergroup relations often study social identity to better understand human behaviour in diverse social contexts. But most of all, what is increasingly evident is that we are actually seeking to live with dual or multiple identities, even at the physical level.

Psychological Identity

Our self-concept, self-image, and self-awareness are how we define and understand ourselves. Psychological identity includes all these aspects. The key here is how we perceive ourselves, what our beliefs are about who we are, and the roles that we see ourselves playing in society and in relation to others. Self-concept includes physical attributes, personality traits, values, beliefs, and abilities. We develop our self-concept through self-reflection and feedback from others. Self-esteem is crucial as it relates to the overall evaluation or judgement that we have about ourselves. High self-esteem means we have a positive self-image and feel valuable, capable, and confident. Low self-esteem means negative self-perception and self-doubt. This can lead to disastrous outcomes like isolation, depression, and, in the long term, mental and physical ailments. We commonly frame our self-perception within the context of personality, which is then influenced by social and cultural dimensions.

We tend to differentiate ourselves from others with a unique combination of characteristics and experiences. This includes personal history, life experiences, and our individual sense of continuity over time. These experiences are largely influenced by our internal awareness of existence, even though we do not introspect enough to understand it that way. Instead, we remain caught at the level of personality and identity.

Psychology encourages the appreciation of this "I", which it terms to be the expression of a personality. The personality is the combination of our emotional, attitudinal, and behavioural response patterns. It can be defined as a dynamic and organized set of characteristics that we possess. These characteristics influence our understanding, emotions, behaviour, and motivations in all kinds of situations.

Many of us derive a significant part of our psychological identity from our social and cultural group memberships, based on ethnicity, nationality, religion, gender, and other sociocultural categories. Group memberships give a sense of belonging to shared values, traditions, languages, and customs. They influence how we see ourselves and others, particularly in multicultural societies. The aspects of the role we play are context-driven, such as being a parent, a student, an employee, or a friend. This is in addition to the roles we play or the personas we have in our professional worlds. We often adopt different roles in different situations, and these roles contribute to and influence our overall sense of self.

Psychosocial theories, such as Erik Erikson's "stages of psychosocial development", suggest that identity formation is a key developmental task throughout human life. We go through stages where we explore and define our identities, ultimately striving for a cohesive and stable sense of self. The eight stages that Erikson defines is the structure of the psychosocial journey from infancy to death.

Identity and Personality

The word "personality" has originated from *persona*, in Latin, which means a mask! The term, when used, did not imply a disguise or a deliberate ploy to be what one is not. It apparently represents a "character". Therefore, the term personality is that particular trait or combination of traits that helps to define ourselves as individuals and points to our uniqueness, our identity, or our character. Does the personality signify the body, the mind, or something else, the spirit, the soul, or ultimately, consciousness? Where does the limit of the physical body end and the jurisdiction of the mind start? Is the mind exclusive to human beings? Don't animals and all other living beings have minds? Can inanimate, man-made machines also have minds? This enquiry must necessarily direct us to the very nature of consciousness and what exactly it stands for. Is consciousness limited to the functions of the brain, as many scientists are determined to prove even today? Does it "stop" or cease to be at any point in time? In deep sleep, are we unconscious? Finally, does consciousness end with the death of a living being?

The terms identity and personality are often used interchangeably, but there is a nuanced or subjective subtle difference between the two terms. Our identity is what we give to ourselves, and it represents what we believe in, physically and mentally, and what we stand for in terms of values, ethics, and who and what we are. Our identity is dynamic and can change and evolve throughout our lives. Personality is how we portray our identity. Our personality develops early in life. On the one hand, it is a genetic hand-me-down, as we inherit some traits from our parents. On the other hand, our surroundings and events as they occur in our lives, our relationships, and how we interact with family and friends affect how our personality develops.

The personality is projected through the pattern of thoughts and behaviour. Personality traits are considered relatively consistent over time and across different situations unless deliberately developed to blend with the identity. Both personality and identity play significant roles in shaping our thoughts, feelings, behaviours, and interactions with others, but they operate at different levels of psychological analysis. The two terms are very similar and together give us the attributes of uniqueness.

Cardiologists Meyer Friedman and Ray Rosenman introduced a personality theory that identifies personality traits and represents them in types A, B, C, and D

- Type A: People placed in this "type" are achievers and go-getters, proactive and goal-driven professionals. Consequently, their personalities are projected as being competitive to the point of being driven by a sense of urgency. They tend to be self-critical and even impatient.
- Type B: In contrast to type A, these people are more laid-back, relaxed, and therefore more flexible and easy-going. They are far more patient and have the ability to adjust and adapt. They may not be as high on the achievers' chart as compared to Type A, but they are certainly not driven by any stress, pressure, or sense of urgency.
- Type C: Almost like an outcome of the two extremes that types A and B seem to indicate, this personality is cautious with the capability to give great attention to detail. A cautious person also tends to be reserved and quiet, at times withdrawn. In such a case, the person, while efficient and conscientious, would lack the ability to be assertive and therefore would be unable to handle stress.
- Type D: The introverted, negative, pessimistic, and anxious person is slotted in the Type D category. Avoiding social

interaction is a natural tendency, and this personality is triggered by a great fear of rejection or disapproval.

Personality Disorders

We all have unique personalities where diverse traits play a pivotal and crucial role in shaping our thoughts and feelings. These, in turn, direct our behaviour and interaction with others. If identity and personality are in sync, we can navigate the complexities of who we are and how we fit into the world around us. Having said that, personality *per se* is not as simple as it may seem. As indicated by the types of personalities we have discussed, we can appreciate that rigidity or inflexibility can become a problem. When that happens, cases of personality disorder emerge, when human beings are incapable of dealing with situations. Excessive anger, frustration, and trust issues become predominant. Mental imbalance, depression, and even extreme acts of suicide develop.

Having come so far in discussing the variance in identities, from the physical to the psychological, we can sense that we are neither the body nor the mind, where thoughts swirl endlessly and relentlessly. Notwithstanding the identification with the body–mind complex, we recognize that we certainly are not this skeleton covered with everything that comprises the body. Our identity and personality cannot be this mass of bones, flesh, blood tissues, and veins. Paradoxically, this body, which remains the central point of focus, is a mass of matter of which we are not even aware and of how intricately it functions. It is mind-boggling to comprehend that millions of cells are born and die within the body with every passing moment. Therefore, do we really believe that we are this ever-changing, decaying, and dying matter? Most certainly not, we would say, because intrinsically, recognizing the nature of the body, we do not identify with it and acknowledge instead that we "have" this body and we "function" through it.

We should be mindful that *we are not* this body, but that we *have* a body. Then we may speculate that perhaps we are the inner vital energy, the *prana* as referred to in ancient Indian psychology and philosophy. But that vital energy also wanes and waxes. With the age of the body, the vital forces diminish and change. Again, the mind protests and says, "No, I am not that either." Then "I" must be the mind, or better still, the intellect, because all my thoughts and emotions swirl in the mind, and the capacity for analysis, understanding, reasoning, and reflection emanates from the intellect. Yes, as a human being, therefore, I am the mind or the intellect. Think again, says my ability to reason; that same ability changes. Emotions rise and fall, and thoughts flit past, mostly uncontrolled and unnoticed. Focused thought also has its variance, as our ability to focus changes almost with every passing moment.

All the activities of the body, the vital forces, the mind, and the intellect are always in flux and change as the senses play with these varying attributes of the psychophysical system. The identification that we as human beings carry—call it the personality, the persona, the ego—is within the limitations of time, space, and causation and proves to be limited and variable, and that is not "I". We have seen the complexity of identity. Despite all these analyses, we think and believe that we are unchanging! We also feel that intrinsically, we cannot be this ageing, material, complexly constituted entity. Finally, the fear and denial of death come from the conviction that we cannot be this friable, destructible "person". Despite all the changes, therefore, can we say that the sum-total of all the variables constitutes a mass, which is a "constant" quantity? There is the "impersonal me", of which all these manifestations form, as it were, parts. This is what we need to understand and reason out to really know "Who am I". The positive assertion of who I am is taken up here with the analysis of *who or what I am not*. If I believe that I cannot be the ageing and changing physical body; if I am not the vacillating, restless, and wavering mind; and if I am not the decaying or degenerating intellect, then "Who am I?"

There are interesting stories found in the Upanishads and the epics about the quest to find the answer to the question, "Who am I?" One such narrative is about King Janaka, who was a wise and just ruler, known for his deep philosophical inquiries and patronage of spiritual learning. One day, while reflecting on his life and responsibilities, he gazed into a mirror and pondered, "Who is the king? Is it the one adorned with this crown?" He proceeded to remove his crown, royal robes, jewels, and all symbols of his worldly status, each time asking, "Now, who is the king?" With each change he realized that none of these external adornments constituted his true identity. He found the answers through a dream, as was explained to him by a learned sage (Box 6.1).

Experiential awareness of "Who am I?" is not limited by simple labels or external roles. We have to know our inner essence, values, and purpose. We have to be constantly mindful of our thoughts, emotions, and actions and reflect on how they align with our sense of self.

Religious Identity

The confluence of science, the vast span to be covered from the physical, biological, and psychological, inevitably culminates in the religious and spiritual identities of human beings because religion has been the mainstay of humanity over the centuries.

Spread across the length and breadth of the world are several religions that are widely practised. They are not always in harmony and not always peacefully because of the rigid fundamentalism of human minds, but the essence of religion is the belief in a higher power, in God. Religion is an organized realm of beliefs, rituals, and ways of life where like-minded people have faith in a common God and collectively follow a path and lifestyle of worship. Religious identity is quite naturally influenced by organizations and religious institutions.

According to Theistic religions, all living beings are finite and limited and must depend entirely on one Supreme Being, God, or the Ultimate Reality. According to Judaism, Christianity, and Islam, the belief is in the duality concept that there is God the Creator, who created the world and all living beings. This concept is also especially appealing to most human beings. And that is primarily the reason, seeing it from a layperson's point of view, for the divide between religion and science. But the belief system does not give clear answers to the quest of "Who am I". God becomes unnecessary in science. Theism seeks, in various ways, to bring

Box 6.1. King Janaka's dream

King Janaka of Mithila once dreamt that he had been defeated by his enemy in war and driven out of his kingdom. He experienced extreme hunger, despondency, and helplessness. He sat among the beggars outside a temple, waiting anxiously for some food that a kind devotee was distributing. He grabbed his meagre share of food and went eagerly to a corner to appease his hunger when a dog came and snatched away the few pieces of bread that Janaka was holding on to. He cried out in despair, only to be woken up by his concerned guards, who wondered why their king was crying out in his sleep! Janaka awoke and saw his opulent chamber where he had been sleeping all the while. Thinking of those hunger pangs he wondered, was *that* true or is *this* true?

King Janaka now sought the counsel of the sage Ashtavakara to go deeper into the question of "Who am I?" The young sage, who had eight severe deformities (literally, *ashtavakara*) in his body, simply said, neither was *that* true nor is *this* true. Our identity is the only truth, the rest are superimpositions (veils of Maya). Ashtavakara led King Janaka through a series of teachings and meditations to help him discern between the transient and the eternal, between the body, mind, and the true self. Essentially, the sage succeeded in shattering King Janaka's ego by teaching him that we cannot be identified merely by physical appearance, position, and status. We have to get to this understanding through our intellectual capabilities, which can guide us to the experiential awareness of "Who Am I". For this experiential awareness, we must dive deep into our thoughts, beliefs, experiences, and relationships with sincere introspection and reflection.

With the help of these profound teachings, Janaka understood that his identity as a king was merely a role played by the body and mind, subject to the vicissitudes of life and time. The true king, the true self, was the Witness Consciousness, the Atman, which observed all changes but remained unchanged as the innermost essence, that is one with Brahman. This realization led him to a state of enlightenment, where he saw himself and all beings as expressions of the same divine reality.

King Janaka attained *moksha* and continued to rule his kingdom with detachment, performing his duties with compassion and equanimity, guided by the wisdom of his true self. This narrative explains the nature of the Self and urges us to look beyond the superficial layers of identity and discover the eternal, unchanging reality within. It teaches that true knowledge of the Self transcends all worldly distinctions, and is the key to knowing "Who Am I".

our relationship to God into closer involvement with the way we understand ourselves and the world around us.

With the belief in a creator God, we as human entities "belong" to that creator, and thus "Who am I?" becomes a rhetorical question. According to the Bible, when Moses asked to see God, all that he could be assured of was that God was real and was bound to be: "I am who I am," Moses was told. On the other hand, in the throes of this humbling and staggering experience, Moses began to learn what was expected of him and how his people should live and be led. God is somehow one with whom we can "talk". With the implicit belief in God's omnipresence and omnipotence, "I" become a chosen one, and thus there are codes of conduct for how to live in God's world.

Over the years, religious identities have multiplied as belief systems have new interpretations. The prominent denominations in Christianity are Roman Catholics, Protestants, and Eastern Orthodoxy, and subsequently, there are further divisions based on what they believe in and want to practice. As denominations increase, religious identities multiply in a similar proportion. Islam believes that humans are created by Allah and have a special purpose and role in the world. Muslims believe that their identity is primarily a matter of religion and that they should follow the teachings and guidance of Allah as revealed through the Quran and Prophet Muhammad. The Islamic view of identity is rooted in the belief that all humans are Allah's creation. Buddhism functions as a religion because it has a community of followers of the Buddha. There are codes of moral conduct, meditation, and forms of worship in monasteries. But it doesn't involve a creator God. The goal is to reach enlightenment or *nirvana*.

Hinduism, the oldest and most diverse religion, has a vast and varied collection of scriptures, teachings, and practices. Hindus see themselves as indestructible souls, integral parts of the supreme soul, Brahman, and who undergo cycles of birth, death, and rebirth, according to their *karma*.

With the focus on religion being the key to conveying or displaying "Who am I", gradually, dress or other visible symbols have taken on great significance. As it is, over centuries and generations, the dress code has very often conveyed an identity. We wear certain clothes to specifically project ourselves as part of a particular cultural, social, and professional identity. Lawyers wear black coats, kings wear crowns, and judges wore white wigs for the longest time.

With twists and tweaks, we have always tried to project our uniqueness in clothes and appearance as part of who we are. In Christianity, we have seen cassocks, robes, and other religious attire. Accessories such as rings, pendants, emblems, headgear such as hats and skull caps, and the hijab and burqa of Muslim women are very particular identifiers. The *bindi*, or *tilak* (a mark worn on the forehead) of the Hindus is significant, especially the shape and size of the tilak. Three horizontal lines are displayed by the followers of Shiva, the Shaivites, while the Vaishnavas, the followers of Vishnu, have either one vertical line on the forehead or a U-shaped tilak. In traditional Sikhism, men and women do not cut their hair; consequently, the men wear turbans and have beards.

Spiritual Identity

This is the last vestige of identities as discussed here, but it is the most profound and enduring aspect that delves into deep-seated inquiries into the very purpose and significance of existence. There is often some confusion between spirituality and religion. While in some contexts, the two might be used interchangeably, there is a significant difference. We human beings are essentially spiritual, as we seek the meaning of life, how we should ideally live it, and the need to sustain higher values to keep progressing. Spirituality is an individual pursuit and quest. The philosophy of both spirituality and religion is the quest for the ultimate understanding of acquiring self-knowledge and, thus, knowing "Who am I". Once we understand the concept of spiritual identity only then can the quest of knowing who we really are take a natural turn inwards, to the study of the self. We have touched upon the thought process of existence, but that has a very wide spectrum as it includes what life itself is and how and why we exist.

At its core, spiritual identity is an intricate combination of our innermost beliefs, values, and experiences that contribute to our understanding of our place in the world and our connection with something greater than ourselves. Our beliefs range from traditional religious affiliations to more eclectic and personally tailored spiritual philosophies.

One of the central functions of spiritual identity is to address existential questions that have always puzzled us. These inquiries include the nature of life and death, the origins of the universe, the existence of a higher power or purpose, and the moral and ethical principles that should guide our behaviour.

Religion, Spirituality, and Peace

Religion and spirituality are parallel paths to knowing who I am, and they are meant to lead us to inner peace and well-being. Science and philosophy also must strive for these core human goals, so that there are sustainable interconnected disciplines. The MIT World Peace Dome in Pune, India, perfectly represents global interconnectedness. Eastern and Western philosophies and religions merge with a holistic vision. There is a 3,000-seat prayer hall, where statues of saints, philosophers, and scientists stand for unity and the pursuit of peace as major religions are symbolically represented. The World Peace Library further emphasizes the interconnectedness of existence with 108 imaginatively arranged pillars. Figure 6.2 is representative of the Dome and religious symbols united in peace.

Fig. 6.2. Religious identities in the world.

Identity Crisis

In recent times, the issue of identity has become increasingly significant, giving rise to the "identity crisis". Erikson coined the term during his study of the psychological development of human beings. The crisis refers to inner conflict and exploration when we are compelled to question our sense of being, our values, beliefs, and purpose in life. Simply put, an "identity crisis" is being uncertain of who or what we are. It primarily occurs during adolescence, although it can surface at any stage of life. It is our struggle to establish a clear and stable identity. Various factors can trigger this process.

- **Personal Reflection**: As and when we mature and accumulate life's experiences, we often engage in introspection and self-assessment. This involves re-evaluating our goals, desires, and aspirations. For example, we may seem to have a clear career path in mind in our early youth, and we might choose to reassess our priorities and interests as we grow older, leading to a paradigm shift in our identity and life choices.

- **Biological Alterations**: Adolescence is a period of significant physical and hormonal changes. These changes can affect our self-image and self-concept. As adolescents, we may grapple with questions related to our physical appearance, gender identity, and sexuality, which can impact how we see ourselves and our place in the world.
- **Major Life Events**: Sometimes, significant life events are catalysts for an identity crisis. For example, relocating disrupts our "comfort zone" and established social and cultural environments, prompting us to reconsider our identity in the context of new ones. Stepping out of a secure home for higher studies, changing careers, or facing the loss of a loved one can lead to the need to re-examine our values, beliefs, and direction of life.
- **Migratory World**: Increased global migrations lead to a fascinating but complex phenomenon: confused identity in the children of migrants. Some such children, often referred to as "second-generation", find themselves caught between their parental heritage culture and the dominant culture of their new home. This cultural tug-of-war can lead to a sense of not fully belonging anywhere. For instance, children of South Asian descent born in America face this struggle sometimes. In informal terms, they are called "America-born confused desi" (ABCD). These children, and Gen Z in general, face several challenges due to confused identities. They navigate between traditions and languages at home that differ from the dominant culture. This can lead to feeling like an outsider in both environments. They might experience subtle prejudice or insensitive comments related to their ethnicity, further emphasizing their "otherness". The Public Religion Research Institute (PRRI) reported in 2019 that Gen Z adults are facing higher rates of discrimination based on race, ethnicity, and sexual orientation compared to the older generation. Interestingly, there is another aspect to this same situation. There are several "second generation" Americans of Indian or South Asian origin who develop a very broad-based healthy combination of the mix of cultures, having imbibed the best of both. They rise

as even more successful and open-minded than perhaps their parents or people of an earlier generation. Therefore, it is evident that while a confused identity can be a challenge, it is also an opportunity for growth. Open communication with parents, supportive extended families from the parental native country, supportive friends, and even therapists can help navigate this complex journey. The goal is not to reject either culture but to find a comfortable blend that represents unique experiences as children of migrants growing up in a globalized world.

- **Exposure to New Ideas**: Learning about different cultures, religions, philosophies, or lifestyles can expand our worldview. But exposure to these diverse perspectives may need re-evaluation of our beliefs and, consequently, our identity. For example, new ideas or cultural practices may challenge our previously held ideas and prod us to reflect on our values and identities in a broader context. Our inability to make this shift could create a "crisis".
- **Social and Cultural Influences**: Pressures from society, family, and peers, as well as cultural expectations, can significantly impact our sense of self. These influences may force us to conform to certain roles, beliefs, or values merely because that is expected of us. The conflict between conforming to external expectations and holding on to our personal beliefs can lead to an identity crisis because we have to deal with the conflicts and contradictions.

An identity crisis can arise from a combination of internal and external factors. Personal reflections, biological changes, major life events, exposure to new ideas, and social and cultural influences. This process of self-exploration and redefinition is a natural part of personal development, and navigating it successfully can lead to a more authentic and stable sense of the self.

Knowledge of the Self

The knowledge of the Self or "Self-knowledge" is at the core of Vedanta philosophy. But it was also regarded as the central point in human understanding by Western philosophers like David Hume and Immanuel Kant. They acknowledged that all human experience is accompanied by a sense of it pertaining to the person, like it is his, her, or my experience. Although it was necessary to understand who was "experiencing". According to Hume, "When we introspect, we do not find a substantial self or ego but rather a stream of constantly changing perceptions, thoughts, and sensations." He famously wrote, "When I enter most intimately into what I call myself, I always stumble on some particular perception or other, of heat or cold, light or shade, love or hatred, pain or pleasure. I never can catch myself at any time without a perception and never can observe anything but the perception." He also pointed out that the self is not an empirical experience to which attributes can be given. This identity of the self, also referred to as "ego", is a field that holds together all aspects of the variety of experiences, providing an underlying unity of cognitive experience, the common "I" when one experiences different things like, "I see, I like, I eat or I feel". Immanuel Kant did not deny the existence of the self but rather emphasized its transcendental and necessary role in human cognition. He argued that the concept of self (or "I") is a fundamental aspect of rational thought and experience. According to Kant, the self is not directly observable or knowable through empirical means but is rather a necessary condition for the possibility of experience and understanding.

The 16th-century French philosopher Rene Descartes concluded that thought cannot be separated from the individual; therefore, I exist, and thus he most famously quoted "*cogito ergo sum*" (which means, "I think, therefore I am"). Descartes confirmed that he can be certain that he exists because he thinks. But in what form? Do we assume that we cease to exist the moment we stop thinking? Are we dead in deep sleep when we apparently cease to think? We perceive our body through the senses, but then what about the functions apart from the physical body, such as thinking, judgement, and reasoning, or those that enable us to have subjective awareness and intentionality towards our environment, to perceive and respond to stimuli with some kind of agency? Descartes determines that the only indubitable knowledge is that he is a *thinking thing*. Thinking is what he does, and his power must come from his essence. Descartes defines "thought" (*cogitatio*) as: What happens in me such that I am immediately conscious of it? Thinking is the one activity of which we are immediately conscious. Even random, uncontrolled thoughts are part of our awareness. We are aware of the faculty of thought. To be able to grasp the nature of different things around us, we have to utilize the faculty of thinking and not be limited to the physical senses. Taking the example of wax—the fact that it is the same substance that appears to change its state when heated and turns into a translucent liquid—requires a faculty of judgement, which is in the mind.

As we have already seen, from the physical gross body, we move to the level of thought and mind, and deeper still to the intellect. Thus, we become aware that we must be an "entity" separate from the physical body, separate from the five senses or their perception, and therefore we are perhaps the mind, or maybe something other than the mind. The mind is debated in philosophy and used in clinical psychology, and it generally refers to consciousness plus autobiographical memory, personal identity, a sense of personal agency (voluntary control over actions), accurate introspection, and the ability to control our thoughts. The mind, as we have already discussed, enables us to think, feel, remember, create, and be self-aware. The mind is influenced by our culture, language, values, beliefs, identity, and behaviour. Our mind is subject to change, to both development and decline, and thus can alter our ability to focus and be aware. The mind is a source of knowledge, a medium of expression, and a vehicle of action. It is important to nurture, stimulate, and expand our mental attributes and activities, as they are the source of our intelligence and creativity. But it is most important to realize that *I am not my mind*.

Prominent atheist Richard Dawkins claims that human beings are just "throwaway survival machines" whose only purpose is to survive and replicate genes. Otherwise, the theory goes, there is very little point to our lives. He further clarifies, "When I say that human beings are just gene machines, one shouldn't put too much emphasis on the word 'just'. There is a very great deal of complication, and indeed beauty in being a gene machine." In *The Selfish Gene*, Dawkins says that life evolves through the differential survival of replicating entities. He articulates a gene's eye view of evolution. According to Dawkins, the processes of reproduction, mutation, and selection are not guided by any sentient being. He contends that religious faith is a delusion and that a supernatural creator does not exist.

Stephen Hawking was not religious, either. He said in an interview, "I regard the brain as a computer, which will stop working when its components fail." "There is no heaven or afterlife for broken down computers; that is a fairy story for people afraid of the dark." Examining the history of scientific knowledge of the universe, he further asserted, "Before we understand science, it is natural to believe that God created the universe. But now science offers a more convincing explanation. What I meant by 'We would know the mind of God' is, we would know everything that God would know, if there were a God, which there isn't. I am an atheist." Hawking said in *The Grand Design* that a creator was "not necessary" in the narrative of how the world was created. For years, he warned that humankind was facing extinction from threats such as climate change, nuclear wars, and genetically engineered viruses. According to him, life on earth may have only 100 years left. The 17th-century Dutch philosopher Spinoza stated that there is only one substance in the universe, and that substance is God. Like Spinoza, Albert Einstein saw the universe itself as the embodiment of an entity called God. He remarked, "I believe in Spinoza's God who reveals himself in the orderly harmony of what exists, not in a God who concerns himself with the fates and actions of human beings." He did not believe in a personal God. In his 1931 essay "The World as I See It", Einstein said, "I cannot conceive of a God who rewards and punishes its creatures or has a will of the kind we experience in ourselves. Neither can I nor would I want to conceive of an individual that survives his physical death."

Thus, this life factor, although present in all living matter, presents itself with a very unique characteristic in human beings. It comes along with the capacity or awareness of being "me" and that the mind is undoubtedly the instrument to gain this experience of "I-ness". This brings us to our next question, agreeing that this "awareness of being alive" is an abstract concept, almost impossible to explain on empirical grounds.

Even from before the Common Era, in the voice of Socrates, who is acclaimed as the founder of Western philosophy, up until now, the question "Who am I?" has remained an enigma. To answer the question, we first look within and discover the layers of our identity and how they interact and influence each other. Then we look at the world outside, and then we also need to look beyond, transcend the limitations and illusions of our identity, and realize our true nature and our true potential. Different cultures, civilizations, and religions have answered the question, "Who am I", with their worldview, values, beliefs, and practices.

Who Am I and the Spiritual Identity

The question "Who am I" is first addressed with negative proofs of who I am not. From the various aspects discussed so far, we do understand that we are not this body, although we have a body; we cannot be the material and decaying body. We are not the mind, although we have a mind. We cannot be the ageing, vacillating, and unpredictable mind that rides on waves of emotions. Anything perishable, we cannot be, and the entire psychophysical system is "perishable matter". The body–mind complex is our instrument for functioning in this world. In the process of identifying with this instrument, we lose track of our real identity.

That "I" exists is never in doubt. Thus we come to the understanding of being the changeless, eternal, immortal Self as we go beyond the body and mind and from the dual to the non-dual "one-ness", where there is only one reality and everything else is a projection or manifestation.

The venerated Indian sage Ramana Maharishi devoted his entire life to discovering "Who am I?" His analysis was a simple understanding of who or what I am not. He said, "When one turns within and searches. Whence this 'I' thought arises, the shamed

'I' vanishes, and wisdom's quest begins." An interesting anecdote connected with Ramana Maharishi is narrated thus: One devotee went to the sage and said, "You talk of the non-dual idea of Who am I, but I am a worshipper of God in form, so will I not go to heaven?" "Yes, you will," said the sage. "Will I see God in the form in which I worship him?" "Yes, you will," was the response. "Will God talk with me?" the devotee asked eagerly. "Yes, he will." "What will he say, I wonder," pondered the enquirer. He will ask you, "Do you know who you are?"

Every scripture, every religion, and every philosophical or theological perspective accepts the journey from birth to death. Every individual knows and accepts, although not by choice, the termination of life. But if this were the only and final truth of life, then we would not be asking, what is life? Who am I? And furthermore, why am I even here? Almost every scripture, which has been dealt with in-depth by all spiritual teachers, commentators, and philosophers, and any enquiry by intellectuals of any denomination, is primarily seeking the answer to the question, "Who am I?"

"Who Am I" According to Vedanta

The most ancient sages or seers of the East, residing in the forests, expressed their most profound thoughts and beliefs, *based on direct experience*, through the Upanishads. There is a difference between experience by reasoning and logic or by inference and suggestion, and direct, unmediated experience of going beyond the senses and body consciousness. According to Advaita Vedanta (the non-dual perspective), which is the final understanding as per these scriptures, the self, which is referred to as the soul in Western theology, although it does not mean the same as the Atman of Vedanta, is our real identity as a human being. The Atman is not the body, not the mind, not the senses, not the ego, or the social role; rather, it is the essence of our being, the pure consciousness, the unchanging witness. It is not separate or different from Brahman, the Universal Consciousness, which is the imperishable, indestructible one without attributes. If the quest for self-realization is kept alive, then we, the "limited" beings, can realize our intrinsic nature and be free from bondage, free from the multiple personas and identities. In that self-realized state, we will find *moksha* or freedom from the cycle of birth and death. Our perishable bodies, born from matter, are destined to die. In contrast, Atman, our unchanging essence, is eternal. By recognizing what we are not (the impermanent body), we gain undeniable knowledge of who we truly are (the Atman). This profound understanding becomes the unshakeable foundation of our real identity. And this realization reverberates with Om.

The Upanishads, thus, give a very emphatic insight into the question, "Who am I". They seem to have encapsulated the entire teachings into four primary Great Statements, the *mahavakyas*:

Tat Tvam Asi—That Thou Art
Aham Brahmāsmi—I am Brahman
Prajnanam Brahma—Consciousness is Brahman
Ayam Atma Brahma—This Self, the Atman, is Brahman

The entire concept of the Vedanta philosophy focuses on getting to the core understanding of the Self. One of the main questions that the Bhagavad Gita addresses is: Who am I? Krishna explains that the human being is the Atman, a part of Brahman, who is the source of all existence. As stated in the Bhagavad Gita, "This soul is never born, nor does it die; weapons cut not, fire burns not; it is unmanifested, unchangeable...." The body-mind (this self) is a manifestation of that which is eternal (the Self). The process of Self-realization is an inward journey. As M. Scott Peck said in this widely acclaimed book *The Road Less Travelled*, "The miracles described indicate that our growth as human beings is being assisted by a force other than our conscious will."

Swami Vivekananda has proclaimed, "Each soul is potentially divine. The goal is to manifest this divinity by controlling nature, external and internal. Do this either by work, or worship, or psychic control, or philosophy—by one, or more, or all of these—and be free. This is the whole of religion. Doctrines, or dogmas, or rituals, or books, or temples, or forms, are but secondary details."

Once we know what life is and who we are, then it is only natural to wonder why we are here. *We are not here by accident.* We have evolved to this stage after much struggle over aeons. The struggle continues because the one driving force for the human race is a goal to be achieved and a clear-cut path to be cut out to reach that goal. In whichever way we have understood who we are—either the realized souls who need to rediscover that identity and be free from the limited identities bound within time and space, or are ordinary biological evolutes whose purpose of life is to ensure the continuation of the species—there has to be a purpose to life, and we have to realize that goal.

CHAPTER 7

Why Am I Here? The Purpose of Human Existence

The unexamined life is not worth living.

—Socrates

We have come a long way and have realized at least this: from among the miracles of nature, as the last of the evolved species, as God's creation, whatever we may believe, we are the finest and most advanced of all living beings on this planet. And until we can discover or find a higher-level intelligent being from somewhere else in this universe, this will remain the current reality! We have discussed the journey through creation, witnessing incredible innovations, getting a grasp of what life is, and getting a glimpse into the mental makeup of our existence. But the next question that seeks answers is, now that we are here, what is the purpose? It might sound paradoxical, and in fact, it may not even rise in the minds of those of us who are moving speedily through life, either driven by some immensely motivating goal or too caught up in the basic struggle of staying alive! But for us, this question is important as we transcend from modern science to meta-science. Our quest now is to understand the purpose of human existence. What we are seeking is not just an academic analysis; it is a personal exploration into the heart of what it means to be human. We wish to reflect on our place in this vast universe. The journey ahead is as much about the collective human experience as it is about self-discovery.

In the Darwinian perspective, if we have reached so far as a consequence of evolution and consequent to the survival of the fittest, then the purpose would be to continue to survive. "First and most basically, modern Darwinism tells us that the primary goal of life is genetic survival. This principle borders on the self-evident. All living things exist only because all of their ancestors passed their genes on successfully, so the inherited nature of all organisms is to strive for genetic survival." A very popular belief is that, in any case, we are the results of a miracle—the fact that we are on this planet! Having evolved to be here, we must continue to evolve. It is widely stated that we evolved so that we could live. But we can also say with greater emphasis that we live to evolve. Evolution prompts organisms to survive and thrive. Humans and every living animal or plant are said to owe their existence to it. Our purpose is to "evolve" during our lifetime because that is consistent with our evolutionary purpose. Thus, *an* answer to the question "What is the purpose of life?" is that we are here so that we can continue to live, adapt, learn, and grow. The purpose of life, and our purpose, therefore, is to continue to evolve.

Continue to evolve? That in itself sets off a plethora of rhetorical questions: how, why, and with what purpose? The most obvious and logical explanation is to have set goals and a direction for our lives. Our journey through life is undoubtedly very deeply personal and constantly evolving. It involves introspection, self-discovery, and an ongoing search for meaning. As we proceed, we have the opportunity to shape our story so that it aligns with our values, passions, and aspirations. Whether we succeed or not is an aspect that needs introspection.

Many of us think of the purpose and meaning of life as the same thing, but they are slightly different. The difference between the meaning and purpose of life can sometimes be very subtle and can vary depending on individual perspectives. This often refers to the significance or interpretation we give to our existence. It involves questions about why we are here, what our experiences signify, and what gives value or importance to our lives. The meaning of life often involves contemplating existential questions about the nature of being, our place in the universe, and understanding life's fundamental truths or principles. The search for meaning can vary greatly among individuals, and it involves philosophical, spiritual, or existential inquiries. Finding the purpose of life involves identifying goals, aspirations, or intentions that give direction and drive to our actions and choices. Purpose is striving towards personal fulfilment, contributing to the well-being of others or society, or aligning with a greater cause or belief system. More specifically, when we have a defined purpose in our lives, we can assume that we feel good about the way we are living our lives. We would have logical and ultimate reasons for our actions and a sense of satisfaction and comfort that we were contributing in some way to the world. This sense of satisfaction and connectedness can help us reach higher levels of well-being.

While the two concepts are related and often intertwined, ultimately, both play significant roles in shaping how we navigate our existence and find fulfilment.

The Purpose of Human Life

Exploring the purpose of life not only reveals the depth and diversity of our existence but also invites a deeper understanding of what it means to be human. Undeniably, the sole purpose and quest of life for every human being is to be happy, content, satisfied, and free from suffering and sorrow!

Philosophers have grappled with the question of life's purpose from the point of view of various schools of thought. The Socratic philosophy emphasizes wisdom and self-realization through self-examination and knowledge. Existentialists like Jean-Paul Sartre suggest that life lacks inherent meaning, proposing that it is up to each of us to define our own purpose through choices and actions, thereby making it entirely personal. Classical philosophical traditions like Stoicism emphasize living virtuously and in harmony with nature as a way of flourishing. This suggests that our purpose is to cultivate wisdom and ethical behaviour. On a social level, many of us find purpose in contributing to the well-being of our communities. This could be through simple acts of kindness, sharing and caring, and being empathetic to our fellow beings. For some, taking care of animals becomes a life's purpose. This shows that many of us realize that our lives are interconnected and that we have great potential for making an impact.

Psychologically, the purpose is often linked to fulfilment and well-being. Positive psychology research shows that purpose correlates with heightened happiness and life satisfaction. This is also highly personal and may be rooted in our passions, interests, or creativity. Economically, of course, everyone's purpose is associated with financial stability and career success. The pursuit of economic prosperity is the primary focus in modern society, and the "rat race" emanates from the mindset of survival of the fittest. This raises questions about whether material wealth is sufficient for lasting fulfilment. Running parallel to this pursuit is the newly acquired role in environmental stewardship, which has become increasingly relevant. With growing environmental challenges, a purpose that includes caring for our planet and advocating sustainable practices is becoming more recognized as an integral part of our existence.

Maslow's "Hierarchy of Needs" is a theory in psychology that proposes a pyramid-like structure of human needs. According to this theory, our most basic needs, like food and water, must be met before we can focus on higher-level needs. Only then can we strive for self-actualization, the pinnacle of the hierarchy, which represents fulfilling our potential and reaching our full capacity (see Fig. 7.1). These various aspects—philosophical thought, social contribution, psychological well-being, economic achievement, and environmental responsibility—create a complex structure for the purpose of human life. Each of these elements contributes to our broader understanding of our existence and our place in the world.

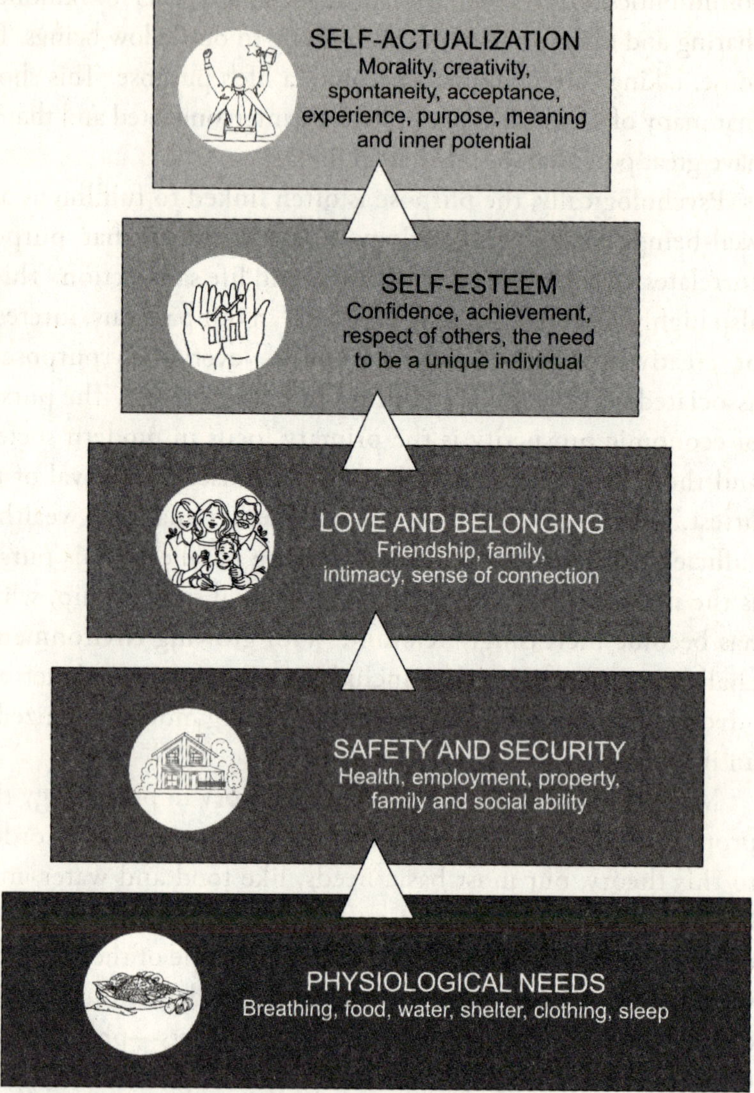

Fig. 7.1. Hierarchy of needs based on Maslow's model.

The Quest for a Purpose

Albert Camus, the French philosopher, found life meaningless, like Sartre did, and therefore challenged us to embrace existence fully if we seek a purpose in life. In his work *The Myth of Sisyphus*, Camus wants us to acknowledge our freedom. One of the key lessons we can glean from the myth of Sisyphus is the importance of resilience and perseverance in the face of adversity. Sisyphus was condemned to an eternity of ceaseless toil, yet he continued to push the boulder up the hill, despite it rolling back down each time (see Fig. 7.2). This relentless determination, despite the futility of his efforts, speaks to the human capacity to endure and persist in the face of seemingly insurmountable challenges.

Fig. 7.2. Sisyphus pushing a boulder up the hill.

But how do we create meaning? Friedrich Nietzsche believed that we needed to cast off universal truths and establish our own values. He believed that meaning comes from our mental states, not external facts. Martin Heidegger emphasized accepting our freedom and responsibility. Recent research emphasizes that people who perceive a sense of purpose tend to live longer. If we are purpose-driven, we will have less stress. It acts as a buffer during adversity, helping us bounce back from negative experiences. We have far better emotional resilience.

"Ikigai" is a very meaningful Japanese concept, which explains that true fulfilment and purpose in life come from finding balance and harmony among four key elements: we must know what we love; what we are good at; look beyond ourselves and see what the world needs; and finally relate to our profession, or what work we are paid for. Sustaining ourselves financially is important, but knowing our strengths, following our passion, and contributing to society are all vital for knowing the purpose of life and finding meaning in even the simplest moments of existence.

At a young, dynamic age, everyone has dreams, ambitions, and goals. Some of us reach and attain; some of us try but fail. Yet, with our inbuilt resilience and ability to "change course" we do carry on with a renewed purpose. If we cannot find the perfect combination of the four key elements, we certainly strive to strike two or three out of four. But somewhere along the journey, we may lose balance and get caught in a whirlpool of uncertainties. Even then, finding meaning in anything that we do helps to keep us focused. As we get older, if we have found some answers to who we are, then knowing the purpose and why we are here evolves seamlessly, and we can remain motivated and inspired. Otherwise, purpose is lost in the anxiety of ageing and fear. This is when age-related physiological changes also make a crucial difference.

Knowing Purpose Helps

There is a deep relationship between purpose in life and resilience against age-related brain changes, particularly in the context of Alzheimer's disease and advanced age. Research shows that higher levels of purpose in life reduce the deleterious effects of AD on cognition at an advanced age. Having a sense of purpose in life may help protect against memory deficits associated with higher depressive symptoms. The present findings emphasize the sense of purpose to promote cognitive reserve in older adulthood, allowing individuals to maintain cognitive performance in the face of accruing neurological insults. At the Century Summit 2022, Bob Waldinger of Harvard University stated, "What are the best predictors of longevity? We thought it was going to be their cholesterol level. We thought it was going to be their blood pressure. It turned out to be the quality of their relationships." A strong sense of purpose is a buffer against the detrimental effects of neurodegeneration, allowing us to maintain cognitive function and independence as we age. In a 2019 case study on whether a purpose-driven life helps us to live longer, the researchers said that life's purpose is defined differently by different people. But in general, it indicates that we have an aim in life and defined goals. This purpose, the study authors said, helps make it more likely that we will engage in behaviours that are good for our health. Some studies have simply asked people what gives them a sense of purpose in life. Growing evidence indicates that a higher sense of purpose is associated with a reduced risk of chronic diseases and mortality. Another study indicates that yoga and transcendental meditation may help students and physicians improve their own well-being, prevent burnout, and improve patient care.

Spirituality as a separate entity (compared to religion) applicable to the medical field needs to be understood. The Indian Council of Medical Research (ICMR), India, and FORTE (the Swedish Research Council for Health) held a joint workshop on ageing and health in 2016. ICMR identified geriatric research as a full-fledged programme to encourage researchers to generate and test scientifically proven methods that can be implemented for active and healthy ageing. Several NGOs and research organizations in India implement community-based interventions to improve well-being and health. These interventions may include components related to finding purpose and meaning in life. HelpAge India has incredible success stories of very elderly people who are motivated and purpose driven. The goals are community welfare, as everyone is motivated to look beyond the self and to live in communal harmony. There are successful self-help groups with a focus on ecology, health, and, most of all, living happy and meaningful lives.

Using functional neuroimaging techniques, researchers have found that individuals with a higher sense of purpose have greater functional integration of the default mode network, particularly the dorsal component (dDMN). The default mode network refers to those areas of the brain that are activated when we are not engaged in any specific cognitive work but are letting our minds wander at rest. A purposeful life is more likely to promote greater functional connectivity within the dDMN, which could reflect a form of brain reserve that protects against age-related cognitive decline.

This increased connectivity may enable more efficient information processing and cognitive flexibility, allowing individuals to adapt to cognitive challenges.

By recognizing the role of purpose in sustaining brain health and cognitive function, healthcare professionals can encourage a focus on purposeful living so as to help healthy ageing and increase resilience against neurodegenerative diseases. By appreciating the protective and positive effects of purposeful living on the brain, we can develop positive approaches to ensure healthy ageing and try to preserve proper cognitive function in our later years.

Purpose of Good Health and Well-being

Good health is not just the absence of disease, and well-being is not just happiness; it is the vibrant foundation upon which we build a meaningful life. Good health and well-being are often taken for granted until we let them slip away and no longer enjoy them. The United States Veterans Administration (VA)'s Whole Health approach also acknowledges that a sense of purpose is vital to health and well-being. By incorporating practices that nurture the spirit, the VA hopes veterans can connect with a deeper sense of purpose, leading to a more fulfilling life.

The United Nations' Sustainable Development Goal (SDG) number 3 proposes "to ensure healthy lives and promote well-being for all at all ages". The aim is to prevent needless suffering from preventable diseases and premature death by focusing on key targets that boost the health of a country's overall population. Regions with the highest burden of disease and neglected population groups are priority areas.

Whatever the purpose of life, good health is necessary to attain it, and it is also the underlying motivator. Good health gives us positivity, energy, and the drive to pursue our goals, thus the purpose of life. Energy allows us to chase dreams; the resilience that carries us through challenges is the quiet confidence that comes from knowing that our bodies are capable instruments. Prioritizing health isn't about chasing a fleeting ideal but about cultivating a wellspring of vitality. It translates to a life brimming with possibility. We have the energy to explore new hobbies, connect deeply with loved ones, and experience the world with a sense of adventure. Essentially, good health and well-being depend on four core determinants: nutrition, lifestyle, environment, and genetics. When one or more factors are compromised, we lose health. Healthcare and medicine are therefore considered allied determinants.

Generally, health promotion can happen effectively at individual, family, and community levels. While allopathic medicine and surgery can be lifesaving, local health traditions play a major role in prevention and primary care. Ayurveda, Yoga, Unani, Siddha, Sowa Rigpa, naturopathy, and homeopathy are together known as "Ayush" in India. Various traditional medicine systems are prevailing in the world, including Chinese, Japanese, Korean, Tibetan, and many others. The current trend is an evidence-based integration of allopathic and traditional medicines.

The Indian systems of medicine, Ayurveda and Yoga, as a body—mind approach are exemplars for good health and well-being. We have discussed them in Chapter 1. The beauty of Ayurveda and Yoga lies in their synergy. Yoga strengthens the mind and body, making them more receptive to the benefits of Ayurveda. These practices reinforce each other, creating a harmonious approach to health that goes beyond the physical. They cultivate self-awareness, a crucial component of maintaining well-being. By understanding our body's signals and emotional triggers, we can make informed choices about diet, exercise, and lifestyle. This empowers us to take charge of our health and become active participants in our well-being. We can build a strong foundation for a vibrant life brimming with energy, resilience, and the freedom to serve the purpose of life and chase our dreams. This energized and purposeful life also makes us good participants in the welfare of the community, and we can become active contributors to the betterment of society and fellow beings.

Ayurvedic principles can guide our yoga practice, ensuring that we choose postures that benefit our specific prakriti. Understanding one's prakriti or *dosha* composition is essential in Ayurveda for maintaining health and well-being. We can tailor our diet, lifestyle, and remedies to create a personalized plan that supports specific individual needs. Imagine aligning our sleep schedule with natural rhythms, adjusting our diet with the changing seasons, or incorporating stress-reduction techniques to ensure harmony within the body. This focus on balance goes beyond treating symptoms; it aims at preventing imbalances before they manifest as illness.

Universal Well-being as Purpose

Research shows that having a sense of purpose may help us live longer, making us resilient to stress. Universal well-being, however, goes beyond personal happiness or fulfilment. It includes evaluative, hedonic, and eudaimonic well-being, all of which contribute to a meaningful and purposeful life. It includes the well-being of everyone, both within society and in the wider ecosystem. It means being deeply aware of our interconnectedness and interdependence with one another and with the planet; being responsible enough to create conditions that promote social justice, economic equity, and environmental sustainability; and ensuring that everyone has access to the resources and opportunities needed to thrive. We must cultivate empathy, compassion, and solidarity towards all beings and recognize the inherent dignity and worth of every individual.

For Aristotle, eudaimonia, the nature of well-being, was the ultimate goal of human existence. It is described as a state of "flourishing" by living virtuously and realizing our highest potential. Eudaimonia means cultivating moral virtues such as courage, wisdom, and justice and engaging in activities that are in sync with our true nature and purpose in life. It means living a life of moral integrity and compassion with an attitude of service to others. Evaluative well-being not only includes positive experiences and achievements but also challenges, setbacks, and personal growth. Hedonic well-being looks at the immediate experiences of pleasure and pain, happiness, sadness, anger, stress, and other emotional states. It focuses on pursuing pleasure and avoiding pain, seeking out enjoyable activities and experiences that bring momentary and immediate gratification. While it is an important aspect, it is transient and entirely dependent on external circumstances; therefore, it is less reliable as a long-term source of fulfilment. Universal well-being guides us towards a meaningful life with fulfilment, contributing to the greater good. It is not a mere utopian ideal but an imperative one for building a more just, equitable, and sustainable world for our future generations.

Working towards the universal good is selfless service to others. This ethos goes beyond personal gains, embracing kindness, compassion, and altruism with a focus on positively impacting others and contributing to the greater good. Those of us committed to the universal good will recognize our role in addressing societal challenges, promoting justice, and alleviating suffering. This sense of responsibility extends to efforts towards creating a more equitable and just world. The pursuit of universal good means having a global outlook and acknowledging the interconnection of all life and the environment. It means tackling global issues like environmental preservation, poverty reduction, and upholding human rights, recognizing the broader implications of our actions. Contributing to the well-being of others and societal betterment can deepen our sense of purpose and self-realization. We are encouraged to explore and cultivate our inner selves while contributing to humanity's and the planet's welfare. Embracing these principles not only leads to personal fulfilment but also plays a crucial role in creating a more compassionate and harmonious world.

Quest for Happiness, Success, Wealth

At the heart of all human endeavours lies the quest for happiness—a state of contentment, joy, and well-being. The pursuit of happiness is the defining characteristic of human history, a universal trait transcending cultural, temporal, and geographical boundaries, and central to human aspirations. Where is happiness to be found? We reach out into this world with our senses and seek happiness in success, wealth, relationships, physical and material possessions, and comforts. The ideal formula appears to be that relationships should be meaningful and fulfilling; experiences should be enriching; goals have to be ambitious; personal growth has to be satisfying;

and finally, sorrow and suffering must be avoided at any cost. Unfortunately, “should”, “have to”, and “must” do not work. The cycle of life has unpredictability as its core structure.

Western cultures, and increasingly now Eastern cultures, have been primarily influenced by individualism and the pursuit of personal happiness and fulfilment. The American Dream, for instance, has prosperity and success as a central purpose, emphasizing individual achievement and self-improvement. The quest for excellence is unrelenting. Great landmarks have been achieved, and human beings, with overarching goals and ambitions, are “over the moon” with their successes.

Success includes personal achievements, professional milestones, or societal recognition, that is, name and fame. This drive for success reflects our deep-seated desire to realize our potential, achieve goals, and create a lasting impact. Success, therefore, is not just about reaching a destination; it is also about fulfilling aspirations and achieving both personal and collective milestones. Along with these, the accumulation of wealth is inevitably the expected result. Equally persistent is the pursuit of wealth, traditionally seen as financial prosperity and material gain. One aspect is acquiring wealth in a righteous way. The other aspect is that the quest for wealth is the primary goal, driven by a desire for comfort, security, and an improved quality of life. It is the key motivator behind economic development, trade, and innovation, significantly influencing the evolution of societies and individual choices.

Science and Technology's Influence on Purpose

The swift advancements in science and technology have significantly reshaped our understanding of life's purpose, introducing an era with vast possibilities and unique challenges. This transformation reveals the intricate relationship between human innovation and our quest to comprehend existence.

Advancements in science and technology have, in the larger scheme of things, deepened our comprehension of the universe, ranging from the intricacies of quantum physics to the vastness of cosmology. This enhanced knowledge defines our role in the cosmos and the meaning of our existence. In the personal scheme of things, the pursuit of comfort, pleasure, and happiness in this age is intricately linked with technological advancements. This pursuit has become a defining aspect of existence for many, shaped by the conveniences and innovations that technology provides. Technology has transformed our daily lives, offering unprecedented conveniences. Smart devices and home automation systems exemplify this trend, simplifying tasks and enhancing comfort. The desire for an effortless existence drives the continual development of innovative gadgets and services. The digital revolution has broadened entertainment options, catering to our need for enjoyment. Streaming services, video games, and virtual reality offer instant entertainment, while social media encourages connection and self-expression, contributing to our sense of joy and community.

Technological progress plays a pivotal role in the quest for economic success and material well-being. E-commerce and online career opportunities have revolutionized how we earn money and spend it, impacting our overall quality of life by providing easier access to a myriad of products and services. Technology has enabled unparalleled personalization. Algorithms and AI now tailor content and services to individual preferences, enhancing satisfaction and happiness. From streaming recommendations to customized health apps, technology caters to our unique needs and desires.

However, if we look around, the ratio of happiness appears almost inversely proportionate to the drive for success, comfort, and satisfaction! Happiness is a state of mind, and what is the measure of happiness in today's world? There is an ongoing tussle between the focus on material comfort and pleasure and a deeper sense of happiness. The greater the accumulation of wealth in the hands of a few, the greater the abject poverty in the hands of many. The more challenging the goals, the greater the success, and the more is the stress, anxiety, dissatisfaction, and distrust. The more we acquire material wealth, the less we seem to garner in terms of inner joy and peace. Wars, communal clashes, distrust, and hatred are rampant today. Suffering is widespread. Where lie the solutions?

The materialist will say this is how it is. This is the world, such is life, the fittest will inevitably survive, and finally, with death, it all ends. Therefore, till then, carry on trying to make the best of what we have. Our senses reach out to every nook and cranny of experience, and we try to grasp everything with both hands. But we reach a stage when these struggles in life are an aimless or fruitless exercise, and thus both the meaning and purpose are lost. This is when, increasingly, we seek answers in spiritual and philosophical teachings or in psychology and psychiatric treatments. We seek inner fulfilment and ethical living.

This current scenario displays the complexity of defining purpose in a technology-driven world. This disparity is very significant as the mercurial rise of science and technology brings forth ethical challenges, especially as we gain the ability to alter our environment and ourselves. These developments demand careful consideration of how technology aligns with our moral values and the broader implications of our choices. Technological progress has raised urgent environmental concerns. The need for sustainable practices and environmental stewardship is now integral to discussions of purpose, urging a re-evaluation of our responsibilities towards the Earth and its ecosystems. The technological pursuit of comfort and pleasure raises significant concerns. Issues like digital addiction, mental health impacts, environmental effects, and a potential shallow focus on immediate gratification need to be faced and acknowledged. Balancing the advantages of technological progress with a deeper understanding of purpose and fulfilment is crucial.

Spiritual Perspectives on Purpose

In exploring the purpose of human existence, we face a fascinating intersection, and sometimes conflict, between scientific and spiritual perspectives. Both domains offer profound insights, yet they can sometimes present divergent views on the meaning and purpose of life.

Science typically focuses on empirical observations and the quest to uncover natural laws. While this reductionist approach has explained much about the physical world, it can seem at odds with spiritual perspectives, particularly when reducing complex experiences like emotions or consciousness to mere neural activities. This can diminish the depth of spiritual experiences. Scientific viewpoints often align with materialism, according to which everything arises from physical matter, including consciousness. This is directly opposite to spiritual beliefs that talk of a transcendent reality beyond the material. Conflicts arise about the existence of a soul or divine essence, where spiritual perspectives emphasize transcendence and science focuses on empirical evidence.

Despite these conflicts, it is important to acknowledge that science and spirituality need not be adversaries. Many seek their integration, arguing that together they can offer a more complete understanding of human existence. For example, exploring consciousness can provide a meeting ground for scientific and spiritual insights, offering a holistic comprehension of life's purpose.

Transcending Materialism

Trying to understand the purpose of human existence often leads to the necessity of transcending materialism and striving for higher, more profound goals. This concept encourages a shift from focusing solely on material wealth to seeking deeper, more fulfilling aspects of life. The focus on material wealth often falls short of offering lasting fulfilment. Transcending materialism means looking beyond these tangible assets to find true meaning and purpose. This shift encourages exploring aspirations that go beyond the material domain, focusing on personal growth, self-realization, and societal contribution. Central to transcending materialism is the pursuit of self-actualization. This includes realizing our full potential, developing personal talents and virtues, and discovering our authentic self.

Higher goals often involve making a positive impact on others and on society. Acts of kindness, service, and altruism become crucial, emphasizing the importance of contributing to the greater good. Many of us turn to spirituality and philosophy in trying to transcend materialism. This gives us deeper insight into life's mysteries and reality and helps us connect to the universe, giving us a sense of spiritual unity. The quest for wisdom is a lifelong endeavour, extending beyond intellectual pursuits to include practical wisdom and ethical living.

Moving beyond materialism does have its challenges. It often involves questioning societal norms and re-evaluating personal values. Balancing higher aspirations with everyday life demands careful consideration. The allure of material comforts draws us back into materialistic pursuits. The concept of transcending materialism to seek higher goals resonates with those seeking a deeper purpose in life. It prompts a reflection on the essence of our existence and the impact of our actions. Navigating the complexities of modern life with this mindset can lead to a more meaningful and purposeful existence, enriching our understanding of the human journey. There are several perspectives that give us a worldview on the purpose of life.

Across Cultures and Philosophies

Throughout history, philosophers and thinkers have provided profound insights into the goals and purposes of human life. Their diverse cultural and intellectual backgrounds provide wisdom, guiding those who seek meaning and purpose in life.

Confucius, the Chinese philosopher, focused on moral virtue and li (propriety) as key to a meaningful life. He advocated ren (benevolence) and fulfilling social responsibilities, aiming for a harmonious, just society through ethical personal development. Immanuel Kant emphasized duty and moral action, saying that we should act as if our actions were to become universal laws. He believed fulfilling moral duties based on universal laws led to moral satisfaction and a purposeful life. Viktor Frankl, an Austrian neurologist and psychiatrist, founded logotherapy, emphasizing the search for meaning as the primary human drive. He believed in finding purpose through meaning in suffering and life experiences, influenced by his survival in the Holocaust.

Everyone emphasizes the importance of ethics, virtue, and personal responsibility. Their teachings encourage self-reflection on values, choices, and actions, guiding us towards a meaningful existence. Some philosophical perspectives, such as hedonism, propose that the purpose of life is to seek pleasure and avoid pain. Happiness and pleasure are the ultimate goals. Utilitarianism, a philosophical theory associated with thinkers like Jeremy Bentham and John Stuart Mill, proposes the purpose of life as maximizing happiness or well-being for the greatest number of people. It emphasizes actions that lead to overall benefit or utility.

Humanism talks about the inherent value and dignity of human beings. The purpose of life is to cultivate human flourishing, emphasizing reason, ethics, and compassion towards individual and collective well-being. Nihilists believe that life lacks intrinsic meaning or value. While some may find this liberating, others may find it bleak or depressing. Several spiritual traditions suggest that the purpose of life is to transcend the limitations of the material world and achieve union with a higher reality or divine consciousness.

These perspectives give us an insight into human thought regarding the purpose of life. Each of us may find resonance with one or more of these perspectives, or we may choose to follow our own understanding based on personal experience, values, and beliefs. Despite their differences, these perspectives have some common ground, such as the importance of ethical behaviour, personal growth, and self-discovery. Virtues like compassion, empathy, and wisdom are recurring qualities, highlighting their universal appeal. Some advocate for blending philosophical and spiritual insights, suggesting a holistic approach for a more comprehensive understanding of life's purpose.

Ancient Cultures and Religions on the Purpose of Life

Religions and cultures around the world provide their own interpretations of the purpose of human life. Many cultures and civilizations throughout history have developed their own explanations for the purpose of life, often rooted in their religious, philosophical, and ethical beliefs.

In ancient Egypt, the purpose of life was often tied to the belief in an afterlife and the journey of the soul. To achieve this, Egyptians engaged in various religious rituals, funerary practices, and burial customs aimed at providing for the deceased in the afterlife. These practices included mummification, burial with grave goods and offerings, and the construction of elaborate tombs and monuments such as pyramids and mastabas. Central to the purpose of life in ancient Egypt was the belief in the judgement of the soul.

In ancient Greece and Rome, the purpose of life varied among different philosophical schools and cultural contexts. Stoicism emphasized living with reason and virtue. The purpose of life was to cultivate wisdom, courage, self-discipline, and resilience in the face of adversity. Epicureanism taught that the purpose of life was to seek pleasure and tranquillity while avoiding pain and suffering. The Epicureans advocated for a simple and modest lifestyle focused on friendship, virtue, and the pursuit of knowledge.

Neoplatonism, influenced by the teachings of Plato, emphasized spiritual ascent and the pursuit of union with the divine. The purpose of life was to transcend the material world and attain a higher state of consciousness or unity with the ultimate reality.

Scepticism questioned the possibility of attaining certain knowledge and advocated for suspension of judgement. The purpose of life was to cultivate intellectual humility, open-mindedness, and tranquillity by suspending belief in dogmatic assertions and embracing uncertainty. These philosophical perspectives reflected the diversity of thought and values within ancient Greek and Roman societies.

African cultures and traditions are incredibly diverse, with a wide range of beliefs, practices, and worldviews. The purpose of life is often tied to ancestral spirits, community values, and rituals. It is difficult to generalize about the purpose of life across all African cultures, but many cultures emphasize the interconnectedness of all living beings and the natural world. The purpose of life is to maintain harmony and balance within the community and with the environment. In many African cultures, the purpose of life is closely tied to the well-being of the community. Individuals are expected to contribute to the collective welfare, support one another in times of need, and uphold shared values and traditions. The purpose of life thus involves fulfilling social roles and responsibilities as defined by these rites of passage.

Native American traditions view the purpose of life as interconnectedness with nature, spirituality, community, and the ancestral lineage. The purpose of life may involve cultivating a deep spiritual connection through rituals, ceremonies, prayer, and communion with the natural world. It means preserving and honouring cultural identity, heritage, and traditions passed down through generations. It includes the well-being of the community, supporting one another in times of need, and upholding shared values and traditions.

In Japanese culture, the purpose of life is a combination of spiritual, social, and personal dimensions. It is influenced by indigenous beliefs, Shintoism, Buddhism, Confucianism, and modern societal values. It often involves finding harmony with nature, fulfilling social responsibilities, contributing to collective well-being, pursuing personal growth, and finding meaning and fulfilment in all endeavours.

Influenced by Shintoism, the indigenous religion of Japan, the purpose of life also honours the spirits of ancestors and kami (deities). In modern Japanese society, work and career play a significant role in shaping a sense of purpose and identity. It may involve finding meaning and fulfilment in professional pursuits, contributing to the success of the organization or company, and achieving personal and professional growth.

Confucian values have played a significant role in shaping Japanese society, emphasizing the importance of fulfilling social roles and duties within the family, community, and society, upholding traditional values, respect for authority, and loyalty to superiors. Influenced by Buddhist teachings, mindfulness, compassion, and the cultivation of inner virtues as means to attain enlightenment and liberation from suffering are important.

Theistic religions, with their belief in a divine or transcendent power, offer distinct perspectives on human purpose.

Christianity and Judaism believe the purpose of life revolves around honouring God. In Christianity, this means accepting Jesus Christ's sacrifice for humanity's sins and living a righteous life to achieve salvation and eternal life with God. Judaism emphasizes following God's laws (mitzvot) and repairing the world (tikkun olam) to create a holy life and deepen the connection between God and creation.

Islam emphasizes submission to the will of Allah and striving for righteousness. Muslims believe that they were created by Allah to worship him and to live according to his guidance as revealed in the Quran and the teachings of the Prophet Muhammad. The purpose of life is to attain closeness to Allah, to seek his pleasure, and to strive for paradise in the afterlife.

Jainism and Buddhism, the followers live by ethical codes to achieve liberation. Jainism's Panchsheel, meaning "five virtues", outlines five core principles: non-violence, truthfulness, not stealing, non-possession, and celibacy (interpreted differently by individuals). Buddhism's Noble Eightfold Path, often pictured as a wheel with eight spokes, provides a broader framework (see Fig. 7.3). It includes right view (understanding reality), right intention (positive motivations), right speech (kind and truthful communication), right action (avoiding harm), right livelihood (ethical work), right effort (cultivating positive states), right mindfulness (awareness of thoughts and feelings), and right concentration (mental focus).

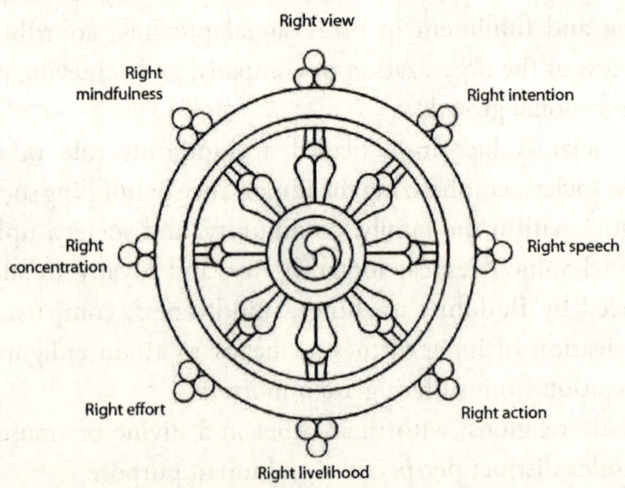

Fig. 7.3. The Noble Eightfold Path wheel.

While both religions emphasize non-violence and good conduct, Panchsheel focuses on avoiding harm and letting go of material things. The Noble Eightfold Path offers a more comprehensive guide for ethical living, mental development, and achieving enlightenment. Ultimately, both codes aim to lead followers on a path of peace, mindfulness, and right action.

In traditional Chinese philosophy, particularly influenced by Confucianism, Taoism, and Buddhism, the purpose of life is often seen as aligning oneself with the natural order and cosmic harmony and fulfilling social roles and responsibilities within the family, community, and society. Influenced by Buddhist teachings, the purpose is also understood as attaining spiritual enlightenment and liberation from suffering. The purpose of life is to transcend worldly concerns through spiritual practice, meditation, and the cultivation of wisdom and compassion.

The Vedic Approach to the Purpose of Life: Sanatana Dharma

Hinduism, or Sanatana Dharma, meaning "timeless wisdom", emphasizes universal values and spiritual growth and has an inclusive and holistic approach to life and existence. It is a framework for understanding reality and the purpose of human life, and it directs us to the path to ultimate fulfilment and liberation. It emphasizes the existence of eternal truths that transcend time, space, and individual belief systems. It acknowledges the diversity of paths and approaches to truth and encourages seekers to explore and realize these truths through personal experience and spiritual practice. Dharma, righteousness or duty, is a central concept in Sanatana Dharma. It implies moral and ethical principles that guide us in living harmoniously with the universe and fulfilling our responsibilities in various roles and relationships.

In the ancient Indian scriptures, there exists a comprehensive framework for understanding the human journey and the purpose of life. This framework builds on the goals or principles, known as Purushartha, and the stages of life or existence, known as the

Ashramas, which provide a timeless guide for every individual seeking meaning, fulfilment, and spiritual growth.

The Four Principles of Life: The Purushartha are interconnected to take us towards a fulfilling life. These are four essential principles that we are encouraged to pursue. They justify living a life of love and happiness, motivated by the quest to accumulate wealth, but the underlying strength is pursuing each goal with righteousness, keeping to our codes of ethics and morals (see Fig. 7.4).

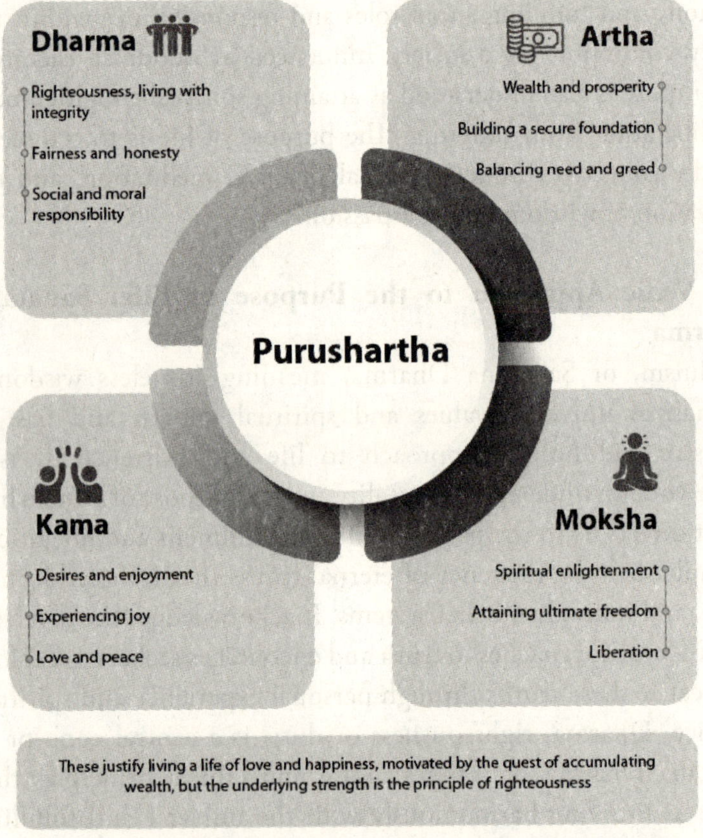

Fig. 7.4. The four core objectives of human life (Purushartha).

The Four Stages of Life: The Ashramas delineate the four distinct stages of an individual's life, each with its own unique purpose and responsibilities, starting from the learning stage, the householder stage, the reflection stage, and finally, the renunciation stage (see Fig. 7.5).

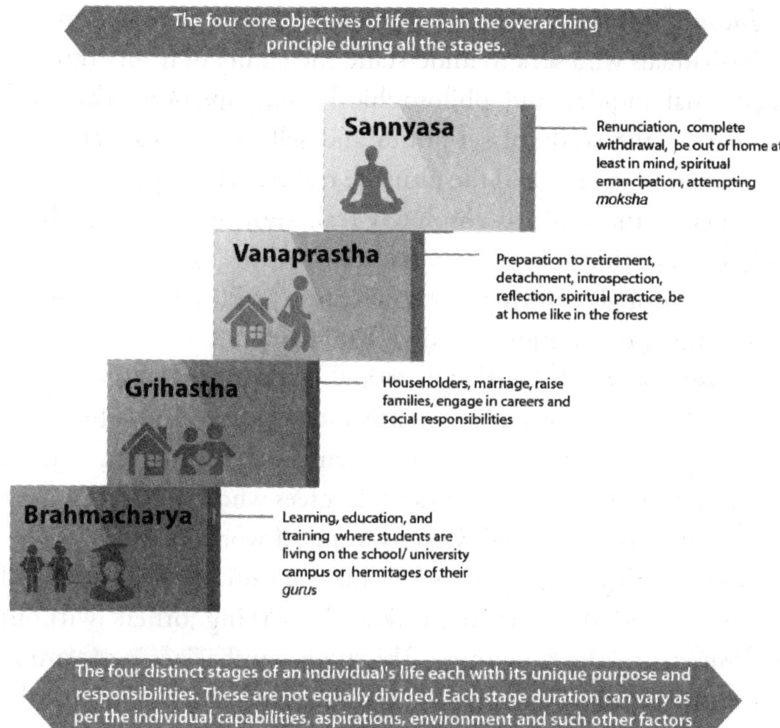

Fig. 7.5. Ashramas: The four stages of life.

Path of the Four Yogas

The purpose of life is to realize the unity of the individual self with the universal and to transcend the cycle of birth and death—to find liberation from the cycle of reincarnation and attain *moksha*. The purpose is directly connected to self-realization and spiritual liberation. According to Swami Vivekananda, who introduced the concept of the four yogas to the world, this liberation can be achieved through the path of spiritual knowledge (*jnana yoga*), devotion (*bhakti yoga*), selfless action (*karma yoga*), or spiritual practice. (*raja yoga*). Vedanta teaches that the ultimate goal of human existence is self-realization (see Fig. 7.6). Swami Vivekananda himself was a prime example of the unification of the four paths. His teachings emphasized the importance of attaining knowledge and self-realization, meditation, ethical living, and mental discipline, along with selfless work and unwavering devotion.

Jnana Yoga, the path of wisdom and knowledge, is epitomized by individuals who seek to understand the nature of reality through intellectual inquiry and philosophical contemplation. This path emphasizes the study of scriptures and self-inquiry to realize the true nature of the self and the ultimate reality. Self-enquiry prompts us to negate the world as absolutely real; know that it is transitory and a projection of our consciousness. The goal is to attain freedom through realizing our true nature as Pure Consciousness, transcending the illusion of individual identity.

Bhakti Yoga is the path of devotion—of belief and faith in God. The single-minded unwavering faith and devotion, and submission to God's will is the path to transcendental realization. This is beautifully embodied by saints and devotees who express their deep love for the divine through various forms of worship.

Karma Yoga, the path of selfless action, is demonstrated by those who dedicate their lives to serving others without attachment to the outcomes. This path emphasizes performing our duties and actions selflessly, without attachment to the results, as a means to purify the mind and attain spiritual liberation. The path leads to freedom by transcending the cycle of *karma*, purifying the heart through selfless service, and working without seeking the fruits thereof. Karma Yoga is increasingly recommended for this age because the human mind is super-active and restless. Dedicating the mental and physical energy in the service of others can bring inner fulfilment and peace.

Raja Yoga is the royal path of meditation and psychic or mind control. This path emphasizes inner discipline, meditation, concentration, and control of the mind-body as the means to attain self-realization and union with the divine. This includes both physical and mental endeavours like breath control (*pranayama*), and concentration exercises to still the mind. The ultimate aim is to achieve self-realization and freedom by gaining absolute control over the mind and transcending material limitations. Patanjali, the sage who compiled the Yoga Sutras, outlined the eight limbs of yoga that guide practitioners towards spiritual liberation.

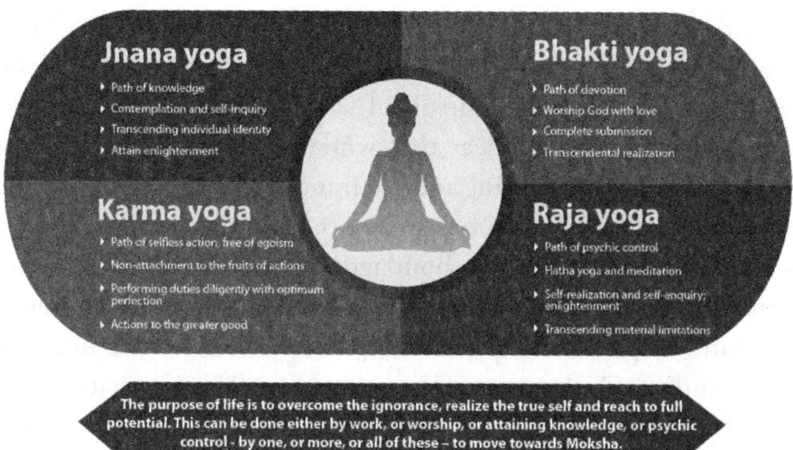

Fig. 7.6. The four yogas.

The beauty of these concepts lies in their interconnectedness. Living with integrity (first principle) through each stage while building a secure foundation (second principle) provides the resources to sustain our journey and allows us to experience joy and love (third principle), but responsibly. Ultimately, this path leads to freedom (fourth principle).

What exactly do we mean by *moksha*, liberation, or freedom? Are we somehow enslaved or bound so that we need freedom? In a manner of speaking, yes, we are bound. Identifying ourselves as this psychophysical system, as the body and mind, our purpose of life is driven by seeking gratification for the body and fulfilment for the mind, all of which is sought and conveyed by the senses in the process of experiencing the world around us. The bondage comes from our delusional belief that the reality of life, or our existence, is connected with the body–mind.

We seek permanence in the transitory, and yet, being aware that this body–mind "instrument" is perishable, we are caught in fear and anxiety and are desperate to hold on to this physical self, which must perish. The process of ageing and dying, therefore, is just not acceptable. This is the bondage, the fear, the despair, and the illusion of seeking permanence in the perishable. And what is liberation? It is the ultimate goal of spiritual liberation and enlightenment. It means transcending the cycle of birth and death and attaining union with the divine. The ultimate freedom is the culmination of a spiritual journey marked by Self-realization and the pursuit of wisdom. The ancient learned thinkers made it very clear that while fulfilling the short-term purposes of life in pursuing and attaining wealth and a pleasurable and comfortable life, ultimately we need to look towards the final goal of freedom from the bondage of the limitations of time, space, and causation and move towards liberation. The moment we can recognize, accept, and experience that we are not this body and mind, that we are the underlying Pure Consciousness, which is undying, we are free that very moment. This paradigm shift in perspective and understanding of the purpose of life takes us from "Genome to Om".

While the interpretations of various cultures, religions, and philosophies may seem disparate, there are common threads that run through them. Compassion, altruism, and the pursuit of wisdom are emphasized as integral to the purpose of life. The evolving global consciousness has given us a more interconnected perspective on purpose, emphasizing the importance of empathy, cooperation, and environmental stewardship. This helps us to recognize that the purpose of human life extends beyond individual and cultural boundaries to look at the well-being of the planet and all its inhabitants. Ethical living with an emphasis on moral behaviour is a universal statement. Compassion, honesty, respect, and justice are central to a purposeful life. Spiritual development is also a common goal. Practices such as prayer, meditation, or self-reflection are emphasized as they help to deepen the divine connection and personal virtue. Devotion to a higher power is common. Through various rituals and prayers, believers express reverence and seek guidance and fulfilment.

Many religions stress the importance of communal bonds and compassion, advocating acts of charity, service, and a sense of interconnectedness. Most cultures and religions believe in an afterlife. The very contemporary idea "we have only one life; make the best of it" is the root cause of extreme materialism, the rat race, and the urgency of over-achieving despite the stress and anxiety it creates. The belief in an afterlife or the goal of attaining freedom provides hope and shapes the view that the purpose of this life is a preparation for what lies beyond or ahead.

Self-realization

To understand the purpose of life, two deeply intertwined concepts stand out: the need for Self-realization and the dedication to universal good. Self-realization means diving deep into our inner world to uncover our true nature. This involves introspection, self-reflection, and recognizing personal strengths, weaknesses, and values, leading to a profound understanding of the higher Self. Self-realization is also about tapping into and nurturing our full potential. It's a process of personal growth, ultimately allowing us to reach our highest potential and contribute uniquely to our life's purpose. It ensures an authentic life where choices and actions are aligned with personal values and beliefs. Authentic living is characterized by purpose, inner peace, and genuine contentment.

Taking the highest understanding of Self-realization is knowing and acknowledging, through experience, that we are free. It is a process of self-discovery and spiritual awakening that can lead to a deeper understanding of ourselves, of others, and of the nature of reality. We go beyond suffering, pain, and sorrow. Not that the body–mind (the self), does not continue to feel the pain and sorrow of disease, the emotional loss of loved ones, or other psychophysical issues. Those have to be borne, but the Self can be the witness of this process and not suffer. The Buddha's *nirvana* and Vedanta's *moksha* convey the same purpose: going beyond suffering. The difference lies only in the background. Buddhism sees life as entirely transitory, from moment to moment of suffering. Vedanta conveys that the Universal Reality is eternal, and we as individuals are that same reality as soon as we can experience that one-ness as expressed in the Great Statements. *Nirvana*, *moksha*, and *kaivalya* are all concepts of liberation from this world and from the karmic cycle, but with subtle differences. *Nirvana* extinguishes suffering through detachment, *moksha* liberates from rebirth and unites with the divine, and *kaivalya* isolates the pure Self from the mind–body complex.

Self-realization and self-discovery are similar but essentially different. Self-discovery is about uncovering our personalities to know who we are. Self-realization is about discovering our true nature and achieving a state of being in one-ness with the Universal Consciousness.

Mindfulness, Meditation, and Self-Discovery

Mindfulness, meditation, and self-discovery are key practices in the journey towards self-realization. These techniques help us with deep inner exploration and personal growth. Mindfulness focuses on cultivating an acute awareness of the present moment. This non-judgemental attention to thoughts, emotions, and physical sensations gives us a profound understanding of our inner self and brings clarity to personal motivations and desires.

By encouraging introspection, mindfulness helps us to examine our core beliefs, values, and motivations. It ensures effective emotional management. Recognizing and accepting emotions without judgement helps in dealing with challenging feelings and developing emotional resilience, which is essential for personal well-being and empathetic interactions with others.

Meditation is deeply rooted in various cultural and spiritual traditions, and it is a profound tool for self-discovery and gaining clarity on life's purpose. Regular meditation practice ensures inner peace and mental clarity. This tranquillity of the mind is conducive to self-discovery and gaining personal insights. Deep meditation leads to self-realization. It allows us to explore our consciousness, uncover hidden aspects of our personality, and enhance self-awareness, which is essential for an authentic existence. This inward journey allows us to delve into our psyche, uncover intrinsic values, and deepen our understanding of our role in life. Meditation helps to filter distractions, allowing clearer discernment of purpose, and provides clarity in setting goals and intentions that resonate with our purpose. Meditation leads to moments of profound insight, guiding us to understand our true self and purpose.

Self-discovery is exploring our true identity, free from societal pressures. It involves deep questioning about personal values and leads to a more authentic lifestyle. This process encourages uncovering hidden talents and potential, driving personal growth, and contributing to a sense of purpose in life. Understanding our vulnerabilities and desires through self-discovery enhances our empathy for others. Integrating mindfulness, meditation, and self-discovery into life is transformative. These practices not only help us to understand our self but also encourage a compassionate and altruistic approach towards others and contribute significantly to the collective welfare of humanity and the planet.

Transcendental meditation (TM) was developed by Maharishi Mahesh Yogi in 1955, and it became a global method of meditation with the silent repetition of a mantra or a sound. This is also a non-religious movement that promotes inner awareness, helps to relieve stress, and promotes self-development, and committed practitioners are known to attain higher states of consciousness.

Swami Rama's methodology of yoga became immensely popular in the 1970s. His way was through the voluntary control of involuntary states.

Vipassana meditation, as taught by the late S.N. Goenka, is a type of mindfulness meditation that involves observing your thoughts and emotions without judgement. It's based on the teachings of the Buddha and is said to help achieve enlightenment.

All such methods, practiced worldwide, are focused on moving away from materialism and going inward to another level of being and awareness. Whatever helps, and in whichever way, the goal is clear: Self-realization is the purpose of life.

Finding Purpose and Moving On

Perhaps we are still hanging suspended, "a mote of life on a pale blue dot in the midst of the cosmic ocean". Carl Sagan's description evokes a profound paradox. Gazing upon the vast indifference of the cosmos, we can't help but contemplate our own insignificance as mere specks. We are just one of the countless species in a vast, uncaring universe. And yet, we are different; we are unique. In the innermost core of our existence is the universal and unyielding quest for meaning and purpose in life. This transcends cultural, temporal, and situational boundaries and presents a defining characteristic. It continually drives us to seek answers to profound existential questions, shaping our collective and individual journeys. Our purpose, therefore, cannot be mere survival and reproduction to keep the species surviving. Each of us cherishes a powerful desire to reach for the skies and attain our full potential. This isn't a singular, grandiose dream of achievement; it is a vibrant desire for exploration and growth.

We have been drawn to question the reasons behind our existence, making it an essential element of our experience. Historically, seeking purpose has fuelled remarkable feats of exploration, discovery, and creation. It prompts us to probe the cosmos, unravel the mysteries of life, and create art, literature, and philosophy that reflect the depths of our psyche. This quest for a purpose sparks our creativity and paves the way for significant scientific, technological, and artistic achievements. In challenging times, the search for meaning gives us resilience. We find significance in strife; it prompts us to relentlessly search for opportunities and encourages us to emerge stronger from adversity.

This journey of life is not just a journey towards a specific destination but an ongoing, evolving process that continually enriches our understanding of what it means to be human. Philosophers, throughout time, have grappled with this drive. The ancient Greeks spoke of "eudaimonia", a flourishing life lived with purpose. The Vedas prescribed the four stages and goals of human life. Modern psychology echoes this sentiment, highlighting self-actualization—the blossoming of our talents across intellectual, social, moral, and spiritual dimensions. Science demonstrates the cognitive benefits of a curious mind, the well-being fostered by strong social bonds, and the resilience gained from ethical action.

The accumulation of material wealth and satisfying the ego can be seductive traps, and these can add fuel to the relentless chase for "more", which offers only fleeting satisfaction. But a life well lived is above and beyond this limited pursuit. Life is about timeless learning, nurturing meaningful connections, and behaving with compassion and empathy. This is where true fulfilment lies. But this journey does not end here. As we confront our seeming insignificance, ancient wisdom traditions offer a bridge to a deeper reality. "Om" represents interconnectedness, a super-cosmic consciousness with limitless potential. We can cultivate focus, self-awareness, and inner peace—tools that not only propel our personal growth but also allow us to tap into this vaster consciousness, giving a new dimension to purpose and goal. Om is just one example; various spiritual traditions offer practices that can complement our diverse pursuits of a meaningful life. The essence is that reaching our full potential is not just about external achievements but also about inner growth and a sense of connection to something larger than ourselves. As we strive to become the best versions of ourselves, we leave a legacy that inspires generations to come, a testament to the potential that resides within each of us.

In this ongoing journey, cognizant of the ripples of positive impact that we can leave on the world and in the connections that we forge and nurture, we discover the essence of a life well lived. Physically, this life has to end. Death ends this cycle of our life as we live it now within time, space, and causation. The body is rendered as dust unto dust, but with death, we transcend into a hitherto unknown realm.

CHAPTER 8

Death and Beyond: The End or a Beginning?

Death is not extinguishing the light; it is only putting out the lamp because the dawn has come.

—Rabindranath Tagore

Death is universal and inevitable, and therefore it has been a subject of fascination, fear, and philosophical inquiry since time immemorial. It is not just a biological event but a complex phenomenon that touches on the existential, philosophical, spiritual, and societal aspects of human existence. Death is an integral and inseparable part of life, yet it heralds its finality, announcing the cessation of living. It is the one certainty that binds all of us on Earth—the final common and "unifying" denominator among all living beings. But our approach to it and how we try to understand it is very diverse across different cultures, religions, and scientific disciplines.

Even the earliest *Homo sapiens* must have been perplexed by its occurrence and questioned death, just as we do. How and why are we here? How and why does it all end, or does it really end with an absolute finality? What happens to the body? What happens to the person we knew and loved but who is no longer with us? Does something survive? If so, how? Where? And in what form? The answers given by science in different cultures, by all religions and philosophies, by psychologists and regressionists, and finally by those who have had out-of-body experiences (OBE) and near-death experiences (NDE), cover a very wide spectrum of possibilities. By examining death through these various lenses, we wish to explore the widespread concepts of death, beginning from its nature—that is, from the biological processes that culminate in the cessation of life—to the diverse cultural and religious interpretations of what lies beyond, as well as the intersection of contemporary scientific achievements with age-old spiritual insights into immortality. We hope to gain a deeper understanding of its nature, its inevitability, and how it shapes our lives and beliefs. We now want to understand how science and spirituality intersect in the discussion of death and immortality.

Communities and belief systems play a crucial role in shaping our responses to death. Reconciling with mortality often means oscillating between fear, denial, and acceptance. Culturally sensitive palliative care and open conversations about death can help find comfort and meaning in the face of life's ultimate certainty. Supportive communities can offer a sense of belonging and comfort, while philosophies and belief systems provide frameworks for understanding death and what, if anything, lies beyond. Whether through religious practices, spiritual beliefs, or secular philosophies, these systems offer narratives that help us make sense of death and approach it with less fear and more acceptance.

Death: The Scientific Approach

In different fields of modern science—biology, psychology, mathematics, and medicine—much has been written about death. It is often considered in matter-of-fact terms and treated analytically, without regard for the need for compassion or human consideration. The journey to death involves several key processes at both the cellular and systemic levels, including apoptosis, necrosis, and the shutdown of major organ systems.

Lifespan of Cells

Our bodies are marvels of constant renewal, with different cell types regenerating at vastly different rates, maintaining the smooth functioning of this most vital instrument of our existence. Beginning from the outermost layer, skin cells that must face and deal with the mercurial changes of the environment have the shortest lifespans, fourteen to twenty-eight days. Next is the intestinal lining, our processing centre, which sheds and replaces cells every three-to-four days to ensure efficient nutrient absorption. There are some cells that nature has built for a long existence. The heart muscle cells, which support this tireless organ that keeps us alive, can last a lifetime. Liver cells, which take care of the consistent detoxification of this physical system, regenerate every 300–500 days with incredible resilience. Our personal filtration system (renal function, the kidneys) has a unique workforce. Some cells last for decades, while others that line the tubules get replaced with more regularity. The brain has neural stem cells in the memory centre, for example, and they generate new neurons throughout our lives. This process, however, slows down with time, explaining the "normal" loss of memory with age. It is interesting to look at the life span and functions of different cells in our body (see Fig. 8.1). The skeleton, the basic frame of our physical existence, has bones that are continuously remodelled! Specialized cells "break down" old bones and replace them with new ones over some years. Osteoclasts are responsible for the breaking down of old or damaged bone tissue through bone resorption. Our teeth constantly undergo repairs.

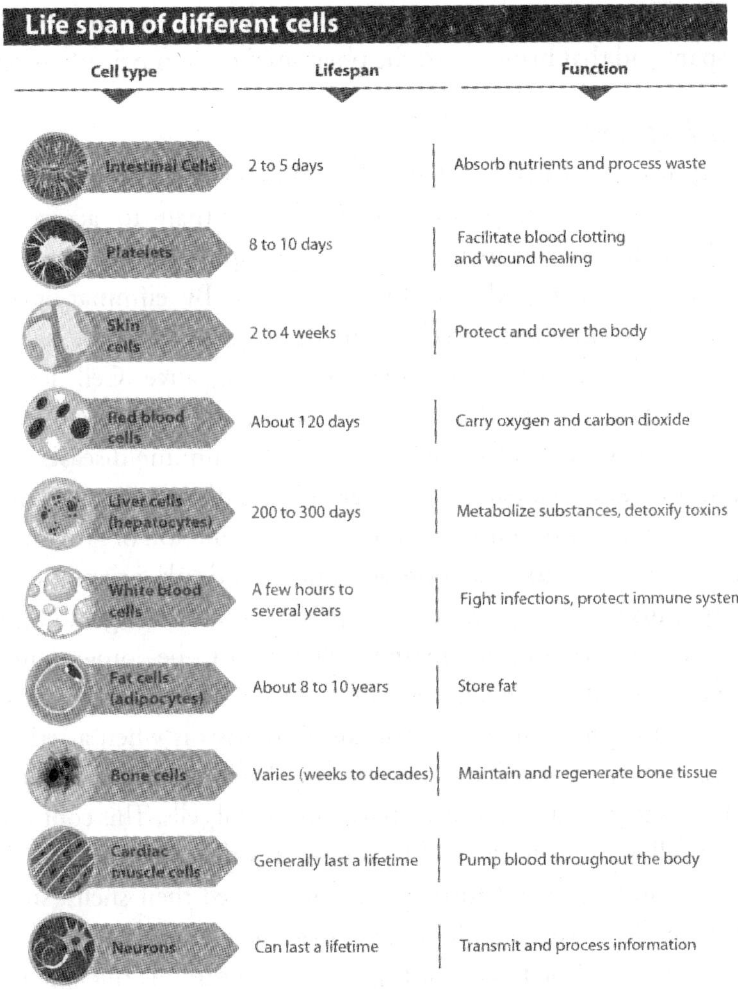

Life span of different cells

Cell type	Lifespan	Function
Intestinal Cells	2 to 5 days	Absorb nutrients and process waste
Platelets	8 to 10 days	Facilitate blood clotting and wound healing
Skin cells	2 to 4 weeks	Protect and cover the body
Red blood cells	About 120 days	Carry oxygen and carbon dioxide
Liver cells (hepatocytes)	200 to 300 days	Metabolize substances, detoxify toxins
White blood cells	A few hours to several years	Fight infections, protect immune system
Fat cells (adipocytes)	About 8 to 10 years	Store fat
Bone cells	Varies (weeks to decades)	Maintain and regenerate bone tissue
Cardiac muscle cells	Generally last a lifetime	Pump blood throughout the body
Neurons	Can last a lifetime	Transmit and process information

Fig. 8.1. The life span of different cells.

This ongoing process of regeneration alongside the inevitable degeneration is the fascinating story of this structure with which we are so inseparably identified that the thought of giving it up or ceasing to exist is unacceptable! Nonetheless, understanding the different cellular lifespans helps us to appreciate the importance of healthy habits that are crucial for optimal cellular renewal. As we age, this natural process slows down, and preventive and corrective measures become even more critical to maintaining a healthy body. The crux of the matter here is the review of the lifespan of cells.

We undoubtedly know and accept that there is an end to a lifespan, and that brings us to the phenomenon of the death of cells.

Death of Cells

The death of cells is inevitable. It is simpler and easier for us to understand and accept the death of cells than to accept the death of our entire organism, this psychophysical system. Cells die, regenerate, and the organism survives. By eliminating old, damaged, or unnecessary cells, cell "death" maintains the balance and integrity of tissues and organs to keep us alive. Cell death is vital when infected or abnormal cells need to be eliminated so that the body is protected from cancer and auto-immune diseases. Cell death is an essential part of life itself. The death of some part of an organism is an important part of the normal function of growth and survival. This is classified as programmed cell death, or apoptosis.

Apoptosis is "programmed cell death", crucial in regulating our organism's health and development. It is said to be "programmed" because it is a deliberate, genetically regulated process that occurs in response to specific signals or stimuli. Apoptosis is when a series of molecular steps in a cell lead to its death. This is one method that the body uses to get rid of non-essential or abnormal cells. This controlled process allows cells to die and be removed without causing harm to the surrounding tissue. Turtles and lobsters shed their shells; snakes slough off their skins; deciduous trees shed their leaves in autumn and conifers steadily shed needles. Especially interesting is the apoptosis administered to maturing egg cells in the ovary. How small a fraction of the original millions survive to puberty, even less to old age, and how very few mature—the rest are weeded out if their condition falls from perfection. And then apoptosis in maturing brains: how could we afford to lose nerve cells? Yet, apoptosis somehow eliminates cells deemed superfluous as the brain matures.

Apoptosis is essential for normal development, maintaining tissue homeostasis, and eliminating damaged or diseased cells. However, when the balance between cell death and cell production is disrupted, it can lead to various health issues, including cancer and autoimmune diseases. The process of apoptosis gets blocked in cancer cells.

Does apoptosis have a paradoxical phenomenon? Does a cell contain the necessary mechanisms to trigger its own demise? Yes. A cell can trigger its demise through a highly regulated and controlled process. Unlike necrosis, which is a form of cell death resulting from acute injury or trauma, apoptosis is an orderly and controlled process that involves a series of biochemical events that dismantle and remove a cell.

In the 1 April 1997 issue of the EMBO Journal, an international team of scientists working in Ann Arbor, Michigan, discovered a gene encoding survival-promoting proteins associated with apoptosis. They proposed a Japanese name for their gene, "harakiri" meaning ritual suicide by the traditional sword. They identified the key mechanisms responsible for apoptotic function. These studies revealed that special "death receptors", are crucial in regulating apoptosis. They are important in recognizing and destroying malfunctioning cells. Failure of this function is a precondition for cancer to develop. However, many cells are programmed to die.

Necrosis is another form of cell death, but unlike apoptosis, it is uncontrolled and often results from external factors such as injury, infection, inadequate blood supply, or exposure to toxins. Necrotic cell death releases cell contents into the surrounding tissue, causing inflammation and damage to the neighbouring cells. Necrosis can cause deterioration of bodily functions and be instrumental in the progression towards biological death, especially in the case of severe injury or disease.

With **autoimmune diseases**, the system mistakenly attacks healthy cells and tissues in the body, causing inflammation and tissue damage. The immune system creates an "enemy within", which does not recognize certain cells or tissues as "self" and reacts against those cells or tissues. As it is, our bodies are in a constant state of controlled demolition and reconstruction. While cell death is a natural and vital process, when it goes rogue, it can lead to the tragic consequences of autoimmune diseases. In these conditions, the defence system that guards against invaders becomes the enemy. It launches a relentless attack on healthy cells and tissues, causing chronic inflammation and, ultimately, tissue death. This conflict leads to the slow erosion of vital organs and tissues.

In rheumatoid arthritis, the joints are damaged as the cartilages get worn down due to an inflammatory process. Lupus, an autoimmune condition, attacks various organs, leaving a trail of fatigue, pain, and dysfunction. Even the body's ability to regulate blood sugar can be compromised in Type 1 diabetes as the immune system mistakenly targets insulin-producing cells. Autoimmune diseases highlight the delicate balance between life and death on a cellular level. While cell death is a necessary part of our body's renewal process, uncontrolled cell destruction leads to a slow and progressive form of cellular death, ultimately impacting entire organs and tissues.

HeLa Cells: An Undying Legacy of Science

Every cell can make life-or-death decisions. Normally, cells in the body grow, divide, and die in a regulated manner as part of the body's natural process of "living". But cancer occurs when this system breaks down and cells begin to grow and divide uncontrollably.

HeLa cells are a paradox in the world of science. Henrietta Lacks was the woman whose remarkable cells, although not truly immortal, continued to hold the potential for future discoveries. In 1951, Henrietta was diagnosed with cervical cancer. Without her consent, tissue samples were taken from her tumour during treatment. While she died as a result of the disease, her cells became one of the most important tools in biomedical research. These cells, named HeLa (after the first two letters of her first and last names), became the first and, for many years, the only human cell line able to reproduce indefinitely in a lab setting. HeLa cells became instrumental in numerous scientific discoveries and medical breakthroughs. They were crucial in developing the polio vaccine. They have been used to understand cancer cell growth, metastasis, and potential therapies. HeLa cells even played a vital role in mapping the human genome, helping us to understand the very blueprint of life. Rebecca Skloot's book

The Immortal Life of Henrietta Lacks raised very pertinent probes into the ethical issues surrounding HeLa cells and their impact on Henrietta's family.

The story of HeLa cells is a very important reminder of the delicate balance between scientific progress and ethics. Henrietta Lacks unknowingly contributed to revolutionizing the course of medical research, but the importance of informed consent and patient rights are crucial. HeLa cells, although not truly immortal, continue to offer a glimpse into the complexities of life, disease, and the dynamic and enduring power of scientific inquiry.

Prolonging Life or Postponing Death

Healthy Ageing and Delaying Death

Ageing is a complex process influenced by genetics, lifestyle, and environment. Gerontology and biotechnology try to explain and understand the mechanisms of ageing with the hope of possibly slowing or even reversing age-related decline or degeneration. Telomere biology, stem cell therapy, and regenerative medicine can extend a healthy lifespan and help combat diseases that we traditionally associate with ageing. Telomeres protect against the deterioration of the ends of chromosomes. Stem cell therapy uses stem cells to repair, replace, or regenerate damaged or diseased tissues and organs within the body. Regenerative medicine, as the name indicates, restores, regenerates, or replaces damaged or diseased tissues and organs in the body. Longevity is not merely about adding years to life but about enhancing the quality of those additional years. Changes in lifestyle, consciousness about suitable physical exercise, appropriate nutrition, and a focus on public health have collectively increased the average life expectancy and changed how we think about ageing and mortality.

Healthy ageing and thus delaying death is quite the focus of modern medicine. Advances in genetics and lifestyle research have helped us better understand the ageing process and have identified ways to enhance the quality of life as we age. These developments not only promise to extend lifespans, but also to ensure that additional years are lived in good health. But however utopian it may sound, while on the one hand "80 is today's 60" as is the popular belief, many in their 40s and 50s succumb to heart attacks, cancer, and stress-related deaths far more than was known before. However cliched it may sound, a healthy, long life is a welcome extension, but death is the inevitable and natural end. Death with dignity is as much a human right and need as is the immense desire to live long and well.

With the possibility of living significantly longer, we might see our place in the world and our contributions to society differently. We could learn to lead a more fulfilling life and have more time to influence the world around us. Extended lifespans can affect personal relationships, with more generations coexisting simultaneously and extended networks of friends and acquaintances. While this can enrich social support systems, it can also pose challenges in maintaining long-term relationships and adapting to social changes over time. The focus on extending lifespans also raises important questions about the quality of life. Ensuring that the added years are marked by good health and well-being is crucial, as is addressing the psychological and emotional aspects of living significantly longer. As science advances, societies must navigate the ethical, social, and economic impacts, ensuring that the benefits of a longer life are accessible to all and contribute to the well-being of individuals and communities.

Twelve hallmarks that drive the process of ageing are proposed by Spanish researchers in an interesting article published in the reputed scientific journal *Cell* (see Fig. 8.2). It is possible to delay the ageing process and promote longevity by modulating these hallmarks through lifestyle and diet modifications, as well as with the help of specialized therapeutic interventions.

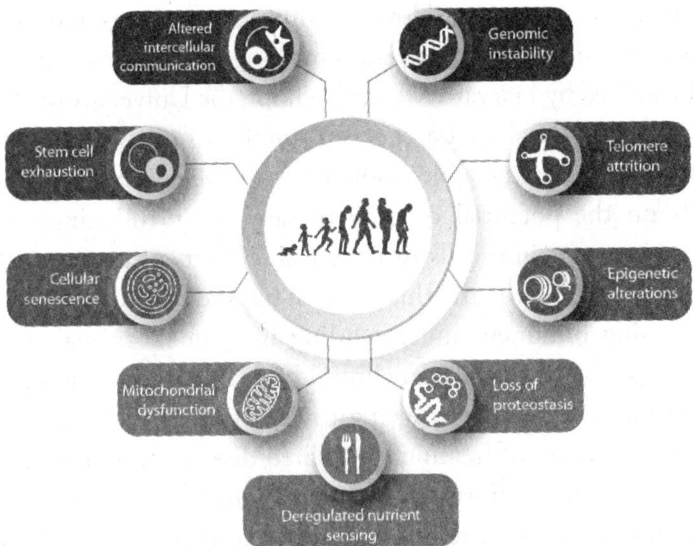

Fig. 8.2. Hallmarks of ageing.

Recent research has opened up new possibilities for extending life and enhancing the quality of our later years. Researchers at the University of California, San Diego, have engineered a synthetic gene oscillator that slows cellular ageing by preventing cells from reaching normal deterioration levels related to ageing. This innovative approach to engineering longevity in cells is a significant leap forward in understanding and potentially extending human lifespan.

Rapamycin has emerged as an interesting candidate in the fight against ageing. It was discovered in the early 1970s by the *Streptomyces hygroscopicus* bacteria from Easter Island. It was meant to be an antifungal drug before its value in addressing ageing was known. Research suggests Rapamycin may mimic calorie restriction, which can slow down certain cellular processes. It also triggers autophagy, a cellular housekeeping process that cleans out damaged proteins and organelles. This cellular rejuvenation is thought to be a key factor in Rapamycin's lifespan-extending effects observed in yeast, worms, and even mice. While these pre-clinical studies are promising, research on the drug's impact on human ageing is still in its early stages. Its potential benefits lie in potentially delaying the onset of age-related diseases like diabetes and neurodegeneration. However, more research is required to endorse its safety and effectiveness for long-term use by humans.

A team led by Harvard Medical School, the University of Maine, and MIT has developed a chemical method for reversing cellular ageing, which could enable whole-body rejuvenation. This method builds on the potential of gene therapy by introducing specific Yamanaka genes into cells to reverse the effects of ageing without uncontrolled cell growth. This research has shown promising results in reversing blindness and extending lifespan in animal models, with preparations for human clinical trials underway. It is a pivotal shift from merely slowing ageing to the possibility of reversing it, offering new ways of treating age-related diseases and injuries.

These scientific efforts make a significant shift in our approach to ageing and longevity, moving from the inevitability of ageing towards the potential for significant lifespan extension and improved quality of life. As we continue to unravel the complexities of biological ageing, the dream of whole-body rejuvenation and the effective treatment of age-related diseases inches closer to reality, promising a future where the boundaries of life are redefined. But despite all endeavours, death of the physiological and biological matter is inevitable, and then we move on to face the idea of what lies beyond death.

Cellular Senescence

Senescence is the age-related decline in an organism's ability to survive and reproduce. It is the age-related death of cells, unlike apoptosis. Cellular senescence is fascinating, where a cell essentially hits the "pause" button on being divided and enters a permanent arrested state. This is a double-edged sword. On the one hand, it can stop damaged cells from replicating. On the other hand, it plays a helpful role in healing wounds. But, as we age, senescent cells accumulate in tissues. Although they no longer divide, they become metabolically active and secrete inflammatory molecules that can damage surrounding tissues. This chronic inflammation, called the senescence-associated secretory phenotype (SASP), contributes to various age-related diseases like diabetes, heart disease, and

Alzheimer's. Senescence itself may not directly cause death, but the accumulation of senescent cells and their harmful SASP causes organ failure, which ultimately leads to death. This research is very important. By developing appropriate drugs, scientists hope to improve health, potentially delay some of the negative effects of ageing, and bring about healthier and potentially longer lifespans.

Regeneration: Nature's Repair Mechanism

While cell death is a natural and necessary process, our bodies aren't passive bystanders. Our psychophysical system has the remarkable ability of regeneration—the power to repair and replace lost cells—keeping tissues functional and our bodies youthful. This capacity varies across different tissues, with the fascinating ability of healing. Skin cells are constantly dividing and differentiating, replacing old, worn-out cells with fresh ones. This is our continuous and consistent barrier against the environment. The liver has an impressive ability to regrow after injury or surgery. Bones and muscles have their own regenerative capacity. Fractures mend through the formation of callus tissues, eventually remodelling into bone. Muscle tissue can repair minor tears and grow in response to exercise. This ability diminishes with age, slower healing times, and eventual irreparable degeneration. The central nervous system, unfortunately, has very little regenerative capacity. Damage to the brain or spinal cord is often permanent. But research in this area continues as scientists explore ways to stimulate neural regeneration.

Stem Cells and Regenerative Biology

Regenerative biology has opened new avenues for cellular renewal. Stem cells have the remarkable ability to transform into various cell types, and this gives phenomenal hope in the treatment of conditions that have been so far untreatable. Researchers are exploring the use of stem cells to regenerate damaged tissues in heart disease, Parkinson's disease, and even spinal cord injuries. After a heart attack, the damaged heart muscle tissue cannot regenerate effectively. Researchers are injecting stem cells directly into the heart muscle to promote the growth of new heart cells and improve their function. In neurological disorders like Parkinson's and Alzheimer's, where specific nerve cells are lost, stem cell therapy can help to replace these lost cells and restore their function. For spinal cord injuries where regenerative capacity is extremely limited, stem cells bring a glimmer of hope. These remarkable cells are found in the bone marrow, in the blood of the umbilical cord, and in other areas, and they have incredible superpowers of self-renewal. This versatility gives them the wonderful ability and advantage to be used for regenerative medicine.

Even in orthopaedics, stem cells can help treat osteoarthritis, where the cartilage deteriorates. Researchers are investigating the use of stem cells to help with cartilage repair and regeneration. This could provide an amazing alternative to joint replacement surgery. However, stem cell therapy has its hurdles. Scientists must minimize the risk of the formation of tumours or rejection after transplants. The source of stem cells is very important. Embryonic stem cells, which are most effective, trigger ethical concerns. Most importantly, delivering stem cells safely and efficiently to the body is crucial.

System cell therapy will revolutionize medicine as long as the related perils are recognized and taken care of. By understanding the process of regeneration, scientists can enhance natural repair, develop biocompatible materials that can support and guide tissue growth, and revolutionize regenerative medicine. Modern regenerative medicine relies on manipulating cellular processes at a much deeper level, often involving stem cells or targeted therapies. In ancient India, *kayakalpa*, or literally "transformation of the body", was an Ayurvedic approach to rejuvenation, longevity, and vitality. While its roots lie in ancient practices and anecdotes, the perspective on cellular rejuvenation that modern science offers today potentially fulfils some aspects of the *kayakalpa* approach (Box 8.1). Understanding how cells repair and regenerate themselves is crucial for understanding the potential of *kayakalpa.* While *kayakalpa* might not be a shortcut to cellular regeneration, it is a bridge between ancient practices of rejuvenation and the fast-evolving field of modern medicine. It is potentially a promising research area for biomedical scientists.

Box 8.1. *Kayakalpa:* The science behind ancient rejuvenation practices

Ayurveda, the foundation of traditional Indian medicine, is replete with *Kayakalpa* practices. These include herbal remedies, dietary modifications, yoga, and detoxification, aiming to achieve a balanced state of mind and body, essential for longevity. Certain herbs and dietary supplements have important antioxidant and anti-inflammatory properties, and they contribute to overall health and well-being, the essence of *Kayakalpa's* philosophy.

Kayakalpa is a specialized therapy in a controlled environment under the supervision of medical experts. The participant undergoing the procedure needs to be isolated in an especially built structure for an extended period, typically ranging from several weeks to months. *Kayakalpa* utilizes *Panchakarma* as a crucial first step to pave the way for its rejuvenating therapies. *Panchakarma* is a step-by-step detoxifying and purifying process. This includes therapeutic oil massage, fomentation, and steaming followed by five therapeutic interventions, which include purgation, medicated anaemias, controlled emesis, nasal oleation, and bloodletting. These processes help to get a person physically and mentally ready for the *Kayakalpa* procedure. The participants are kept under the strict supervision of experts where a controlled diet, yoga, and meditation are prescribed. The treatment known as Rasayana, utilizing various immune modulating and rejuvenating medicines has potential benefits such as enhanced resilience, improved mental clarity, and reduced stress, although there is limited scientific evidence to support these claims.

The Scientific Pursuit of Immortality

The idea of conquering death and achieving immortality is a driving force in human civilization, shaping most myths, religions, and, more recently, science. However, despite significant advancements in medical science and technology, we are in no way close to "conquering death". Research into extending lifespan has made some progress; studies on yeast, worms, and mice show that lifespan can be extended through genetic manipulation or dietary restriction, but none of these can grant immortality!

One radical idea to achieve immortality is the concept of a whole-body transplant, also known as a head transplant. This would involve transplanting the head of one individual onto the body of another. In theory, this could allow a person to continue living after the original body has failed. According to many, the concept of head transplantation is an extraordinary but impossible surgical procedure. But today, the idea is not as far-fetched as it might seem. There have been several successful head transplants performed on animals, and advances in surgical techniques and immunosuppressive drugs have made the prospect of human head transplants seem increasingly feasible. This innovative surgery could offer a life-saving procedure in cases such as those suffering from a terminal disease but whose head and brain are healthy.

Recently, the first cephalosomatic anastomosis (CSA), in simple terms, a head transplant, in a human model was successfully performed, confirming the surgical feasibility of the procedure. CSA has never been attempted before in man, as the transacted spinal cords of the body donor and body recipient could not be "fused" together successfully. Recent advances have made this possible. Hypothetically, one of the primary challenges is reconnecting the severed spinal cord and ensuring that the nerves of the head are properly integrated with the nerves of the donor body. The surgical procedure itself would be incredibly complex and would require a team of highly skilled surgeons and specialists with very advanced technology and extensive pre-operative planning. Assuming this is successfully done, there is a real risk of the immune system rejecting the transplanted head or body, leading to organ failure or death. Immunosuppressive drugs would most likely be necessary, but they come with their own risks and complications.

Undoubtedly, there are huge ethical concerns connected to the concept of head transplantation, with relevant questions about identity, autonomy, and the potential exploitation of vulnerable individuals. For example, who would be eligible to receive such a transplant, and who would provide the donor bodies? While the scientific community continues to explore the boundaries of life and death, conquering death and attaining immortality is, for now, largely in the realm of science fiction. As we continue to push the boundaries of what is scientifically possible, we must consider the ethical implications of our actions. The pursuit of immortality raises basic questions about the nature of life, identity, and what it means to be human. Most importantly, this concept has far-reaching spiritual ramifications, as it would be taken as interfering with the natural order of life. But efforts to increase longevity without compromising the quality of life have to be the main consideration in all such endeavours. Despite these challenges, researchers and medical professionals are continuing to explore the feasibility of head transplants as a last resort for individuals with severe medical conditions.

Transhumanism talks of enhancing human capabilities and traits through technology to transcend current human limitations. It encourages using scientific advancements, particularly in biotechnology, nanotechnology, artificial intelligence, and genetics, to improve human physical, intellectual, and psychological capacities. This may include enhancing physical strength, endurance, and longevity, as well as cognitive abilities such as memory, intelligence, and creativity. According to transhumanists, technology is a way of accelerating human evolution and overcoming biological constraints. With the use of technology, we can shape our evolution and achieve a post-human condition. The significant focus is to extend the human lifespan and achieve immortality through advancements in medical technology, regenerative medicine, and the potential transfer of consciousness to artificial substrates. Some transhumanists even speculate about uploading human consciousness or mind to a digital substrate, such as a computer or virtual reality environment, thereby generating a form of digital immortality. They see technology as a potential way of overcoming death. While not achieving true immortality, it looks into a future where advancements significantly delay or redefine what it means to die.

As a private effort, Bryan Johnson, a tech entrepreneur, has embarked on an ambitious project called "Project Blueprint" to slow his biological ageing. He implements a data-driven regimen that includes caloric restriction, intermittent fasting, supplements, a strict sleep schedule, and frequent testing. This approach has yielded promising results, reducing his biological age to thirty-six compared to his chronological age of forty-six. Project Blueprint verifies Johnson's commitment to pushing the boundaries of longevity, and it could have significant implications for our understanding of ageing and potential future interventions.

The philosophical question of conquering death prompts us to contemplate not just the possibility of escaping mortality but also the potential consequences and the very meaning of existence in a finite world. Perhaps the true conquest of death lies not in avoiding it altogether but in living a life so meaningful and impactful that it transcends the limitations of our physical lifespan. The body–mind complex is matter, and matter that gets created or caused must degenerate and die. Immortality is only conceivable outside of time, space, and causation. The fascination with immortality finds a vibrant expression in the Indian concept of chiranjeevi, which literally translates to "long-lived" or "immortal". This transcends mere physical longevity and signifies instead a life with purpose and service (Box 8.2).

Death Drive

In direct contrast to the quest for immortality, Sigmund Freud introduced a controversial concept in his 1920 work *Beyond the Pleasure Principle* challenging his earlier theories centred around pleasure-seeking instincts. The death drive, or Thanatos, is a psychological state of a human being driven by a desire for self-destruction, aggression, and the return to an inorganic state. The death drive is a force pushing us towards a state of rest and stillness, with the desire to return to the state before life began. Individuals unconsciously recreate negative experiences and even the existence of civilization itself. This is a highly controversial concept in psychoanalysis. Some find it too dark and pessimistic a view of human nature. However, it reflects on the complexities of human behaviour, the coexistence of creation and destruction, and the forces that drive us beyond just seeking pleasure.

Freud initially introduced the concept of the death drive in contrast to the life instinct, or Eros, which prompts us towards

Box 8.2. The Concept of Chiranjeevi

The fascination with immortality finds a vibrant expression in the Indian concept of Chiranjeevi, which literally translates to "long-lived" or "immortal". This transcends mere physical longevity, and signifies instead a life with purpose and service. The most common understanding of Chiranjeevi revolves around the Ashta Chiranjeevi, the eight "immortals" connected with the epic Mahabharata—Ashwatthama, Bali, Vyasa, Hanuman, Vibhishana, Kripacharya, and Parashurama are the seven death-defeating beings. They are counted along with Sage Markandeya, the eighth Chiranjeevi. Each embodies virtues such as strength, wisdom, and unwavering adherence to righteousness. It is believed that these figures aren't passive observers but actively uphold the cosmic order and guide those in need.

However, the Mahabharata warrior Ashwathama, blessed as he was with immortality, became a curse when he violated all ethics of morality in murdering the five Pandava sons in their sleep, at the end of the great Mahabharata war. Sri Krishna cursed him—that although he was immortal, he would roam the universe plagued by illness and guilt, known more for his heinous crime than for his erstwhile valour.

The devoted monkey god Hanuman is blessed to remain immortal as long as the name of Lord Rama echoes in the hearts of human beings. Hanuman exemplifies boundless devotion. Vibhishana, also a key figure in the epic Ramayana and extolled in the Mahabharata, displays the courage to stand by righteousness even against family.

Besides these characters from the epics, the concept of Chiranjeevi represents individuals who achieve exceptional longevity through healthy living or those whose stellar contributions to society leave a lasting impact. Philanthropists, environmental activists, and social reformers are the modern-day Chiranjeevis, their legacies outlasting their physical presence. Chiranjeevi is a reminder that true immortality is not in escaping physical death but in leading a life that leaves an enduring mark. The Chiranjeevi ideal inspires us to live with purpose, uphold righteousness, and try to make the world a better place.

self-preservation, reproduction, and seeking pleasure. According to Freud, the death drive operates alongside the life instinct and is responsible for destructive behaviours and tendencies that contradict the instinct for self-preservation. Freud saw these two drives locked in an eternal struggle within the human psyche. Whether it is a literal biological force or a metaphor for the darker aspects of the human psyche, it is an intriguing concept.

Nature of Death

What Is the Process of Death?

Biological death occurs when the complex systems that keep an organism alive stop functioning. In humans, this can result from the failure of critical organs, the cessation of cellular activities, or catastrophic injury. The countless cells in a body are programmed to play specific roles within certain lifespans. Programmed cell death is a natural part of the body's maintenance and development. However, when cells can no longer replicate or repair the damage due to age or disease, the system itself begins to fail, leading to the organism's death.

Medical science distinguishes between clinical death, biological death, and brain death. Clinical death is when the heart stops beating, breathing ceases, and there is no brain function. Clinically, death is the irreversible cessation of these vital functions. Today, cardiopulmonary resuscitation (CPR) techniques can sometimes reverse clinical death if successfully administered at a crucial moment. Biological death, or irreversible death, occurs when oxygen deprivation and other factors cause irreparable damage to the body's cells and tissues, leading to the irreversible cessation of all vital functions. Brain death is the complete and irreversible cessation of brain activity. It is a key criterion for death in many legal contexts because it signifies the absence of consciousness in the body and the cessation of all brain functions, making recovery impossible.

Despite the advancements in medical science that delay death, it is an inescapable, inevitable, and universal occurrence. Death is the great equalizer, sparing no one regardless of wealth, status, or achievements. Its universality prompts us to reflect on the meaning of life and our mortality. It forces us to confront our finiteness and can motivate us to live our lives more fully, with greater appreciation for the moments we have. Accepting the inevitability and nature of death encourages a compassionate approach to end-of-life care, emphasizing the quality of life for those approaching death. It brings attention to the importance of palliative care and the ethical considerations surrounding life support and euthanasia.

Therefore, while death is an inevitable part of life's cycle, our understanding of it—both scientifically and philosophically—shapes how we live our lives, how we approach the end of life, and how we perceive the concept of mortality itself. This gives us a deeper appreciation of life and of the bonds that connect us to each other.

Shutdown of Major Organ Systems

The shutdown of major organ systems is the final stage of biological death. It can occur gradually due to chronic illness or ageing, or it can happen rapidly in cases of acute medical conditions such as a heart attack or severe trauma. The failure of one system triggers a domino effect, compromising other systems and ultimately resulting in the cessation of all vital functions.

- Cardiovascular System Failure: The heart's inability to pump blood causes decreased oxygen and delivery of nutrients to tissues, accumulating waste products and causing multi-organ failure.
- Respiratory System Failure: Inadequate respiratory function in the lungs results in insufficient oxygen levels and a sharp build-up of carbon dioxide in the blood, impairing the functioning of other organs.
- Renal System Failure: The kidneys fail to filter blood and remove waste products. This causes toxicity in the body, affecting other organs and systems.
- Neurological System Failure: Severe brain injury or degeneration disrupts the regulatory mechanisms for vital functions and leads to the cessation of life-sustaining activities.

These and other related processes cause biological death. Understanding these processes is crucial in medical science, not only to treat and manage diseases but also to make informed decisions about end-of-life care.

The Stages of Dying

There are several known and subtle stages of dying. But they usually begin with changes in patterns of sleep, decreased intake of food and liquids, withdrawal from people and surroundings, and far less movement. The physiological changes are the body's natural ways of preparing to shut down. Vital signs may begin to fluctuate, and the person may have periods of confusion or disorientation. In the active stage of dying, the person is less responsive, and there is difficulty swallowing, irregular breathing, changes in skin colour, and a drop in body temperature. There is a marked decrease in the urinary output, and the body's ability to regulate temperature also decreases. The digestive system slows down, with a lack of interest in food and drink.

Causes of Death

Where death is concerned, two questions arise in the mind: how and why? Most ironically, death is not seen as a natural consequence of birth, despite the fact that it is built into "life". Aside from the "natural" cause of old age and consequent physical degeneration, catastrophes or disasters perpetrated by nature, the causes are either disease, accident, failure of timely medical aid, or deliberate acts like murder or suicide.

Over the centuries and decades, the reasons why people die have changed or shifted significantly. Infectious and other virulent diseases such as smallpox, malaria, cholera, and tuberculosis were among the most commonly known killer diseases. But cures or preventions were discovered in medical science that changed the outcomes. The new waves brought higher incidences of heart diseases, diabetes, respiratory or lung-related diseases, liver and kidney-related afflictions, and cancer. Not that these are "new" in occurrence; in earlier days, people died "suddenly" and that was not necessarily classified as an attack, failure, or infarction of the heart. Similarly, any organ failure was not necessarily even detected in good time.

Some "new" conditions have also developed due to changed choices in lifestyle. Increased consumption or use of tobacco, alcohol, and fatty-rich foods and the explosion of the "junk" and "fast-food" industries have certainly made a difference to the reactions and responses of the human body. Air, water, and food pollution take their toll.

According to the World Health Organization (WHO), non-communicable diseases (NCDs) are said to be the leading causes of death worldwide today, accounting for over 70 per cent of all mortalities. Heart conditions are the top killers. Among cancers, lung cancer stands out with close to 2 million deaths annually. While chronic diseases dominate today, mortality due to infectious and viral diseases skyrocketed with the pandemic of 2020–2022, and it continues to prevail.

Interestingly, in a rich and scientifically advanced country like the USA, the third leading cause of death is reported as "medical errors" or "unintentional injury". This paradoxical situation highlights the need for modern medicine to transcend its current limitations and embrace a more integrative health approach informed by meta-science.

Apart from environmental and constitutional causes, socioeconomic situations have a great impact. In the wealthier nations, chronic diseases like heart disease, cancer, and strokes are more prevalent, triggered by lifestyle, diet, exercise or lack thereof. In poorer countries, infectious diseases like malaria, tuberculosis, childbirth, pulmonary or respiratory conditions, influenza, and diarrhoea-related conditions remain major causes of mortality. These are triggered by inadequate sanitation, limited access to clean water, and, at times, lack of proper healthcare infrastructure.

Declaration of Death

The medical and legal criteria for declaring or announcing death have evolved with advances in medical technology and our understanding of biological processes. While specifics can vary, there are generally two widely recognized criteria for declaring death: brain death and circulatory death. These criteria reflect the dual understanding of death as either the irreversible cessation of all brain functions or the irreversible cessation of circulatory and respiratory functions. Legally, the declaration of death is usually done by a qualified healthcare professional. The precise moment of death has significant legal implications for issues such as organ donation, cessation of life support, inheritance, and the issuance of death certificates. Therefore, the process is regulated and involves strict protocols to ensure ethical integrity and accuracy.

The criteria for declaring death, especially brain death, have ethical and practical implications, particularly for organ donation and end-of-life care. The determination of death often requires a multidisciplinary approach, which will include healthcare providers, ethicists, and legal experts to balance the medical evidence, ethical considerations, cultural beliefs, and legal requirements. As medical science continues to advance and societal values change, the criteria and procedures for declaring death are also being continually debated and refined. It is important for all persons connected to be informed about these criteria and engage in open discussions about preferences and beliefs regarding end-of-life care and death.

The Finality and Universality of Death

Death is the ultimate equalizer, a fundamental aspect of life that unites all living beings in a shared destiny. Despite the vast differences in how we live our lives, regardless of status, wealth, power, culture, and belief systems, death is the one inevitable outcome for all. The inevitability of death is central to the meaning of life, and its universality is the profound truth about existence: we all face the same end. Death transcends all socioeconomic barriers. Up to a point, wealth and power can provide access to better healthcare and potentially extend life or improve the quality of life, but death still prevails. This should make us realize the importance of the intangible aspects of life, such as relationships, kindness, and personal fulfilment, over material wealth and external achievements. Socrates to Heidegger pondered death's significance and argued that awareness of mortality can be a motivating force to live more authentically, ethically, and meaningfully.

Throughout history and among all cultures around the world, there are beliefs, rituals, and philosophies regarding the end of life and, in many instances, the possibility of an afterlife or rebirth. These provide comfort, offer explanations, and establish moral or ethical guidelines. Accepting the inevitability of death is what many philanthropists or thinkers propose in the legacy that they leave behind. Whether it is material contributions to society, the impacts of their lives on the lives of others, or through their descendants. The desire to be remembered, to know that we have made a difference, is a natural face-off to the finality of death, offering a way to transcend the temporal boundaries of finite existence.

Death puts into perspective the transient nature of life. It prompts us to reflect on what truly matters, encouraging us to live with awareness, nurture relationships, and focus on the legacies that we wish to create. In facing the universality of death, we are reminded of its interconnectedness and the shared journey through life. We have a greater appreciation for life's experiences and the commonalities that bind us.

Preparing for Death

Ramana Maharishi very graphically described his experiment at the age of sixteen, "I seldom had any sickness and on that day there was nothing wrong with my health, but a sudden, violent fear of death overtook me. . . . I just felt, 'I am going to die,' and began thinking what to do about it. It did not occur to me to consult a doctor or my elders or friends. I felt that I had to solve the problem myself, then and there. The shock of the fear of death drove my mind inwards and I said to myself mentally, without actually framing the words: 'Now death has come; what does it mean? What is it that is dying? This body dies.' And I at once dramatized the occurrence of death. I lay with my limbs stretched out stiff as though rigor mortis had set in and imitated a corpse to give greater reality to the enquiry. I held my breath and kept my lips tightly closed so that no sound could escape, so that neither the word 'I' or any other word could be uttered, 'Well then,' I said to myself, 'this body is dead. It will be carried stiff to the burning ground and there burnt and reduced to ashes. But with the death of this body am I dead? Is the body 'I'? It is silent and inert but I feel the full force of my personality and even the voice of the 'I' within me, apart from it. So I am Spirit transcending the body. The body dies but the Spirit that transcends it cannot be touched by death. This means I am the deathless Spirit." During this period of intense introspection, Ramana Maharshi experienced what he described as a spontaneous and profound spiritual awakening. He realized that the true "I" transcends the physical body and the limitations of individual identity. He recognized that the essence of his being was eternal and unaffected by physical birth or death.

This realization led to a profound shift in consciousness for him. The realized state of inner peace, clarity, and self-realization remained with him for the rest of his life. Following this awakening, he left his family home and travelled to the sacred mountain of Arunachala in Tiruvannamalai, where he lived in solitude and continued his spiritual practice. Ramana Maharshi's experiment with death was a catalyst for his spiritual awakening and the development of his teachings on self-inquiry and non-duality. He shared his insights with seekers from all walks of life, guiding them on the path to self-realization and liberation from the cycle of birth and death. His teachings continue to inspire spiritual seekers around the world to this day.

Cultural and Religious Perspectives

Many cultures have traditions that grant the right to die to those who are mentally prepared for death when it is time. In Jainism, *santhara* is the practice of giving up food when death is imminent. It is permitted for ascetics and householders. The concepts of *karma* and reincarnation are two steady pillars of Vedanta. The belief that our actions in this life determine the circumstances of future lives suggests that exercising the right to die or taking one's life amounts to interfering with the natural course of life and death and can disrupt the karmic cycle. But even in this system, there are instances of "withdrawing" consciously by giving up food and water.

The practice of *prayopavesha* is a voluntary fast to attain spiritual liberation before death. Sages and ascetics were known to adopt this practice, to give up their bodies with dignity. The ancient choice of householders to move to the forests was motivated by the need to take the first step towards final liberation from the body. Sanjeevan samadhi is a yogic meditative practice of "remaining absorbed in meditation till the end of the mortal body". Swami Vivekananda gave up his body in this manner after having decided the appropriate date and time according to the almanac.

Nirvikalpa samadhi is the highest meditative level, when the person attains an elevated level of losing complete body consciousness. In such a meditative state, the choice of physical termination is entirely voluntary. Whether such people stay in that state to give up the body or "bring" themselves down to a world-conscious level depends on their desire and ability to "come back". Pramahansa Yogananda, another realized monk and yogi, explained that sleep is a partial life-force control, while death is the complete withdrawal of life-force from the body. In Hindu culture, the god of death is called Yama. Ecstasy is "*Yama* by your own will". The highly advanced yogi learns to "die" (or leave the body) at will in meditation by consciously stopping breath and heartbeat, after which one can freely return to the physical frame.

Honouring Death

There have been some very remarkable and unbelievable exemplars of people who have stoically faced and accepted death and used their approach to sharing the death experience. The neurosurgeon Paul Kalanithi wrote his "ultimate moving life-and-death story" in *When Breath Becomes Air*, just as he succumbed to lung cancer at the age of thirty-seven. Reflecting on what makes life worth living in the face of death was his very brave and memorable contribution.

Randy Pausch, a computer science lecturer, said in the introduction to his *Last Lecture*, "While for the most part, I'm in terrific physical shape, I have ten tumors in my liver and I have only a few months to live." He was forty-seven years old. His goal now was to teach his children what he would have taught them over the next twenty years! They were too small to absorb all that he would want to share, and that led him to deliver the "last lecture" at Carnegie Mellon University.

Dr Elisabeth Kübler-Ross, a Swiss-American psychiatrist, was a pioneer in the field of death and dying studies. She talked of the five stages of grief in her book *On Death and Dying*. Her research was based on her interactions with terminally ill patients and their families, as well as her own observations and experiences as a physician working in hospice care.

Kübler-Ross identified five stages that we would commonly experience when facing death: the process of preparing for it and facing death, confronting our own mortality, or coping with the impending death of a loved one. The initial stage is disbelief or denial of the reality of impending death. We may refuse to accept impending death or its significance while trying to cope with the shock and fear. As the reality of death sets in, we may experience anger and frustration directed towards ourselves, others, or even a higher power. This anger may come from feelings of injustice, abandonment, or powerlessness in the face of death. Next, we may even attempt to bargain, negotiate, or make deals to postpone death. We may plead for more time, promise to change our behaviour, or seek alternative treatments or interventions. As death becomes more apparent, we may experience immense sadness, grief, and despair. At this stage, we may come to terms with the loss and face or confront the emotional pain associated with an impending death. The final stage is the acceptance of the reality of death. For some of us, this may generate a sense of peace, closure, or readiness to let go. This acceptance does not necessarily mean we feel okay with death, but rather that we come to terms with its inevitability and find a sense of calm in the midst of uncertainty.

Kübler-Ross's work revolutionized attitudes towards death and dying, directing greater awareness and compassion for terminally ill patients and their families. Her insights into the emotional and psychological aspects of death have had a very great impact on healthcare, counselling, support services, and how society approaches end-of-life care.

Accepting Death

The mere word "death" evokes several emotions: fear, sadness, and even morbid curiosity. Several scientists and thinkers have approached the subject from various perspectives, including psychological, philosophical, and scientific angles. Their insights have contributed to a broader understanding of mortality and its impact on our existence. Across cultures and throughout history, philosophies and practices have emerged that challenge this fear and offer a path of not just accepting death but also celebrating it. Some people from several indigenous cultures do not fear death or have any dread of facing it. It is seen as a continuation of existence. It is even an occasion for rejoicing and celebrating in some cultures because death means moving ahead to higher realms. Primitive people perhaps feared life and its uncertainties more than the apparent peace of death. Ernest Becker talked of the denial of death and pointed out that the fear of death is a fundamental motivator for our behaviour. He believed that individuals create symbolic systems, such as religions and cultural beliefs, to manage the anxiety that surrounds mortality. Albert Camus went to the extent of talking about the absurdity of life and the inevitability of death. He questioned whether life itself is worth living, given that death is certain. Sigmund Freud believed that human beings fundamentally fear death, and this fear is deeply rooted in the unconscious mind. According to him, we develop defence mechanisms to cope with this fear by either denying it, repressing it, or trying to displace it and thus protect our conscious mind from having to face it, accept it, or deal with the overwhelming fear of death. Otherwise, it would be impossible for us to function in daily life if we had this constant existential dread in our conscious mind. Freud saw religion and cultural beliefs as means to provide comfort and meaning in the face of death. Carl Jung had a more holistic and spiritual approach to the concept of death. He believed that anxiety could be a transformative experience that could lead to our personal growth and spiritual development. According to Jung's psychology, confronting death and accepting its inevitability can deepen our understanding of the self and our place in the universe. Jung also explored the symbolism of death in dreams, myths, and religious traditions, viewing it as a symbol of psychological rebirth and renewal.

Acceptance of death is a sign of psychological and spiritual maturity. Many cultures and philosophical traditions emphasize the importance of accepting death as a natural way of life. This acceptance ensures greater peace of mind and a focus on living fully in the present. Acceptance might come through personal growth, experiences of loss, or deep contemplation of life's transient nature. In societies where death is viewed as a transition rather than an end, rituals and communal practices help to normalize death, making acceptance a communal as well as a personal journey.

Fear of death (thanatophobia) is a complex subject and the most instinctive reaction when we try to face or accept death. This fear is dread of the unknown, of ceasing to exist, or anxiety about the potential suffering involved in dying.

According to science, our primal fear response is triggered by the unknown, and death's finality certainly qualifies as what happens thereafter is entirely unknown. Philosophers like Epicurus believed this fear stemmed from our attachment to life's pleasures, fearing their absence. On the other hand, existentialists like Sartre point to the fear of losing control and the void that death presents. From a spiritual perspective, some religions focus on the fear of separation from loved ones in the earthly realm, while others emphasize judgement or consequences in the afterlife. It is also the fear of losing our very existence, which is the core of our being. Ultimately, the fear of death includes anxieties about physical suffering, emotional separation, the finality of being, and the fear of the unknown.

Seeing absolute reality in identity and personality prompts the exclusive "I-ness" that we cherish throughout our lives. The thought of its termination is immensely daunting. Often, this fear drives us towards religions, elaborate rituals, and beliefs about the afterlife, as they seem to help us face the fear and provide a structured way to cope with loss. At the individual level, the fear of death can lead to such behaviours as avoiding risks with a focus on health and longevity or even acknowledging existential crises that prompt a deeper search for meaning. While self-preservation and self-care are natural and instinctive, the dread of death and the inability to face it as an eventuality do create extreme emotional crises and trigger emotional reactions such as depression and despondency. All said and done, intellectual understanding alone cannot take away the very deep-seated fear of the unknown, which possibly resides in our subconscious and even the unconscious mind. Here, meticulous self-discipline becomes paramount. With regular meditation and mindfulness, we can quieten the incessant chatter of the mind, face anxieties, and develop a sense of inner peace. These practices refine our awareness, allow us to connect with the present moment, and detach from the anxieties associated with the impermanence of life. This path is undeniably arduous. There will always be moments of doubt, setbacks, and the urge to give up. However, with each dedicated step, the fear of death decreases, and we can replace it with an acceptance of life's impermanence.

Denial of death is another common response, where the reality of death is minimized or ignored. At the level of society, we see denial in the glorification of youth, the taboo on discussions of death, and the constant pursuit of medical interventions to indefinitely prolong life. We are in denial when we avoid or circumvent thoughts about mortality, when we neglect to make end-of-life plans, or when we deliberately disregard the risks of death. While denial can temporarily shield us from the anxiety associated with death, it may also prevent us from fully appreciating life or addressing important emotional and practical considerations related to dying. By living a life brimming with purpose and meaning, we can create a legacy that transcends death.

Palliative Care Approaches

Fear of death often stems from pain and suffering. Palliative care provides comfort and support for the patient and family emotionally, spiritually, and socially, with dignity. Medications, including opioids, help to effectively manage pain without causing undue sedation, allowing patients to be as comfortable and alert as they wish to be.

Palliative care includes supplemental oxygen or medications to ease breathing and address anxiety related to shortness of breath. Psychological support helps the patient and their family cope with fear, grief, and other complex emotions. It is important to address end-of-life wishes, including advanced directives and spiritual or religious rituals that are important to the patient and their family. Ensuring a peaceful and comfortable environment is the best that anyone can do.

Understanding the physiological stages of dying and ensuring palliative care can impact the experience of the patient and their loved ones and provide a dignified and peaceful transition. The Vedic science of Yoga makes clear that the fear of death is one of a human being's main subconscious motivators. Overcoming it is a means to gaining increased freedom of action in this life. Certainly, those who show no fear when faced with the possibility often reap high rewards. The ultimate heroism is often equated with devotion to duty in the face of mortal danger, and crosses and hearts are showered on those in whom it manifests.

The Right to Die: Euthanasia

Throughout our lives, we have heard this clarion call, "the right to live", being broadcast loud and clear, especially from the oppressed and the downtrodden, from those who have indefatigably fought for it. But the "right to die" with dignity and by choice has been a relatively more recent uprising and is finding expression and even acceptance in many ways. Why do we say this is recent? Because, over centuries, the ideas of *karma*, destiny, luck, free will, and fate had a bigger significance than they do today. Many of us argue that just as we have the autonomy to chart the course of our lives, we should also have the right to determine how and when it ends. As long as the mind can discern rationally, and since the body is our property, we should have the right to decide to give it up if living in it becomes an unbearable agony.

Medically or scientifically, the right to die is often connected with euthanasia, the act of intentionally ending a life to relieve suffering. Euthanasia can be voluntary when a patient requests it or involuntary without consent, which is almost always the case when the patient or person concerned is unable or incompetent to take the call. Physician-assisted suicide (PAS) and voluntary assisted dying (VAD) are specific forms where a doctor provides the means for a patient to self-administer lethal medication. Euthanasia is a complex and extremely debatable subject, as there are several implications—humanitarian, ethical, religious, and legal.

The Legal Perspective

The legal status of euthanasia varies very widely. Countries such as Belgium, the Netherlands, Canada, Australia, and some US states have legalized some form of euthanasia, or PAS and VAD, under very strict guidelines. These typically require a competent patient's informed consent if he or she is terminally ill, undergoing unbearable suffering, and is very aware of a justifiable reason to make the call. MAID (medical assistance in dying) is an accepted form of euthanasia in Canada. Although it has been strongly criticized as lacking in safeguards, it is quite an accepted practice. The Swiss company Dignitas offers the option "Live with dignity, die with dignity". Most countries, however, prohibit euthanasia because it is seen as assisted suicide or even murder. Supporters of euthanasia emphasize a person's right to avoid prolonged suffering and to choose to die with dignity. Critics remain concerned about potential abuse and emphasize the sanctity of life. Religious objections focus on the belief that life is divinely given and shouldn't be terminated, as that would be going against the will of God.

Commercializing Healthcare

The protagonists of euthanasia, supporting the right-to-die point to the commercialization of healthcare. People are kept alive on ventilators or in long-term care facilities, which can be financially beneficial for hospitals, nursing homes, and pharmaceutical companies, prolonging the suffering both physically and financially for caregivers and family, besides the patients themselves. This nexus between healthcare providers, insurance companies, and the pharmaceutical industry can create situations where patients are denied a dignified death due to obvious vested financial interests. There are moral and ethical issues on the other side. The patient's well-being must always be prioritized under all circumstances.

A Balance Is Needed

The right to die is a very sensitive issue. It is extremely crucial to balance the right to live with the right to die. The right to die is a complex and evolving concept. While the right to life is fundamental, acknowledging the right to die with dignity under certain conditions must include patient autonomy and a humane approach to end-of-life scenarios. Terminally ill patients have to be given access to pain management and dignified end-of-life care. But most importantly, when faced with the inevitable, their right to choose must be respected within ethical and legal structures. The patient's freedom to impose a "do not resuscitate (DNR)" directive is a "living will" and it is a person's right that it should be honoured, and it is in all countries where such an opportunity exists both medically and legally.

Beyond Death

Across cultures and throughout history, philosophies and practices have emerged that challenge the fear of death and offer a path towards not just accepting death but also celebrating it. There are several cultural and religious points of view on death and beyond, with many beliefs, rituals, and teachings reflecting our diverse efforts to understand and cope with mortality. They not only influence how we view and experience death but also our attitudes towards the afterlife, ethics, and the meaning of life.

Cyclical Nature of Life and Death

Indigenous cultures view death as an integral part of life and nature. These cultures focus on the ancestors' ongoing presence and influence in the lives of the living, seeing death as a transition rather than an end. It is a journey to join the ancestors. Several indigenous cultures believe in the cyclical nature of life and death, seeing both birth and death not as linear endpoints but as integral parts of a continuous cycle that includes rebirth and the ongoing presence of ancestors. These beliefs influence the diverse practices and rituals and the way that individuals relate to the environment and each other. Life and death are seen as part of the natural world's cycles, emphasizing the interconnectedness of all living things and the natural transitions between different states of being.

The concept of rebirth, which is intrinsic to Eastern philosophies, is the next phase, with death as a transition to another form of existence rather than an end. This may involve beliefs in reincarnation, the spiritual presence of the deceased, or the return of the spirit to sacred landscapes. Such beliefs focus on the continuity of life, and death marks a change in existence rather than a final closure. Interestingly, there are known cases of infants having recollections of their past births in the initial few years of their lives. What happens when a toddler is haunted by memories that aren't hers or his? Regressionists like Brian Weiss and Sylvia Browne have narrated many cases of human minds carrying forward very deep impressions from previous lives that affect their present lives in several ways. This continuity, the thread that does not break, is Universal Consciousness, which is represented by Om in this book.

Ancestors hold a significant place in indigenous cultures, as they are believed to be ever active in the community's life as spiritual guides, protectors, and sources of wisdom. The relationship with ancestors is kept alive through rituals, offerings, and communication, highlighting the enduring bond between the living and the dead. Rituals surrounding death and ancestral worship are central to maintaining the balance between the spiritual and physical worlds. These practices can include ceremonies to honour the dead, festivals to celebrate ancestral spirits, and rites of passage that signify transitions in life's cycles. Through these rituals, communities acknowledge the contributions of their ancestors, seek their guidance, and ensure their memory and legacy continue. The indigenous perspectives on the cyclical nature of life and death, along with the reverence for ancestors, can give valuable insights into sustainability, community, and the understanding of existence.

Ancestral worship is also a means of cultural transmission, connecting current and future generations with their heritage, traditions, and the accumulated wisdom of the past. This continuity is crucial for maintaining a sense of identity, belonging, and resilience within the community. These worldviews challenge the more linear and materialistic approaches prevalent in many modern societies, proposing a model of living that emphasizes harmony with the natural world, intergenerational responsibility, and an understanding of life's cycles.

We live in a world facing severe environmental and social challenges. Indigenous beliefs about death and life cycles prompt us to re-evaluate our relationship with nature, the community, and the legacy that we are leaving for future generations. By recognizing the cyclical interplay of life and death, we can have a deeper appreciation for the continuity of existence and the importance of living in a way that respects those who have come before and those who will follow.

Understanding and explaining death and the quest for immortality intersect at the crossroads of science and spirituality. Scientific advancements strive to understand the biological mechanisms of ageing and death, seeking ways to extend life and improve the quality of our years. Spirituality provides perspectives on the meaning and transcendence of death, often focusing on the soul's journey and the possibilities of an existence beyond the physical life. Together, they enrich our understanding of death and our longing for immortality. Several epics, including the Ramayana and the Mahabharata, carry stories about rebirth.

Philosophical Perspectives

Philosophical perspectives on death and the afterlife explore existential questions about the meaning of life, consciousness, and what, if anything, exists or continues beyond physical death. Socrates and Plato discussed the immortality of the soul and the importance of living a virtuous life in preparation for what comes after. Existentialists Jean-Paul Sartre and Albert Camus questioned the traditional notions of the afterlife, focusing instead on the significance of individual choice and the inherent absurdity of seeking meaning in an indifferent universe.

The ancient Egyptians believed in a complex afterlife where the dead embarked on a long journey. They practised elaborate burial rituals to ensure safe passage and a favourable judgement by Osiris, the god of the afterlife. The Book of the Dead, with spells and instructions for navigating the afterlife, was a guide for the deceased.

The idea of conquering death forces us to confront the very essence of life and existence. The definition of death itself becomes a philosophical minefield. Is it the absolute cessation of being or a more nuanced transition? Advances in life support and cryonics further blur the lines, leaving us to question what truly constitutes the end. The finiteness of life has meaning and purpose, with short-term and long-term goals. Would an immortal existence, without the inevitability of death, have the same value? Would life hold the same weight and the same urgency if we did not have the spectre of death looming over us? What meaning and purpose could there be in an endless existence? The knowledge of our mortality motivates us to make the most of our limited time, and we push ourselves to achieve and experience more fully, making the most of the time and opportunities we have. Hypothetically, even if conquering death in the biological sense was possible, an immortal population could lead to resource depletion and societal collapse. An endless lifespan would lead to stagnation and a loss of motivation in the face of unending time. And perhaps, most fundamentally, how would our sense of identity and purpose hold in a world where death is no longer the ultimate horizon?

The desire to "conquer" death can be interpreted as a yearning for transcendence, a bid to escape the confines of our physical bodies. And that kind of immortality is conveyed in several religions and philosophies. Some philosophical schools, like Stoicism, advocate for accepting death as a natural part of the life cycle, urging us to focus on living virtuously in the present moment and finding meaning and purpose within the inescapable reality of our mortality. Existentialism takes a different approach, suggesting that death defines the human experience. We are thrust into existence with no inherent purpose, and it's up to each of us to create meaning in a world ultimately defined by finitude.

Hinduism introduced the concept of the endless cycle of birth, death, and rebirth, driven by the results of our past and present actions. The ultimate goal is to achieve *moksha*, liberation from this cycle, which is attained through living a life of righteousness and spiritual practice. The world we live in is driven by greed, suspicion, avarice, intolerance, and all those negative emotions that human beings trigger and which fester as wounds in the psyche work overtime. The fear of loss, the fear of suffering, and finally the fear of death always hold us in their tenacious grip.

One question persists: why is it like this? The answer, ironically, is very simple. As long as we live in this world as separate entities, identified entirely with this body–mind complex, which we all do, the physical pain of disease, the emotional pain of relationships, the intellectual pain of accepting the limitations of the system, and the sense of loss will persist and prevail. If at all there is anything we can do at this level, it is to reach out and share and care, empathize and understand, hold a hand, and wipe a tear. At least from this micro-level personal, or zoom-lens, point of view, we can change the lens and look at the world from a wide-angle point of view.

There is another perspective that changes the entire scenario. This is the understanding, or at least an awareness, that we can only ask for ourselves and answer for ourselves. Are we indeed *only* this body–mind complex? Because if we are, then there is no question of escaping or getting away from this grief and pain, or from death. Just as we have laughed and enjoyed moments of extreme joy and pleasure with this body and mind, sharing love in relationships, excitement in fulfilling desires, revelling in success, name, and fame, then the pains, griefs, and losses are part of the same deal. The flip sides of that coin called "life" show us all the angles, whether we like them or not! Certainly, if there is birth, there is death. But it is the death of this physical system. It is not the death of the Atman; it is not the end of consciousness. The body and brain are matter, and matter generates and degenerates. Therefore, the death of a body is a natural phenomenon.

Contemporary spiritual movements also accept the longing for immortality, and they blend ancient wisdom with modern metaphysics and New Age beliefs. Beliefs in reincarnation, spiritual evolution, and consciousness suggest that immortality is not just about endless life but about achieving a higher state of awareness and being. Spiritual practices aimed at immortality provide a sense of peace, purpose, and connection, helping us to cope with the fear of death and the challenges of life. Meditation, prayer, and rituals can enhance mental and emotional health. The different spiritual approaches to immortality reflect diverse belief systems and the universal quest for meaning. They also highlight the potential for interfaith dialogue and mutual respect as people explore different paths towards understanding life, death, and what lies beyond.

Humanity's deepest desire is to transcend the limitations of physical existence and to find enduring meaning and purpose. In the face of life's impermanence, various spiritual traditions offer hope, continuity, and the possibility of a reality that surpasses the physical world.

Death and Afterlife

Cultural and religious beliefs about death and the afterlife are extremely relevant in our societies. They deeply impact our ethical behaviour and laws because there is a complex connection between spirituality, morality, and civilization. These influences are evident in the emphasis on ethical conduct, the legal systems, and policies on social welfare and end-of-life care. Across different cultural and religious perspectives, there is a belief in some form of life after death. Therefore, we are always conscious of the moral implications of our actions in this life and in the afterlife. There are elaborate rituals to honour the dead and support the living in their grief. These perspectives give us insight into how different cultures and religions cope with death, and they highlight the universal quest for meaning, the desire for immortality, and how communities come together to support each other in times of loss.

Historically, the moral codes derived from religious teachings have influenced the development of legal systems. For example, many Western legal principles have roots in Judeo-Christian ethics. The Islamic law, Sharia, directly governs the legal framework in some Muslim-majority countries, covering aspects of criminal justice, family law, and business transactions. Religious and cultural beliefs can inspire laws and social policies aimed at protecting the vulnerable, promoting fairness, and ensuring community welfare. Different cultures and religions have varying views on the sanctity of life and the appropriateness of life-sustaining treatments. Beliefs about death and dying influence laws and policies related to end-of-life care, euthanasia, the ethics of organ donation, and shaping healthcare policies and patient rights.

Pluralistic societies must navigate the diversity of beliefs to create inclusive laws that respect individual freedoms while upholding communal values. This balancing act requires ongoing dialogue, respect for human rights, and adaptive legal frameworks that can accommodate changing beliefs and ethical understandings.

Certain forms of atheism, materialism, or nihilism simply propose that there is nothing beyond death—that consciousness ceases to exist and there is no continuation of the self as there is no concept of an afterlife. The Semitic religions—Judaism, Christianity, and Islam—share a belief in an afterlife where souls are judged based on their deeds in this earthly life. There is a strong emphasis on moral accountability. The goal is to live according to ethical principles, such as honesty, charity, and compassion, with the hope of achieving a favourable outcome in the afterlife. Judaism traditionally focuses on the importance of living a righteous life in accordance with God's laws, with varying beliefs about the afterlife ranging from a vague notion of Sheol to more developed concepts of Gan Eden (heaven) and Gehinnom (hell). Christianity teaches about heaven and hell; it talks of the resurrection of Jesus Christ, and that is the foundation for belief in eternal life. The faithful are promised resurrection and eternal life in heaven, a place of everlasting peace and communion with God. In the New Testament, believers see death as a transition to eternal life. Islam emphasizes life after death as very significant, where the Day of Judgement leads to paradise for the faithful and righteous, or hell for the disbelievers and sinners. There is an emphasis on moral accountability. This belief encourages adherents to live according to ethical principles, such as honesty, charity, and compassion, with the hope of achieving a favourable outcome in the afterlife.

Hinduism, Buddhism, and Jainism have the concept of rebirth or reincarnation, closely tied to the law of *karma*. In these religions that advocate the cycle of rebirth, the law of *karma* is paramount; the idea that actions in this life directly affect future circumstances is a powerful motivator for ethical behaviour. Everyone is encouraged to live virtuously, perform good deeds, and avoid harmful actions to ensure a better rebirth.

Moksha means "liberation"; it signifies the liberation of the self from the cycle of death and rebirth. It is rooted in the belief that the individual conscious soul is part of the universal consciousness. The unity of the two is central to seeking and attaining freedom from this cycle. As we discussed in the goal of human life, the four-way path of the yogas includes devotion, knowledge, concerted effort, and selfless action. Each path offers a different approach to realizing one-ness with the Ultimate. It is a state of understanding the true nature of the Self and the universe beyond the limitations of time, space, and causation that bind us to the material world.

Nirvana implies "extinguishing" the fires of desire, aversion, and ignorance that cause suffering and attain enlightenment. Buddhism does not posit an eternal soul. It talks of *anatman*, the no-self; it emphasizes the impermanence of all things. It says that life is essentially full of suffering and sorrow. Enlightenment can be achieved by following the Eightfold Path. This is a middle way to avoid the extremes of sensual indulgence and self-mortification. We can be free from suffering and the cycle of rebirth by overcoming our desires and attachments. It is a state of peace and freedom, beyond concepts and dualities, often described as the ultimate reality.

Nirvana, *moksha*, and *kaivalya* by Patanjali are all concepts of enlightenment and liberation, but with subtle differences. *Nirvana* extinguishes suffering through detachment, *moksha* liberates from rebirth and unites with the divine, and *kaivalya* isolates the pure Self from the mind–body complex.

Integrating Science and Spirituality

The face-off between science and spirituality about death and the desire for immortality is an integration of beliefs, practices, and discoveries. It does not try to blur the distinct lines between empirical evidence and faith but tries to explore how each can inform and enrich the other in answering life's ultimate questions. Blending scientific knowledge with spiritual wisdom can offer a more holistic approach to death, immortality, and the essence of human existence.

Scientific efforts to extend life through medical and technological advances can be seen as complementary to spiritual pursuits of immortality. While science focuses on the physical and biological aspects, spirituality dives into the psychological, ethical, and existential dimensions of living longer lives. Together, they can address the need to navigate the possibilities of extended lifespans. Integrating scientific and spiritual insights raises important questions about the value of life and what constitutes a life worth living. These discussions can help society face the ethical implications of technologies to extend life and the pursuit of spiritual enlightenment, ensuring that advancements contribute to meaningful and ethically sound lives. The integration of science and spirituality offers diverse perspectives on death and the dying process, from palliative care practices that emphasize dignity and comfort to spiritual rituals that prepare the individual and their loved ones for the transition. This can transform the experience of dying into one of meaning, acceptance, and peace. It offers a comprehensive framework for understanding aspects of human existence. By acknowledging the contributions and limitations of both realms, we can accept the ethical, philosophical, and practical implications of extending life and defining immortality. This approach encourages a deeper exploration of what it means to live well, die with dignity, and possibly transcend the known boundaries of life and death.

Consciousness is an area where science and spirituality intersect. Neuroscience and psychology offer frameworks for understanding the brain and mind, while spiritual traditions provide experiential knowledge of consciousness beyond the physical.

Immanence of Consciousness

The immanence of consciousness beyond death is a deeply personal belief. It is a subject of ongoing debate and inquiry within both the spiritual and scientific communities. The concept is deeply rooted in the idea of non-duality in Advaita Vedanta, which asserts the fundamental unity of all existence. Brahman, which is Universal Consciousness, is immanent, all-pervading, and transcendental. The goal is to transcend the illusion of duality and realize our true nature and the ultimate unity of existence. The concept of consciousness persisting beyond death in Advaita Vedanta is the essence of Self-realization. Through spiritual practice, such as meditation and self-inquiry, we can directly experience our true nature as pure consciousness, which is beyond birth and death. This experiential understanding of consciousness leads to *moksha* from the cycle of birth and death. Death is merely the shedding of the physical body, while the eternal consciousness, the true Self, continues to exist beyond the limitations of the physical realm.

While traditional Buddhism talks of the impermanence and the cessation of individual consciousness upon attaining *nirvana*, some also believe in the existence of a subtle continuum of consciousness that transcends physical death. This continuum is sometimes referred to as the "ground consciousness".

Plato considered consciousness an essential aspect of the human soul, which is immortal and exists independently of the physical body. Plato's theory of Forms suggests that the soul's knowledge of eternal, unchanging truths persists beyond the material realm. Various mystical traditions propose the idea of consciousness persisting beyond death as part of a larger cosmic reality or divine unity. These traditions often emphasize direct mystical experience and the transcendent nature of consciousness. Some indigenous and shamanic traditions around the world teach the existence of a spiritual realm or dimension where consciousness continues to exist after physical death. These beliefs often involve communication with ancestors or spirits and rituals aimed at maintaining a connection with them. Within contemporary spiritual movements, such as New Age spirituality and esoteric traditions, there are beliefs in the continuity of consciousness beyond death. These beliefs draw on a combination of mystical teachings, Eastern philosophies, and metaphysical concepts. Beliefs about the immanence of consciousness beyond death vary widely across cultures, religions, and philosophical systems. Different traditions offer diverse interpretations of the nature of consciousness and its relationship to mortality, reflecting humanity's enduring fascination with questions of existence, identity, and the afterlife.

Out-of-body Experiences (OBE)

Among the most telling experiences that seem to endorse the reality of the afterlife are the out-of-body experiences (OBE) of several people from around the world. These are subjective phenomena, and in this state, people feel as though their consciousness or awareness is detached from the physical body and they can observe their surroundings from outside the body. During an OBE, individuals report sensations of floating, flying, or seeing their physical body from a distance.

How exactly people have OBEs is not fully understood, but various theories, including neurological, psychological, and spiritual perspectives, have been proposed. The most prevalent explanation is that OBEs are related to alterations in brain activity and perception.

Henrik Ehrsson, a neuroscientist, utilizes mannequins, rubber arms, and virtual reality to create illusions that manipulate people's sense of self, inducing feelings of out-of-body experiences, foreign body ownership, and even size distortion. This unconventional research, published in *Nature* (2011), challenges our perception of a fixed self and highlights the brain's dependence on sensory information to construct our body image.

Another study published in *Nature* investigated the effects of stimulating specific brain regions, including the pineal gland, on inducing OBEs. The study found that electrical stimulation triggered feelings of detachment from the body and altered self-perception in some participants. This opens exciting avenues for further research into the neural correlates of OBEs and may even lead to the development of techniques for safely inducing and studying these fascinating experiences.

Psychological factors, such as beliefs, expectations, and altered states of consciousness, can also influence the occurrence of OBEs. But for those who experience it, OBE is a sensation where consciousness seems to leave the physical body. People often describe OBEs as floating outside their bodies, with altered perceptions such as looking down from a height or feeling like they're looking down at themselves from above. An uncanny sense of "what's happening?" is very real. OBEs typically happen without warning and usually don't last very long. While some individuals with neurological conditions, like epilepsy, may experience them more frequently, for many people, OBEs are rare—perhaps only once in a lifetime. The ongoing debate is whether an OBE actually occurs physically or whether it is a kind of hallucination. Some studies suggest that during resuscitation, people report a separation from their body and awareness of events they wouldn't have seen from their "normal" or actual perspective. However, there's no scientific evidence that a person can actually travel outside the body.

Near-death Experiences (NDEs)

NDEs invariably start with an OBE. First is the sensation of floating outside the body, observing the surrounding environment from a different perspective, then moving on, travelling through tunnels or to other realms, encountering deceased loved ones or spiritual beings, and meeting with spiritual guides. These are common descriptions and experiences.

From a psychological perspective, both OBEs and NDEs are the outcome of various psychological factors, including dissociation, depersonalization, and altered states of consciousness. These experiences can be influenced by cultural beliefs, expectations, and personal interpretations of unusual or mystical experiences. Psychodynamic theories suggest that OBEs and NDEs may also serve psychological functions, such as coping with stress, trauma, or existential concerns.

Many individuals who have had NDEs see them as profound spiritual or transcendent experiences providing insights into the nature of consciousness, existence, and the afterlife. These experiences are evidence of the soul's separation from the body or the existence of higher dimensions of reality. NDEs offer confirmation of spiritual beliefs and the continuity of consciousness beyond physical death for those who experience them and those who believe them.

Some researchers explore parapsychological explanations and talk of these being extrasensory perceptions (ESP), telepathy, or psychic phenomena. However, scientific evidence supporting these hypotheses is limited, and many mainstream scientists are sceptical of paranormal interpretations of NDEs. While scientific research focuses on the neural and psychological mechanisms to explain these experiences, their subjective nature makes them difficult to fully explain or understand.

Renowned neurosurgeon Dr Eben Alexander gave his account of his near-death experience. In 2008, Alexander fell into a coma due to bacterial meningitis and spent seven days in a state of deep unconsciousness. He wrote about his experience in the book *Proof of Heaven: A Neurosurgeon's Journey into the Afterlife* in 2012. He describes his journey from scepticism about NDEs and the afterlife to a belief in the existence of a spiritual reality beyond the physical world, based on his personal experience.

Alexander's story resonates with many individuals who have had similar experiences or who are interested in exploring questions about consciousness, spirituality, and the nature of reality. This is just one of many reported cases of NDEs, each with its own unique features and interpretations. Scientific research on NDEs continues to explore the underlying mechanisms and implications of these phenomena, but conclusive evidence for the existence of an afterlife or spiritual dimension remains elusive.

Pam Reynolds underwent a unique form of surgery called "standstill" surgery, during which her body temperature was lowered to near-freezing and her heart and brain activity were stopped for an extended period. Soon after, the surgeons began to lower her body temperature to sixty degrees. It was about that time that Reynolds believed she noticed a tunnel and a bright light. She eventually flat-lined completely, and the surgeons drained the blood out of her head. During her NDE, she says her body looked like a train wreck, and she said she didn't want to return to it.

Anita Moorjani's NDE was during a near-fatal illness due to cancer. As she says, "In this near-death state, I was more acutely aware of all that was going on around me than I'd ever been in a normal physical state." Moorjani experienced a transformative NDE characterized by feelings of overwhelming love, a sense of one-ness with the universe, and profound insights into the nature of reality and human existence.

There are innumerable cases and studies of people's NDEs. There are also enough studies based on awareness of previous lives through regression. The mystery of life and death does not have a definitive explanation. Trying to understand death and the possibility of immortality is an ongoing search marked by continuous discovery, reflection, and growth. As science advances and our spiritual understanding evolves, we will undoubtedly uncover new insights and face new questions. Yet, at the heart of this quest is the enduring human spirit—driven by curiosity, hope, and a desire for connection and continuity. In contemplating death and beyond, we are reminded of the richness of the human experience and the boundless capacity for love, creativity, and transcendence. Whether we find answers in the meticulous details of scientific research, the profound teachings of spiritual traditions, or the simple acts of kindness that define a life well-lived, the exploration of these eternal questions connects us to each other and to the mystery and beauty of existence itself.

Dialogue with Death

Finally, we refer to a dialogue with death itself. In the Katha Upanishad, one of the primary Upanishads in Vedic literature, a young boy named Nachiketa engages in a dialogue with Yama, the god of death, after having been sent to Yama's abode by his father. The story of the courageous boy going to Yama's abode is fascinating (Box 8.3).

Nachiketa's third and crucial question is: "Going from here, from death, going beyond, people have this doubt or question—some say he still is, 'it exists', others say he is not, 'He does not exist'. I want to know the truth about this." After trying to dissuade the boy from seeking this very deep and mystifying answer, Yama tells him, "The one who believes that this level is all that there is comes under my power again and again. This belief in the body is so strong that it will be born again and again." Finally, Yama gives Nachiketa the secret. "The Self is immortal. It was not born, nor does it die. It did not come out of anything, neither did anything came out of it. Even if this body is destroyed, the soul is not destroyed." "All people who concentrate, wanting to follow the path of Brahman, I am telling you all that briefly, It is this one word, Om" (Canto 1.ii.15).

Celebrate Death and Beyond

As long as death is inevitable, as long as birth and death remain two sides of the same coin, then understanding and accepting death as a natural consequence of life is the wisest and only way to move forward. For those of us who see life as a one-time finite event with nothing beyond, there is no option but to accept its finality, and then this journey ends here. But for the many more for whom there is an afterlife, the final phase has to now evolve.

What if we reframe death not as a fearful end but as a cause for celebration? Then death would be the culmination of a meaningful life. In stark contrast to fearing death, trying to delay it, or denying it, several cultures have the unique tradition of celebrating death. Taking death as a natural part of life, the next of kin, along with friends and well-wishers, honour the memory of the deceased person with celebrations. Perhaps one of the most well-known celebrations of death is Día de los Muertos, which is a Mexican holiday observed on 1st and 2nd November. Families gather to honour and remember deceased loved ones by creating elaborate altars (ofrendas) and adorning them with photographs,

Box 8.3. A dialogue with death: The story of Nachiketa

Nachiketa was a curious and thoughtful young boy. One day, he watched his father perform a special fire sacrifice (*yajna*). But this astute child noticed something was amiss. His father was gifting away his cows as alms to the priests and the poor, but these were old cows which couldn't give milk anymore. Nachiketa felt this was an unfair or mean offering. Since the sacrifice meant giving away all that belonged to the one performing the *yajna*, Nachiketa asked his father, to whom would he gift his son, Nachiketa himself.

His father, extremely upset with Nachiketa's question, blurted out in anger, "I shall give you away to Yamaraja." Nachiketa took his father's words seriously, and he set off on a long journey to find Yama, the Lord of Death. When he finally arrived at Death's abode, Yama wasn't even home! Nachiketa waited patiently for three days. Finally, Yama returned and apologized profusely for making him wait. To make it up to Nachiketa, Yama offered him three boons for the three days that he had waited.

Nachiketa, wise beyond his years, used his boons cleverly. For the first boon Nachiketa sought peace with his father; next he learnt about the Agnihotra ritual that his father was performing; and finally, he asked the most important question: what happens after death? Where does the person go? Yama tried to dissuade Nachiketa from asking for this ultimate truth, and offered him all the wealth, pleasures, and physical comforts that even the gods in heaven did not have; but the boy did not waver. Finally, Yama, realizing young Nachiketa's determination, wisdom, and maturity, gave him the ultimate knowledge. With Yama's teachings, Nachiketa gained the profound understanding of the Atman, the immortal Self, and the path to liberation.

Moral lessons from the Nachiketa story: Nachiketa's courage to question his father, a respected sage, emphasizes the importance of independent thinking and seeking the truth fearlessly. Nachiketa's unwavering determination in his quest for knowledge, even when facing the Lord of Death, shows his perseverance. Nachiketa rejects fleeting pleasures and material things, thus looking beyond the material gains, focuses on the eternal truth about the Self, the afterlife, and gives the greatest importance to Self-knowledge. The core message revolves around understanding the nature of the Atman, the immortal Self, which we are. This knowledge is essential for self-realization and liberation from the cycle of death and rebirth.

Fearlessness, confidence to seek the truth, and unwavering determination.

candles, flowers, and offerings. They believe that during this time, the departed souls return to visit their families, and they are welcomed with food, music, and festivities. In Ghana, families and friends create elaborate fantasy coffins, abebuu adekai or "proverbial coffins". These coffins are crafted in various shapes and designs, representing the occupations, hobbies, or interests of the deceased. In this way, the living honour the departed souls and celebrate their journey into the afterlife. In Bali, Indonesia, cremation ceremonies are grand and elaborate, celebrating the transition of the deceased from the physical world to the spiritual realm. The ceremonies involve intricate rituals, processions, and offerings, all aimed at ensuring a smooth journey into the afterlife. Irish wakes are traditional gatherings held to mourn the passing of a loved one but also to celebrate their life. Wakes involve storytelling, music, dancing, and sharing memories of the deceased. It is a way to honour the departed and support the grieving family.

The Obon Festival, observed by Japanese communities worldwide, is a Buddhist tradition that honours the spirits of ancestors. During Obon, families gather to clean gravesites, light lanterns, and offer food and prayers to the deceased. It is believed that during this time, the spirits of the ancestors return to visit their living relatives. These are just a few examples of cultures that celebrate death in unique and meaningful ways. In India, the death of a very aged person is especially celebrated as it is the end of suffering, and prayers and good wishes are offered for an easy transition. Families and friends get together to enjoy a good meal and share reminiscences about the departed soul. Across the world, various traditions and customs reflect a deep reverence for the deceased and a recognition of the interconnectedness between the living and the dead. These celebrations serve to honour the memory of loved ones, provide comfort to the bereaved, and reaffirm the continuity of life beyond death.

Celebrating death doesn't diminish the naturalness of grief. A loss will still evoke sadness, a necessary human emotion. However, through dedicated practices and the guidance of a wise teacher, we learn to navigate these emotions with equanimity and understanding. Ultimately, celebrating death becomes an act of courage, proving our ability to overcome a primal fear and transform it into a source of wisdom and growth. It is a transformative journey, demanding yet rewarding, that allows us to embrace life in all its richness and greet death, not with fear but with the serenity of a returning guest at the final grand finale.

Advancing from Genome to Om

As we understand, especially from the wisdom of the Vedas, Upanishads, and Bhagavad Gita, we are infinite conscious beings for whom birth and death are another projection within *maya*, limited within time, space, and causation, and now, being free of all such bondages, we can transcend into a state of bliss and tranquillity. This life is then a transformative journey towards eternal peace, accepting the impermanence of the body–mind complex as part of the journey, with its death as a natural evolution, not a tragic end. The individual self "gives up" or "leaves" the physical frame that it occupies for that "event" within the limited time and space. Thus, it transcends the physical body. Unfortunately, almost all of us remain ignorant of this aspect because we do not know or perceive this understanding in our awareness. But if we had this understanding, it would allow us to view death as a transition, a shedding of the worn-out cloak of our physical form.

As we come to the end of this journey, evolving from modern science to meta-science, death becomes not an ending but a transformation, a graduation from the known into the unknown world with a heart full of acceptance and peace. We are confident that we can enter the state of pure consciousness, often described as enlightenment and blissful liberation. It is a state of complete detachment from earthly desires and anxieties. We can move to this final understanding with a sense of curiosity and wonder.

CHAPTER 9

Finding Unity in Diversity: The Om Way

The wise choose out the good from the pleasant, but the dull soul chooses the pleasant rather than the getting of his good and its having.
—Katha Upanishad, 1.2.2, translated by Sri Aurobindo.

We are now standing at a very crucial crossroad. As we come to the end of this quest of modern science evolving to meta-science, our *Genome to Om* journey is bridging science and spirituality, seeking unity in diversity to achieve our optimal potential, and working towards attaining the goal of universal peace and well-being. As co-partners in the need to nurture and cherish our lives on planet Earth, we must know where we stand as we continue to seek answers to the perennial questions of life. It is up to us to walk this path.

What have we brought with us up to this point in our journey? The ground-breaking discoveries in science and technology continue to astound us as the desire for comfort, the related joys, and the resounding successes with claims to fame are satisfied as an ongoing feature. Unfortunately, we had to hoist several red flags thereafter, flags that are now globally hoisted, and remedial solutions are also being hastily researched. If not on the anvil already, answers need to be found with immediate urgency. Enough scientists, researchers, and thinkers have voiced their concerns. Having seen the rapid journey break all speed breakers, we knew we had to press the pause button and stop. We needed a flashback to the beginning, with the creation of this universe and our existence. Despite the very well-researched explanations and empirical data on how this universe came into being, we are still at a crossroad here as well. We do not know irrefutably: how did the universe come into being? Was it a big bang? Is it in a steady state, immanent and eternal? Or is it cyclic, ever-expanding, or possibly hurtling towards a big crunch? The questions are still being debated and discussed.

This unbelievable reiteration of where we have come from and where we stand today is a very valid question about our ability as the "most intelligent" of living beings. As we discussed, whether through the aeons and epochs we began as the single-celled amoeba and evolved to who we are today, or whether we were created by a creator God "in his own image", we certainly came into being effectively and strongly enough to walk the lands, choose our habitat, use what nature had already generated as flora and fauna, and explore widely to utilize not-so-wisely what we found as our "birth right". Whether by natural selection, whether we survived as the fittest, or whether we are meant to get eliminated as the weakest in the future, we took over and have reigned unchallenged. We know from the earliest scriptures, every written or orally transmitted, the annals of recorded history, and especially over the past four-five centuries, that we have advanced phenomenally. But simultaneously, we are also going around in circles with the basic, in-depth questions of who, why, when, how, and where about our existence. We have looked at the skies and wondered and asked all our questions. And today, the one question we are all compelled to ask is: *Where are we going*?

Perhaps we are now in a mindset that says, however it began, *where do we stand now*? Accepting where we have come from, we recognize the need to go deeper towards understanding ourselves and, through that, our place in this universe. With that journey in mind, we have sought answers to the fundamental questions about the intrinsic nature of the universe and life, the complexities of the mind, and our purpose, and we have tried to understand death because what has a beginning also has an end. The cosmic miracle remains mind-boggling and humbling for us. We seem to know how we evolved into the unique species, *Homo sapiens*, but having taken over this planet, which mysteriously seems to be the only "just right" place for our survival, we have managed to bring it to the verge of extinction!

A Ticking Time Bomb: The Perils and Poisons

The beginning of the 20th century brought unforgettable world wars, the inhumane hands of humanity with the Holocaust, the partition of the Indian subcontinent, and various wars across continents. That trend continued in one way or another, and in this third decade of the 21st century, we are already in the thick of it. Regional wars, large-scale conflicts, religious extremism, and terrorism are devastating entire regions, wiping out humanity in horrendous ways, and we are possibly sitting on the time bomb of another world war!

Today, climate change is the most immediate threat, and for the moment, a live grenade with a human finger on that pin. What we are already experiencing and what lies ahead is no longer an imaginative science fiction fantasy. Climate change will progressively render vast lands totally uninhabitable, and we can expect entire populations or communities to be displaced, which will trigger resource wars. Almost uncontrolled rising temperatures, already melted or in the process of melting and disintegrating glaciers and ice caps, and rising sea levels are triggering sweeping floods on the one hand while we are facing an acute shortage of water on the other; widespread forest fires are burning through our lives. The extreme shortage of food and water could lead to a total breakdown of social structures, and we could face mass starvation.

For decades, we relied on fossil fuels and nuclear energy, and these two factors have exposed us to ecological and radioactive devastation. We stand on the brink of disaster also because the race for excellence has led to rising nationalism, global competition for fast-depleting resources, and the widespread proliferation of weapons of mass destruction, which include the potential for lethal, poisonous, biological, and chemical weapons. This is perhaps the most terrifying threat: nuclear and biological warfare. The devastating potential is an unimaginable catastrophe, wiping out entire populations and rendering vast areas of land uninhabitable. A "nuclear winter" would be the chilling consequence of a large-scale nuclear exchange, and it could plunge our planet into a prolonged period of darkness and cold, with a devastating impact on global ecosystems. The delicate balance of life on Earth could be irrevocably disrupted, leading to a mass extinction event. Biological warfare, whether purposefully unleashed or accidentally released, could decimate populations and wipe out humanity at large. Imagine a world ravaged by a highly contagious and deadly virus, against which we have no defence. Pandemics, epidemics, natural disasters, wars, and resource scarcity have already taken the widespread social unrest to a new peak, warning us of the imminent collapse of the existing geopolitical and social structures. All it needs is a single act of aggression, a miscalculation, or a "mistake".

Learning from the Past

Unchecked technological advancements could lead to scenarios like uncontrollable artificial intelligence, the emergence of engineered pathogens, or even pandemic civil wars. Industrial, atmospheric, and environmental pollution is adding all kinds of poisons to the stratosphere, soil, air, forests, rivers, and oceans. The oceans have been churned, and the poisons are out here now.

A very famous story from Hindu mythology regarding the churning of the ocean (*samudra manthan*) in search of elixir is relevant here. This story appears in ancient Hindu scriptures, including *Vishnu Purana* and the Mahabharata. In this mythological story, the king of snakes, named Vasuki, is used as a rope to churn the ocean. According to the story, the gods and the demons jointly churned the ocean in search of *amrita*, the elixir for immortality. Along with the precious jewels and wealth that came out with the churning also arose the most lethal poisons that threatened to wipe out all living creatures. Lord Shiva, known for his benevolence swallowed the poison, and saved mankind. Today, we have churned nature to the point of bringing up all the poisons of the pollution of the atmosphere, air, rivers, oceans, forests, mountains, and environment. We are all set to send our planet and ourselves to extinction! Who will swallow these poisons? Who is the modern *avatara* of Shiva to save us?

Aside from the symbolism in this mythological story, a very recent article published in *Nature* scientific reports (18 April 2024) is quite intriguing, where researchers found fossils of a giant snake from an early Middle Eocene period (56 to 47.8 million years ago) in Kutch, western India. The estimated body length of approximately 11–15 m makes this new taxon from the Genus Vasuki. Biogeographic evidence suggests that Vasuki represents a relic lineage that originated in the Indian subcontinent. A huge sculpture of *samudra manthan* at the central point of Suvarnabhumi International Airport in Bangkok stands tall as a reminiscence of the mythological tale (see Plate 21).

Beyond the mythology, the epic, and ancient history, the real intriguing questions are: How was the principle of centrifugal and gravitational forces behind churning known several thousand years ago? How was the existence of a giant snake like Vasuki known when mythological stories were composed? How could something like *amrita* as an elixir of longevity have been imagined? How can the same story be part of diverse civilizations, from India to Thailand? To find answers to these questions, we need an open, inquisitive unbiased mind. In fact, this is the primary purpose of scientific research in search of truth.

The Need for a Balanced Scientific Approach

We suggest that mythology isn't merely a relic of the past, nor is science fiction the fantasizing of the future. We can certainly say here that they both can, and do, fuel imagination, creativity, and modern scientific exploration. We are looking squarely at the recent discovery of the Vasuki fossil in India, named after the celestial giant snake in mythology; ESA's Gaia satellite, named after the Greek goddess of Earth, observes our planet with unparalleled detail; and missions like Phoenix and Hermes, named after the firebird of rebirth and the messenger god, respectively, seek life on Mars. Recently, astronomers identified two ancient streams of stars named after Shiva and Shakti. These are believed to be the Milky Way

Galaxy's earliest building blocks. Both Shiva and Shakti were found using the Gaia space telescope. Even within the hallowed halls of CERN, the Nataraja statue—a depiction of the cosmic dance of Shiva—stands as a potent symbol of creation and destruction, mirroring the fundamental forces that scientists' probe.

As science continues its relentless march, potentially towards a meta-scientific future, the scientific community must strive for a balanced approach. The future we see is where scientific advancement empowers us to both predict and respond to challenges, ensuring greater safety and well-being for all. Science fiction has often served as thought experiments, pushing the boundaries of scientific possibility. The film *57 Seconds* (2023) projects a world where quantum crystal technology allows for the prediction and prevention of disasters, rendering accidents, injuries, and even medicine obsolete. This pushes us to ask: Can scientific progress ever catch up to such a utopian ideal?

These are not mere coincidences. They speak to a collective yearning to bridge the seemingly disparate realms of mythology, fiction, and science. Mythology, history, culture, and spirituality, with their rich knowledge, symbols, and archetypes, might act as a wellspring of inspiration. This can provide a framework for understanding as we venture into the unknown, be it the outward leap into the vastness of space and the complexities of nature or the inward dive to understand ourselves, our experiences, and our purpose. We call this the meta-scientific approach, where the empirical methods of science harmonize with the experiential and existential questions that mythology, as embedded in spirituality, asks. As science and technology continue to discover and innovate, and as we take a look at our stories, they give momentum to our quest for knowledge, but all the while we maintain a critical and unbiased perspective to ensure a clear distinction from pseudoscience and blind faith.

The concept of meta-science emerges as a potential driver of this shift. It sees a future where previously siloed scientific disciplines converge, presenting a more holistic understanding of the universe. Quantum mechanics, the study of matter and energy at the atomic and subatomic level, holds immense potential. While manipulating probability fields, as depicted in fiction, might seem far-fetched, it focuses on the transformative possibilities that lie at the intersection of established scientific frameworks. We reiterate that adopting the meta-science approach and learning from mythology, philosophy, and spirituality does not mean encouraging pseudoscientific gimmicks.

For most of us, material well-being assures us the comfort and security that we seek. And we believe that is enough. But it never is. The uncertainties, fears, and doubts triggered by deep-seated insecurities generate high levels of restlessness. But if we progress to an inner awareness, then that spiritual life gives us the strength to deal with all the vagaries of life with equanimity. Within society with its collective quest for materialism and consumerism, is it possible to find a balance between material and spiritual well-being?

Root Causes: Need, Greed, and Ego

Martin Luther King Jr talked of his dream: "I have a dream that one day every valley shall be engulfed, every hill shall be exalted, and every mountain shall be made low, the rough places will be made plains, and the crooked places will be made straight, and the glory of the Lord shall be revealed, and all flesh shall see it together." King's path was nonviolent resistance to achieve equal rights for Black Americans, which earned him the Nobel Peace Prize in 1964. But in 1968, he was assassinated for having this great dream and mission. Mahatma Gandhi's words reverberate poignantly today: "The world has enough for everyone's need, but not enough for everyone's greed." Rabindranath Tagore commented, "The greed of gain has no time or limit to its capaciousness. Its one object is to produce and consume. It has pity neither for beautiful nature nor for living human beings. It is ruthlessly ready without a moment's hesitation to crush beauty and life."

Albert Einstein echoed these sentiments, warning of the dangers of unchecked greed and the need for a shift towards a more peaceful and sustainable way of life. Similarly, environmentalist Rachel

Carson documented the devastating effects of human actions on the environment, urging a course correction towards ecological harmony in *Silent Spring*.

We have crushed, killed, and conquered; we have overtaken and ruled. We have discriminated, hated, loved, and fought to survive. Goaded on by our greed and ego, we are today so deeply entrenched in our belief in our existence as human beings that we question death itself. Which only means that the most intrinsic desire, quest, and motive is to exist, but entirely at our terms. Our existence as a species must not be threatened. But there is one problem here: we want to exist by defying nature. The advancements in S&T have built a kind of hubris that we can conquer and redefine nature. Our egos have inflated disproportionately, and our greed seems to be never-ending.

Why did we do this? Why did we assume that humanity could thrive by destroying nature? When and why did we bring ourselves to this "either-or" situation? This "humanity-versus-nature" condition that we are faced with today has given us the ultimatum. We have no option. Either we exist in harmony or we become extinct. Perhaps the "only outward" direction was wrong, or at least, to some extent, imbalanced?

Our current reality is "after the fact" reactive measures. Earthquakes and floods strike with brutal frequency, and we scramble for solutions. Medicine remains largely focused on treating existing conditions and barely keeping up. Ongoing scientific endeavours do hint at a future where prevention might take centre stage. Whether we will walk on that stage ourselves, we do not know, but for future generations to be comfortable on that stage, it is a foreseeable possibility.

On World Environment Day 2020 the UN Secretary-General Antonio Guterres highlighted the "existential threats" posed by environmental degradation and the urgency of decisive action. The United Nations Development Programme's (UNDP) 2020 Human Development Report, "The Next Frontier: Human Development and the Anthropocene", also points to the critical shift in our relationship with the environment. In the Anthropocene, human progress is inseparable from planetary health. We are not separate entities on a planet; we are, in fact, inextricably interconnected parts of a complex ecological web. If modern society does not take bold steps towards environmental sustainability, both humanity and nature will be at risk of collective decline and potential extinction. This new reality necessitates a fundamental shift in the development paradigm of modern society, transcending towards a matured, humane society. This does not in any way imply abandoning technological progress but rather redefining it within the context of planetary boundaries and universal well-being. We call this a move towards meta-society, as a means towards a state of harmony, peace, and universal well-being. But before we make this move, let's look at societal revolutions.

Evolution and Challenges of Modern Society

Historically, the first social revolution started when hunting was the main purpose and means of livelihood. Then, with the advent of agriculture and animal husbandry, the world entered the second social revolution. The third and fourth social revolutions were marked by the invention of the steam engine and microchip, respectively. This was the beginning of the industrial revolution as well. But today, there is growing concern over all the evils that have permeated global society. Perhaps what we are waiting for is the fifth revolution towards a brighter future, as Society 5.0, where anyone can create value anytime, anywhere, in harmony with nature. This is also the basis of the UN Sustainable Development Goals (see Plate 22).

What are the main challenges in our day-to-day lives? The desires to be part of something bigger and to follow trends and emulate popular culture are very powerful. This in itself is a motivating mindset. But accepting trends, unquestionably, particularly those related to lifestyle, has had a very adverse impact, especially when applied across diverse cultural contexts. The most widely spread change that took place over the past few decades was the "explosion" in Western social culture of fast food, nightlife, casinos, pubs, and bars. The culture of extended partying held a certain glamour. The alluring marketing advertisements of unhealthy lifestyle products, violent, pervert products of the entertainment industry, and all the extending tentacles of social media played a significant role in distorting the healthy social fabric. The rapid growth and outreach of these mediums gave unprecedented impetus to the recreation, hospitality, gambling, liquor, and tobacco industries. So are the powers behind this social explosion to be held solely responsible? If we, as individuals in the "Pied Piper" societies, are ready to follow unthinkingly and grasp everything out here with our unbridled senses, are we really absolved of all responsibility in this collective journey of unthinkingly taking in all that is out there?

Consumerism is a social and economic ideology that focuses on acquiring and consuming various goods and services, even if they are not needed, because it is believed that this is for our well-being and the well-being of society and promises happiness, higher social status, and fulfilment. Consequently, the emphasis is on material possessions. With a simple classification process, various countries are reviewed and assessed to know where they stand in relation to consumerism and minimalism. In reality, minimalist nations have better health and well-being, which declines as consumeristic market forces start dominating the so-called modern lifestyle (see Fig. 9.1).

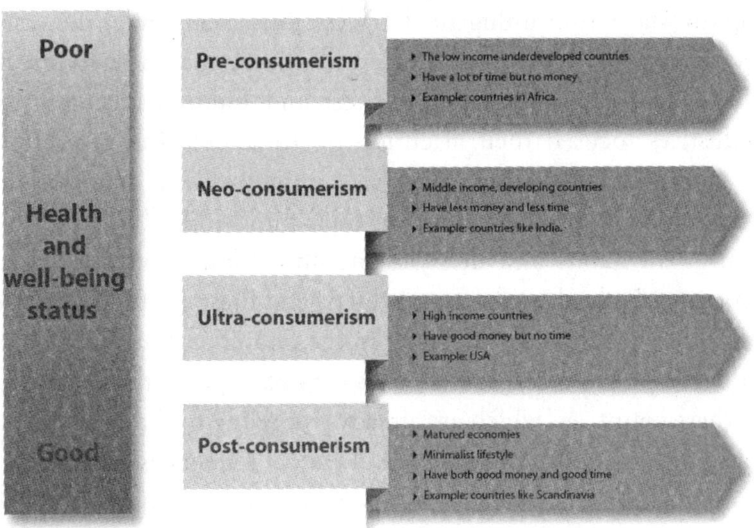

Fig. 9.1. Consumerism, minimalism, and well-being.

This societal trend had a negative impact on public health. Here, the USA can be the exemplar. Studies by the American Heart Association highlighted a link between unhealthy dietary choices associated with late-night eating and increased risk factors for non-communicable diseases (NCDs). The 2021 data from the Centers for Disease Control and Prevention (CDC) pointed to a correlation that drastically increased NCDs such as heart disease, diabetes, respiratory diseases, and obesity. The CDC reported that chronic diseases are the leading causes of death and disability, with NCDs accounting for seven in ten deaths. It is also reported that medical errors and unintentional injury were the fourth cause of death in the US. The 2023 data from the CDC reported that nearly 60 per cent teenage girls in the US felt persistent sadness and hopelessness, and one in three girls might have succumbed to substance abuse and seriously considered attempting suicide. These are just representative examples of a broken society in one of the richest and most powerful countries in the world.

Gradually and to some extent, the educated and conscious public from the West became aware of the negative health impacts, and there was a decline in smoking, consuming sugary and aerated drinks, and eating unhealthy fast food. Markets for these products rapidly started declining in the West. But meanwhile, the rest of world, especially the poor, low-, and middle-income nations, had already chosen to imitate the West and follow suit. And those industries focused their attention on these evolving markets full of gullible populations. Several commonwealth countries still carrying Western dominance on their minds, like India, followed many bad habits and lifestyles blindly as fashionable or socially "correct". Unhealthy foods, products, and lifestyle trends adversely impacted health and well-being in the poorer and underdeveloped or developing nations as well. This led to a triple burden of diseases in these countries, which are already struggling for basic healthcare. Although the realization of the grave consequences has begun to hit home, the correction across the board, if at all possible, takes its own time.

Happiness: Any society, most pertinently, needs peace and happiness. The example of the USA reminds us that just being resourceful, powerful, and rich does not guarantee peace and happiness. There are exemplars. Bhutan, once a constitutional monarchy, is now said to be a successful social democracy. Despite its relatively modest economic conditions, Bhutan is a happy country. Although Finland ranks as the happiest country in the world, Bhutan is referred to as the "Kingdom of Happiness". How has Bhutan achieved this landmark recognition? Probably because it developed different assessment criteria than Gross Domestic Profit (GDP) and introduced the idea of Gross National Happiness (GNH). The Bhutanese people have realized that consumerism, or material well-being, is not the sole parameter for happiness. Bhutan's focus on its national well-being has been immensely successful, translating into the well-being of its citizens. This happiness is not related to material wealth. It is the outcome of a comprehensive approach to development that prioritizes GNH over GDP. The Bhutanese focus on conserving the environment and culture and trying to keep up the vitality of the community and mental health through spirituality. Bhutan has managed to bridge disparities in happiness and quality of life in ways that wealthier nations have struggled to achieve. Undoubtedly, the size of the country is very small in comparison to a nation like the USA, but the important fact is to acknowledge that true well-being extends beyond economic indicators.

The primary lesson to learn from Bhutan is that a society must recognize that the spiritual and emotional health of the people is as important as their physical and material well-being. We are not condemning consumerism. But the greater its hold or impact, the farther a country and its people are from the scale of contentment and collective well-being. Nonetheless, there is some hope when we realize that, while on the one hand, there is a race to "satisfy and fulfil" all desires and wants, on the other, there is a changing trend in this over-the-top consumerism.

Minimalism: The after-effects of extreme consumerism probably led to minimalism as a way of living a mature, though modern, lifestyle. Scandinavian minimalism, or Nordic practicality, focuses on simplicity, using natural materials, and functionality. It prioritizes natural, light-filled spaces that are both simple and comfortable; it creates a design philosophy that extends to a life lived in harmony with nature. Minimalism counteracts the fast-paced, busy lifestyle with movements such as "slow food" and "slow living". It emphasizes mindful consumption, local sourcing, and savouring and enjoying all experiences. These minimalistic trends can certainly promote regeneration and a more balanced approach to life. We can say that they are getting closer to the ideal matured societies, although the world order is not letting them reach there because of migrations and the changing demographic profile which is the dynamism of the Anthropocene.

Having cited the examples of Bhutan and Scandinavia, our eyes are not shut to the over-populated cities where barely educated masses are crowded in, many of whom live in little huts of tarpaulins and torn fabrics, under bridges and underpasses in the buzzing cities, with malls and multiplexes and struggling villages where malnourished women have to walk for miles just to get a bucket of water. But here also there is a determination to face all adversity and seek a better life. And spirituality cannot be taught on an empty stomach! But we cannot hide behind any of this, and it is time to make a start. And changing the mindset is the first step.

To make a start, we must transcend from a consumer-driven, materialistic society, prone to exploitation, to a minimalist one. Instead of constantly striving for more, we have to learn to be content with less. While at a personal level we may even realize momentarily that happiness and fulfilment don't come from material possessions, in our relationships, experiences, and personal growth, we tend to miss the larger good. It is not just about having fewer possessions but also about reducing our drain on the environment. It's about making deliberate choices that are sustainable and eco-friendly and that will contribute to the well-being of our planet. Enough damage has already been done, and there are ongoing global efforts at both the individual and national levels to try and halt, slow down, or, if possible, reverse the progressive damage. All such simple individual decisions, multiplied exponentially, are the countless drops of water that we can bring back into the drying sources.

Modern societies often prioritize individualism and competition. A meta-society emphasizes cooperation, collective well-being, and environmental sustainability. The transition from an individualistic modern society to a future-oriented "meta-society" requires concerted efforts by various stakeholders. Governments, businesses, and the scientific community, and most importantly, we the people, all have crucial roles to play in this transformation, which we can see as a journey to the "Om Way" with a sense of solidarity and unity.

We are aware of how challenging it is for societies and countries to be convinced and ready to look at the growth of their nations and their people as more inclined to implement a mid-course shift of focus towards personal, national, and universal well-being. It is challenging, but not unviable. It is a process of evolution. We trust that the "Om Way" can help in this evolution. We can now see a world in transition, where we move from the Anthropocene, the age dominated by human impact, to what we would like to refer as the "Omcene", an era where humanity thrives in harmony with all and with universal well-being. We call this a meta-society and are convinced that it is not a mere utopian concept.

Meta-science and Meta-society

The pivotal question here is: how can we bring about change? It is no longer just about understanding the physical world, where the struggles, the hurdles, and the ups and downs are unrelenting in their cyclic nature; it is about understanding ourselves, our societies, and our place in the universe. If it is possible for modern science to evolve to meta-science, then the transformational revolution will lead modern society to meta-society. Although the reverse is also possible. Meta-science helps us transcend conventional scientific inquiry and integrate subjective human experience, intuition, and the role of consciousness into scientific discovery. It is a more holistic approach, acknowledging the potential existence of non-material realities and their influence on our physical universe. Even at the risk of repetition, we must reiterate that meta-science goes beyond the boundaries of empiricism and logical reasoning. It stands on the spirit of Karl Popper's concept of falsifiability, which suggests that if a scientific theory is strong, it can be potentially proven false through experiment or observation. Meta-science involves the study of the principles and effects of scientific research itself, including the people who conduct it, the decisions they make, the effects of those decisions, and the ways that research affects the world. Meta-science encourages a holistic approach to knowledge where empirical evidence and spiritual insights complement each other, leading to a deeper understanding of reality. This needs both experience and realization.

What Do We Mean by Meta-society?

The human condition is constantly evolving, both in terms of technological advancement and societal organization. Scientific progress has historically been driven by societal needs and aspirations. Modern societies are primarily focused on individual needs and competition. This emphasis, while it boosts innovation and economic growth, leads to social inequalities and alienation. Therefore, it hinders progress. A meta-society would prioritize the "common good", kindling a spirit of cooperation and mutual respect for diverse viewpoints. This collective orientation will ensure a more sustainable and equitable future.

Throughout history, philosophers and thinkers have grappled with the ideal society. Ancient Greek Stoics like Seneca emphasized the importance of virtue and living a simple life, free from the anxieties of material possessions. Similarly, in Eastern traditions like Buddhism and as stated in the Bhagavad Gita, the ideal is to promote detachment from material desires and achieve enlightenment through compassion for all beings. These philosophies resonate deeply with the tenets of the Omcene.

In this context, Immanuel Kant's moral philosophy of *The Kingdom of Ends* perceives a world where all human beings are treated as ends in themselves rather than mere means to the ends of others. All rational beings act according to laws that imply an absolute necessity. According to Kant, if we are capable of moral reasoning, then we can govern ourselves according to such moral laws, indicating that we are ready to follow ourselves. Therefore, in a meta-society, we would not need laws imposed on us by "another" authority. We would determine them according to our capacity for moral reasoning. Is this a utopian idea? Is it too far-fetched to expect all of us to "do unto others as we would have others do unto us"?

In the *Kingdom of Ends*, all of us "rational beings" are expected to act and behave according to our individual capacity for moral reasoning. In this ideal ethical meta-society, we would recognize each other's intrinsic value and dignity. Therefore, all our actions would be guided by principles of mutual respect and autonomy. Ours would be an ethical community where it would be possible to govern ourselves according to universal moral principles, based on mutual respect and fairness, because we would know the value and worth of every individual.

How is this even possible? These are age-old philosophies, and we may sceptically scoff at them, because they were utopian then as they are utopian now; if they had no impact on our psyche through all these centuries, what are we proposing now? But is it too late for a paradigm shift? Is this "kingdom" attainable? In a world fraught with distrust and dishonesty, with hate and anger, what are we talking about here? In simple terms, it is the realm of acceptance, or tolerance, and of introspection. In the 1970s, Oscar Ichazo offered a valuable perspective on what we are referring to as meta-society. He "mapped" the human condition, which "provides us with the ability and knowledge to recognize and transcend our relative ego process into the Higher States of Mind, which are found in and available to each of us". He believed that all things in the universe were interconnected.

The concept of a meta-society also draws inspiration from an idealized era, such as the Satya Yuga, the first of the four yugas in the cosmic timeline in Hinduism, and what was known as the *Rama rajya*, the legendary period of peace and prosperity. The notion is the simple and universally accepted idea of "common good" across diverse cultures and ideologies. It means a social structure where the primary focus is universal well-being. Undoubtedly, peace and well-being are always at the core of any society. An ancient aphorism, a peace chant (Box 9.1) that prays for balance and universal peace, captures this spirit. SDG-17 echoes these ancient ideals. Because without peaceful and sustainable partnerships between nations, none of the specific goals can be satisfactorily attained.

We want a future characterized by a balanced science, a just society, reciprocity and harmony between people and nations, and a deep respect for our planet. How can we reach this? The choice is ours. Introspection and awakening are the essence of our endeavour. Can we make this shift and look at moving into the Omcene?

Box 9.1. Shanti Mantra (peace aphorism)

ॐ द्यौः शांतिर-अन्तरिक्षसं शान्तिः पृथिवी शान्तिर-आपः
शान्तिर-ओससाध्यः शान्तिः | वनस्पतयः शान्तिर-विश्वेदेवः
शान्तिर-ब्रह्म शान्तिः सर्वं शांतिः शांतिः-एव शान्तिः सा मां
शान्तिः-एधि | ॐ शांतिः शांतिः ||

Om!
Peace is in the sky; peace is in space; peace is on Earth;
peace is in water;
Peace is in plants; peace is in trees; peace is in the gods;
Peace is in Brahman, peace is pervading everywhere;
Peace alone is in peace.
Om! Peace, peace, peace.

Anthropocene to Omcene

The relentless pursuit of material gain ushered in the Anthropocene, an epoch marked by our detrimental impact on the planet. Conflicts rage on, casting a shadow over global well-being. However, a path towards a more harmonious existence beckons—a world built on the cornerstones of compassion, simplicity, and universal well-being. We call this the Omcene epoch.

The transition from the Anthropocene to the Omcene is not at all simple. A concerted global effort is pivotal to tackling climate change. We need global education that prioritizes empathy and environmental consciousness. We want meta-societies where sustainable practices, from renewable energy to mindful consumption, are sustainable. Each one of us has to demonstrate our willingness to engage in international cooperation to address global challenges like climate change. Diverse indigenous communities have always had a deep understanding of local ecosystems. They can offer invaluable insights for sustainable land management practices. At the same time, international scientific collaboration can bring forth innovations in renewable energy and carbon capture technologies.

We can ease geopolitical tensions through dialogue and understanding between different cultures and ethnicities. Sincere diplomacy and well-directed international relations can ensure more creative and inclusive solutions to global challenges.

The journey from "Genome to Om" is not a linear progression but a continuous evolution. The journey represents a fascinating shift in the very fabric of science and society. As we unravel the complexities of the world, we can leverage the power of science and technology to create a society that thrives on compassion and universal well-being. This, in turn, opens the doors to the Omcene, a future brimming with hope and possibility.

The road ahead is not easy. Powerful economic forces and mighty lobbies perpetuate materialism and conflict. However, history is replete with examples of humanity's capacity for positive change. By embracing the values of compassion, minimalism, and universal well-being, we can usher in the Omcene, a world where humanity thrives in harmony with nature and each other. The choice is ours: to succumb to the perils of the Anthropocene or strive for the promise of the Omcene.

The unity at both the macrocosmic and microcosmic levels lies in the one consciousness that pervades everything. Metaphorically, we represent it as Om. We know that to move forward and deal with the diversity in our utilitarian day-to-day lives, unity has to be found in the diversity. And most importantly, to think of Omcene humans as the most evolved and conscious species, we must set a target to reach their maximum potential both as individuals and as society as a whole. We trust that the Om Way can help and guide us in this evolving journey—from modern science to meta-science, from modern society to meta-society, and from the Anthropocene to the Omcene. But before we begin, it is necessary to understand the significance of this journey, "Genome to Om", and what it means to us. We need to dive deep into what Om means to us and, more importantly, what and why Om?

The Significance of Om

In the foregoing chapters, there have been frequent references to "Om" (or A-U-M). Symbolically, it is written as ॐ. In *Genome to Om*, we take Om to represent ancient wisdom. Om symbolizes the interconnectedness of all things, emphasizing the unity of existence. Om is believed to be the sound that reverberated through the cosmos at the moment of creation, symbolizing the birth of the universe. It is that cosmic sound that is said to have emerged from the void before any specific form or change was manifested. Om represents the one-ness, the spiritual merging into one consciousness, and the recognition of a unifying reality of humanity. Om is the key to life, to death, and the means to spiritual liberation and enlightenment. Our focus here is self-realization to realize that we are Consciousness itself. How do we get to this realization? The most powerful method, as was irrevocably discovered and explained by the ancient sages of India, is Om, which is central to the existence of the universe. Om is not a restricted representative of one religion or philosophy. It stands as a symbol of the unity of humanity. Here we go deeper into the concept and explanation of this sacred syllable.

Composed of an interesting trilogy, the three sounds A-U-M represent earth, atmosphere, and heaven; the three aspects of thought, speech, and action; and the three qualities of matter. The Om symbol ॐ transcends a single shape. It is composed of the three letters A-U-M in Sanskrit. It has distinct curves and markings that hold symbolic meaning. The large lower curve represents the waking state of consciousness, our everyday experience; the upper curve signifies the deep sleep state, where awareness is withdrawn;

the middle curve denotes the dream state, a realm between waking and sleeping; and the dot represents the transcendent state of pure consciousness overarching the three other states, the underlying "fourth" which runs through the three states and is their essence. Combined, the shape ॐ symbolizes the entirety of existence, encompassing all states of being and the ultimate reality, Brahman (see Fig. 9.2).

Fig. 9.2. The Om symbol.

"A-U-M" is pronounced as: *aaa* (sound coming from the naval region), *uuu* (sound from the chest), and an extended humming *mmmm* (from the throat), after which there is silence. Thus, "Om" mystically embodies the essence of the entire universe. But what exactly does that mean? What is the significance of this symbol? How can we apply it to our lives? How can it transform our spiritual practice and help us gain enlightenment? We cannot help but wonder: what is the depth of this simple syllable for it to be extolled to such heights and depicted almost as the central theme?

The Primordial Sound

Om is considered the most elemental sound, encompassing all other sounds within it. Its resonance is believed to permeate through every level of existence, from the smallest particles to the vast expanse of space. This primordial sound represents the vibration of the universe and the ultimate reality.

The concept of a primordial, universal sound or vibration is found in various cultures and spiritual traditions around the world. While they may not be direct equivalents, many cultures have similar concepts that symbolize the fundamental, cosmic sound, or universal consciousness. Across diverse cultures and religions, there is a common thread of perceiving a supreme force that is responsible for creation and governance. While the specific sounds or symbols vary, these traditions recognize the principle of symbolizing universal consciousness.

Om is very significant in Buddhism and Jainism as well. Buddhism talks of *shunya* or *shunyata*. Both Om and *shunyata* point to realities beyond ordinary conceptual understanding. Om represents the ultimate reality or essence of the universe, the interconnectedness of all things, emphasizing the unity of existence. *Shunyata* points to the absence of inherent existence in everything. It highlights the interconnectedness and interdependence of all things, as they lack independent existence. In Jainism, Om is taken as a symbol of the essence of the universe, representing the divine energy or the soul. It symbolizes the eternal nature of reality and the interconnectedness of all living beings. Om is auspicious and is often used in prayers and rituals. In Sikhism, the symbol "*Ik Onkar*" represents the one-ness of God and the interconnectedness of all creation.

Both Om and *shunyata* represent aspects of the ultimate truth or reality in Buddhist philosophy. The Buddhist mantra "*Om Mani Padme Hum*" (praise to the jewel in the lotus) is very significant. Om is chanted in meditation practices. It is believed to represent the sound of the universe and the interconnectedness of all things. Om is chanted in Tibetan Buddhism. In Japanese Shingon Buddhism, the syllable "Un" (sometimes spelt as "Aun" or even "Om") has a significant spiritual meaning. Shingon Buddhism emphasizes the use of mantras, rituals, and visualizations to attain enlightenment. "Un" also represents the fundamental sound of the universe and the interconnectedness of all things. It is associated with the three aspects of awakening, the three mysteries or aspects of reality: the physical (body), the vocal (speech), and the mental (mind). These correspond to the trilogy of "A-U-N".

"A" represents the physical aspect, the body of the practitioner, and the physical world. "U" represents the vocal aspect, the speech, and the teachings of the Buddha. "N" represents the mental aspect, the mind, and the ultimate reality. In Chinese philosophy, "Tao" represents the source of all existence, the fundamental force that underlies and unites everything. The concept of yin and yang, representing opposites in balance, is similar to the dual nature of "Om" signifying both the manifest and unmanifest aspects of reality.

Muslims use the term "Amin" or "Ameen" to conclude prayers. It voices the hope of the acceptance of prayers by the almighty God. While it is not a sound, the Hebrew word "Shalom" represents peace and completeness. In Christianity, especially in the Bible, "Amen" is often used at the end of prayers and hymns. God's governance is seen through divine providence and the belief in a purposeful plan for humanity. The divine providence and purposeful plan are like the Hindu concept of "Om" as the reality governing existence.

Explaining Om

The *Mandukya Upanishad* explains the states of consciousness (Box 9.2) and the nature of the ultimate reality with the symbol Om. The sound of A-U-M during pranayama, meditation, and techniques like Bhramari Pranayama goes beyond mere symbolism. The systematic changes in pronunciation of the A-U-M are believed to have a significant impact on both our physiology and psychology. Studies suggest these vibrations can influence metabolism, brain function, and mental well-being. For example, Bhramari Pranayama, which incorporates extended humming of the "mmm" sound of A-U-M, is linked to stimulating the hypothalamus gland. Also referred to as "humming bee breath" it is a calming breathing technique with a humming sound emanating from the back of the throat. It is immensely relaxing, and it lowers stress. Similarly, chanting A-U-M silently with very steady and rhythmic breathing during meditation helps in focusing on an internal point of concentration, which quietens the mind and helps with single-pointed focus

Box 9.2. Om as the unity of consciousness in Mandukya Upanishad
The Mandukya Upanishad, the shortest but the most compact of the Upanishads, is all of 12 aphorisms. It explains the nature of ultimate reality of Brahman with the help of the syllable "Om" (verily all this), which is a composite of three letters: A, U, M. This Upanishad talks of three states of knowing and the corresponding three states of being, which together form the ultimate reality. The nature of this reality is explained by analysing our experience to show the underlying unity of one and the same consciousness running through all the three states of experience such as jagrat (waking), swapna (dreaming), and sushupti (deep sleep).

Corresponding to these states there are three kinds of knowers named as Vaishvanara, Taijasa, and Prajna. Vaishvanara, symbolized with the letter A, experiences and enjoys the material world of gross physical objects as "outside" of him or her. Taijasa, symbolized with the letter U, experiences and enjoys the subtle dream world as "inside" of him or her. Prajna, symbolized with the letter M, does not experience anything either outside or inside but enjoys the sheer bliss, which is undifferentiated mass of consciousness devoid of all dualities. Thus, there is an uninterrupted continuity of consciousness throughout the three states of experience, which correspondingly establishes the identity of three kinds of knowers as one unitary Self.

In addition to these three states, the Mandukya Upanishad mentions without naming the fourth state (turiya) along with but not in addition to the three states of knowing and being. This state is absolutely devoid of any characterization whatsoever but is the essence of consciousness, of oneness of the self, is a cessation of all expression, quiescent, blissful, and without the second. This fourth state is the Self, which has to be realized.

[Courtesy: Prof. Sharad Deshpande, Pune, India]

on Self-realization. Chanting Om and following the technique of rejecting each sheath of this body–mind, as discussed earlier, is a direct path to discovering "Who am I?". Overall, Om is a multifaceted tool for realizing consciousness and improving well-being, working through both audible chanting and silent internal focus.

Om is the reverberation at the beginning and at the end of all Vedic chants. It is integral to the practice of the breathing and meditation regimes in Ashtanga Yoga. The comprehensive path of the four yogas, as explained in chapter 7, gives us the freedom to choose any one or a balanced combination of all four paths to find our way towards self-realization in this journey of "Genome to Om". So, here we need to understand Ashtanga Yoga, which is the practice methodology for Raja Yoga. Regular practice of Ashtanga Yoga ensures that we can be stable, in good health, and motivated to walk the path towards the Om Way.

The Eight Limbs for Well-being: Ashtanga Yoga

Yoga offers a unique perspective on the quest for knowledge through the subtle shades of experience. As opposed to the popular notion, "yoga" is not merely *asana*s or postures, physical exercises, and arduous discipline of the body and mind. Undoubtedly, regular practice of yoga builds strength, improves flexibility, and enhances circulation, creating a strong and resilient physical foundation. But in essence, yoga means "union". It is the union of body and mind, incorporating breathing exercises, meditation, and ethical principles. In ancient Eastern philosophy, as in Jainism, Buddhism, and Hinduism, yoga is often interpreted as the union of the individual self with the universal consciousness, or the union of the individual with the divine. The purpose of yoga as explained in Patanjali's Yoga Sutras, is to realize our true nature. The philosophical significance of yoga, therefore, is not mere physical exercise or relaxation techniques; it means a profound spiritual journey towards self-realization, attaining inner peace, and freedom from suffering.

Performing perfect yoga posture is possible through meticulous practice and experience. Such experience also enriches skills. However, deeper spiritual feeling in the meditative states is beyond experience, which can only be felt through realization. In Vedic system, these levels of experience are known as *anubhava* and *anubhuti,* respectively. The former is linked more to knowledge and skill, while the latter is linked more to wisdom.

The eight limbs of Ashtanga Yoga, outlined by Patanjali, are a roadmap to holistic well-being. These eight interconnected steps progress from outward ethical conduct to inward awareness, and ultimately, to a state of complete union of the body–mind and Atman (see Fig. 9.3).

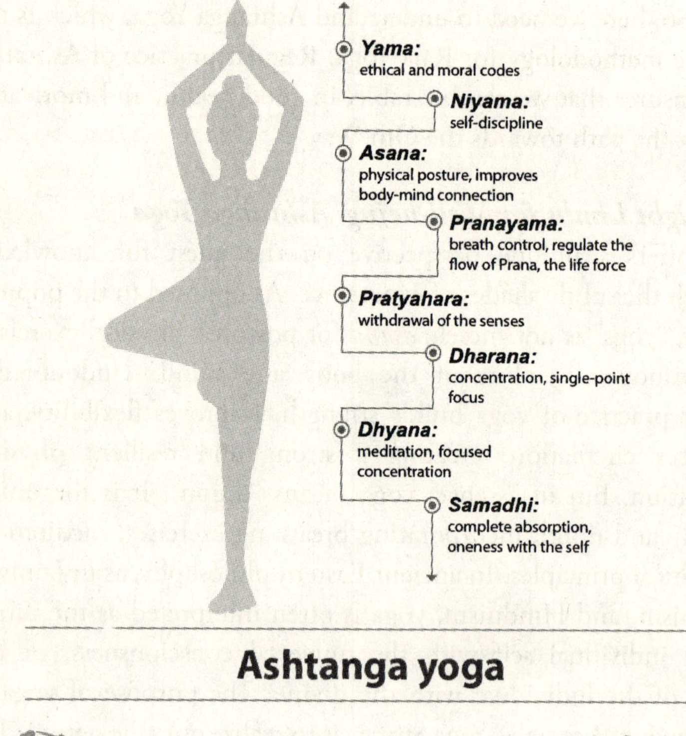

Fig. 9.3. The eight limbs of Ashtanga Yoga.

The first two limbs, *Yama* and *Niyama*, focus on ethical principles and self-discipline. These cultivate a strong foundation for inner exploration. *Asana*, the most well-known limb, is a range of physical postures that improve strength, flexibility, resilience, and mind–body connection. *Pranayama*, the yogic art of breath control, refines the flow of energy and sharpens focus to link the body and mind. Using specific breathing patterns such as inhaling, exhaling, and holding the breath in regimented ways, the practice improves the flow of *prana*, the life force in the body, and gives a range of other wellness benefits. Proper breathing and meditation techniques calm the mind, enhance mental clarity, boost energy levels, and help us manage stress, a major culprit behind many health problems. *Pratyahara*, the withdrawal of the senses, directs attention inward, while *Dharana*, concentration, holds that focus on a single point. *Dhyana*, the seventh limb, signifies a state of deep meditative absorption, where the mind transcends focus on a single point and becomes effortlessly one with the object of contemplation. It deepens the state of inner awareness. Finally, *Samadhi* is the culmination, a state of complete absorption and liberation. While often described sequentially, the eight limbs work together. Each step supports the journey towards a more mindful, meaningful, and integrated life.

This path to inner peace and self-awareness finds echoes in contemporary practices like mindfulness meditation, focusing on present moment awareness; Vipassana meditation, observing fleeting thoughts and sensations; and transcendental meditation, using a *mantra* to achieve deep relaxation. Against this background, we reiterate that yoga and spirituality can be a highway for the evolving journey to meta-science, meta-society, and the Omcene.

Unity in Diversity—The Way Forward

While the challenges are immense, the answer doesn't lie in uniformity. In fact, diversity of ideas, cultures, and scientific approaches is our greatest strength, but only when we can unify it all for a common purpose. Technological advancements will continue, but they must be guided by a united front. The concerns and perils are known. Therefore, ethical considerations, assessments of environmental impact, and robust regulatory frameworks that involve all the stakeholders are crucial. Only then is it possible to ensure responsible development of the technologies of today and tomorrow, like AI and bioengineering.

Are we ready to take it all in stride knowing who or what we really are? Are we ready to embrace the richness of diversity by recognizing our shared humanity? If yes, then what do we need to do? We must be all-inclusive, where we, the people from diverse backgrounds, are welcome. We must discard our discriminating, prejudicial, judgemental, and marginalizing attitudes. Finding unity in diversity is an ongoing process that needs continuous learning, self-reflection, and growth.

This emphasizes the importance of rediscovering the wisdom embedded in our cultural heritage. Ancient Vedic aphorisms like "*Vasudhaiva Kutumbakam*" (the world is one family), "*Ekam Sat Vipra Bahudha Vadanti*" (the truth is one, the wise give it different names) which promotes inclusivity and respect for diverse perspectives, and "*Ano Bhadra Krtavo Yantu Vishvatah*" (let noble thoughts come to us from all directions). We call this an open-minded, rigorous meta-scientific approach.

One hundred and fifty years ago, Sri Ramakrishna said that we all go to take water from the well. It is the same water, but it may have different names. For some, it is water; for others, it is *pani*, *jal*, or any of the countless names (aqua, eau, or any other). For scientists, it is just H2O. But it is the same thing and is accepted as the most vital element of life. The crux is to be free to embrace our own way of being, but also embrace others' ways of being, and hold on to the promise of a more compassionate, sustainable, and harmonious world for all. These principles can promote a healthy sense of global community, a crucial aspect of collective consciousness.

The Overarching Goal

The goal isn't a singular state of perfection for the elite few, but a collective awakening and a shift towards a sustainable future. The challenges will be formidable, but the alternative is extinction. The scepticism towards meta-science, the complexity of expecting changes in individuals and societies, and a conscious discarding of vested interests if they stand as a threat to sustainability are just a few of the several hurdles. Defining "universality" or universal well-being, which means ensuring inclusivity across diverse cultures and perspectives, is in itself formidable.

We know that there are different points of view about what people consider the ideal world for them. We will probably never find a world without suffering and pain. The coin will always have two sides. Perspectives will always be divided. What is perfect for one is incomplete for the other. The world, the macrocosm, is in a constant state of flux, just as is the microcosm, the individual.

Social, economic, political, and environmental systems are completely interconnected today, and that makes it challenging to address all issues simultaneously. The reality on the ground, however, is that if a perfect world may be unrealistic, that does not mean that improvement or progress in that direction is futile. A so-called "ideal or near-perfect world" is one where, despite its imperfectness, we are surrounded by compassionate, caring human beings. We have countless exemplars of individuals, communities, and organizations who work tirelessly towards humanitarian causes, and they face endless social, environmental, and political challenges. The aim of creating a better world for the present and future generations is a universal goal and purpose. For that, we need to be compassionate and cooperative, just and non-judgemental, and most importantly, at peace and harmony within ourselves.

The Ascent of All: A Call to Realize Our Full Potential

In the present challenging times, we must move beyond the "Ascent of Man" narrative, which often celebrates individual perspectives and prowess at the expense of the universal. Let us instead champion the "Ascent of All", where scientific advancement, societal betterment, and individual flourishing are inextricably linked. The ascent of all is only possible if, at the individual level, we are already ascending. This inward journey to transcend upward will require unwavering determination, discipline, and an unshakeable conviction in the transformative power of both science and self-awareness. However, the potential rewards are equally immense. By embracing this grand vision, we can usher in an era of Omcene—a future where humanity leverages its full potential to create a world that thrives, not just survives.

The way forward demands unwavering dedication—none other than the arduous journeys undertaken by spiritual seekers throughout history. The Bhagavad Gita warns of its immense difficulty: "Among thousands who aspire towards spiritual perfection, only a handful try to attain comprehension." Even those who reach such heights find true understanding elusive.

The wisdom of Lao Tzu applies perfectly to the Om Way. "A journey of a thousand miles begins with the first step." This reminds us that even the grandest spiritual pursuit starts with a single act of dedication, like the first firm step in the right direction on the path. However, just like Alice tumbling down the rabbit hole, simply entering the extraordinary world of the Om Way isn't enough. Without understanding its principles and practices, we risk wandering aimlessly, like Alice in Wonderland. Therefore, while the initial spark of curiosity is vital, true progress on the Om Way requires both the courage to take the first step and the wisdom to seek guidance on the path towards enlightenment.

As we now look to the future, we again listen to Yama's teachings to young Nachiketa in the Katha Upanishad. Yama, the Lord of death, points to the choices that define humanity's path. Life brings us two kinds of experiences, and it is up to us to choose. These are *preyas*, with the allure of being pleasing to the senses and immediately gratifying—comfort, material possessions, and the inflation of the ego. Or there is *shreyas*, which represents the pursuit of universal well-being, characterized by compassion, commitment to justice, the spiritual path, and unwavering dedication to the greater good. While *preyas* may offer fleeting satisfaction, *shreyas* paves the way for the future, which is meaningful and purposeful. By embracing *shreyas,* we can choose not to simply survive but to thrive as a species, bound together by empathy and a shared vision for a flourishing world. It is up to us to make a prudent choice.

This is a challenge. Let us strive, both individually and collectively, to create a new dawn—an era where science and spirituality coalesce for the universal good. Seeking unity in diversity is not a vague, distant ideal; it is a beacon of hope, illuminating the path towards a brighter future for all. Even a single step on this path can change the entire meaning and purpose of our lives, leading us towards peace, universal well-being, and liberation. This is an ongoing journey, perhaps as long as humanity exists, towards a future where the objectives of individual progress and the well-being of all life and the planet become a common goal. *It is the road to be travelled.*

We trust that in this evolving journey from science to meta-science and meta-society, every conscious individual will play a key role. Not just to thrive for personal happiness and peace but to reach our full potential, doing true justice to this precious life. And also contribute our bit to the larger cause of moving towards universal well-being. This will be true *moksha, kaivalya,* and *nirvana.*

Glossary of Vedic Terms

Ahimsa, the principle of non-violence that applies to action towards all living beings. In the deepest sense, it is about the intent more than the action itself. It is an attitude of universal benevolence.

Ajiva, non-living matter; inanimate, lifeless.

Anahata, also *anahat*, the unstruck, timeless sound of the universe expressed as Om.

Anubhava, the "experiential knowledge", direct perception or cognition, similar to the concept in modern science, emphasizes observation, experimentation, and analysis. *Anubhava* is experiencing through the senses and the mind. What we perceive through our sense organs and the experience of the perceptions is *anubhava*. It also refers to knowledge or skills gained through doing a job or activity. In Vedanta, it is intuition, or experience of an individual's awareness of knowledge of Brahman, the Absolute. It is the knowledge gained or apprehended through experience, as opposed to that as recalled by memory or *smriti*.

Anubhuti, the "sensory and emotional response", knowledge based on direct perception, inference, and comparison. *Anubhuti* is a subjective or emotional feeling. It is a feeling, an independent phenomenal experience. It is a general insight caused by a particular experience, a realization or a mental experience.

Apara Vidya, knowledge focused on the finite world based on the intellect and senses.

Ashtanga Yoga, the practice of Yoga, as divided into "eight limbs"; each limb is developed to bring the body and mental energy under control. The core of the eight limbs is *ahimsa*, non-harming, which can be considered the foundation of the eight limbs. *Yama* and *Niyama* focus on ethical principles and self-discipline. *Asana*, a range of physical postures that improve strength, flexibility, resilience, and mind-body connection. *Pranayama*, the yogic art of breath control, refines the flow of energy and sharpens focus to link the body and mind. Using specific breathing patterns such as inhaling, exhaling, and holding the breath in regimented ways, the practice improves the flow of *Prana*, the life force in the body, and gives a range of other wellness benefits. *Pratyahara*, the withdrawal of the senses, directs attention inwards, while *Dharana*, concentration, holds that focus on a single point. *Dhyana*, the seventh limb signifies a state of deep meditative absorption, where the mind transcends focus on a single point and becomes effortlessly one with the object of contemplation. It deepens the state of inner awareness. Finally, *Samadhi* is the culmination, a state of complete absorption and liberation.

Astika, from the Sanskrit word, *asti*, meaning there is or there exists; theist, one who accepts the authority of the sacred scriptures. There are six *astika* schools (*darshanas*) of Hindu philosophy: Nyaya, Vaisheshika, Sankhya, Yoga, Mimamsa, and Vedanta.

Atman, a very basic concept in Hinduism, and in the Vedanta philosophy. It refers to the inner and real Self, the core of the living being, the very essence. The Atman is indestructible and inseparable from the Universal Self, Brahman. It is the underlying essence of the individual being, identical to the eternal core of the personality that after death either transmigrates to a new life and a new body-mind complex, or is freed from the cycle of *samsara*, by attaining *moksha*, or liberation from the bondage of existence, merging eternally into Brahman.

Avatara, incarnation, or material appearance. In Indian philosophy, it specifically refers to the reincarnation of a deity or realized soul, as a messenger to help humanity. In Hinduism, it refers to the ten reincarnations of Lord Vishnu.

Bhakti Yoga, the path of devotion. The devotee surrenders completely to God, to his or her chosen ideal and with complete faith and surrender takes everything that comes in life as "God's will". This kind of faith does not question, nor doubt. But it is not blind faith driven by fears or superstitions. This is the path of pure love, and the follower attains *moksha* or 'merges' with his or her chosen ideal and is freed from the bondage of *samsara*.

Brahman, the supreme, unchanging, omniscient, omnipresent, universal reality. It is the eternal principle irreducible core of existence-consciousness-bliss, *sat-chit-ananda*. Brahman is nondifferent from its manifestation or projection, the universe, the world. It is the cause of all creation, maintenance, and destruction. The 'Om' symbol represents the essence of Brahman.

Chakra, the centre of energy in the subtle body as indicated in Yoga. The seven centres, from the *Mooladhara* (at the base of the spine), *Swadhisthana* (in the sacral bone), *Manipura* (at the navel), *Anahata* (in the chest or heart region), *Vishuddha* (in the throat), *Adnya* (the forehead), and *Sahasrara* (top of head) are connected by several *nadi*s, the channels through which the life force or vital energy moves.

Chitta, mind, state of mind or mindset, or to be conscious. In Vedanta, the subconscious and unconscious mind is called *Chitta*. It is considered the unconscious storehouse or reservoir of all impressions. Recognizing the movements of thoughts is imperative to control the flow of random or uncontrolled thoughts in the mind.

Chitta vritti nirodhah, the cessation of the fluctuations of the mind. This is emphasized in the second aphorism of the Yoga Sutras by Patanjali. It points out that Yoga ensures the control of the movements of the mind, and stresses the importance of keeping the mind in control

Darshana, literal meaning, a glimpse of a view. In Hindu philosophy, school or principal system of philosophy: Nyaya, Vaisheshika, Sankhya, Yoga, Mimamsa, and Vedanta.

Dosha, the three fundamental principles or humors in Ayurveda, Kapha, (phlegm) Pitta (bile), and Vata (wind). They correspond to the elements of nature, of water, fire and air, respectively. These *dosha*s are said to be responsible for the physiological, mental and emotional health of human beings.

Ekam Sat Vipra Bahudha Vadanti, the truth is one, the wise give it different names. This statement from the Rig Veda rings all the truer today, as unity in diversity is the goal.

Garbha samskara, derived from the two Sanskrit words, *garbha* (meaning the womb, and thus foetus in the womb) and *samskara*, is educating the mind, dealing with the imprints and impressions. Traditionally, it is believed that a child's mental development starts at conception because it already has brought forward past impressions.

Guna, a tendency of the mind, body, and consciousness; the individual attributes and qualities. There are three *guna*s that are present in every human being in different ratios or proportions, and they indicate behavioural and thinking patterns; they are also connected to mental and physical health. *Sattva guna* manifests as purity and goodness; calm and harmonious; *Raja guna* manifests as passion, hectic activity, and restlessness; *Tama guna* displays as inertia, ignorance, laziness, and tardiness. All these attributes of energy are present in every person in different proportions, often with one *guna* predominating at a time. The mental state of a human being can vary as well between the three states of energy. Foods are also identified as being either *sattvic*, *rajasic,* or *tamasic*, and a prudent choice of appropriate foods is always recommended for good health and well-being.

Jeevo Jeevasya Jeevanam, the life of one being is dependent on that of another living being. All living beings are connected. The link of the food chain has a deep significance and must not be ignored. The web of life concept is rooted in this Vedic concept.

Jiva, a living sentient being. Most *jiva*s are bound to the cycle of *samsara*, to the cycle of birth and death, based on their karmic cycle.

Jivatma, an individual soul, the "self", which identifies itself as a *jiva*, the embodied being, connected as it is to the psychophysical system.

Jiva-mukta, the realized soul or being, the atman, who lives in the world as a realized perfected being, finishing off the final karmic deeds before giving up the physical body to attain final liberation, never to be born again.

Jnana, (also *gyan*) knowledge. In the realm of Indian philosophy, *jnana* is spiritual knowledge or wisdom, the kind of knowledge that is a total experience or awareness of its object.

Jnana Yoga, the path or discipline of knowledge or self-realization, through intellectual inquiry and study of the true Self.

Kaivalya, a state of aloneness, isolation, and solitude. The person who attains this state is completely detached from the material world. This is the prime goal of those practising Raja Yoga. The individual is free of all relationships, devoid of ego and arrogance; and is free from the karmic cycle of birth and death. To achieve this state the disciple performs austerities and follows the practice of yoga very earnestly. He or she must isolate himself or herself from the external world. The person who attains this state of pure consciousness is in a state of *kaivalya*.

Kapha, one of the *dosha*s, characterized by the qualities of earth and water. People with a predominant *Kapha* constitution are calm, compassionate, and nurturing. They have a sturdy build, smooth skin, and slow digestion. Balanced *Kapha* provides stability, strength, and immunity.

Karma, action, deed, also the result of action. The word originates from the Sanskrit verb, *kri*, which means "to do". *Karma* is also driven by the law of cause and effect. Every action has an effect. Whatever actions we perform, with whatever intent, the kind of mental motivation we have while performing any action, and the impressions they leave on our lives and on the lives of those around us contribute to the results of *karma*. Good, selfless deeds generate good *karma*; self-centred, actions motivated by personal benefit, especially with disregard of the impact on the world around us, generate a negative balance in the *karma* count! Nature gives back to us what we give to it.

Karma Yoga, one of the paths of four Yogas, Karma Yoga, is the yoga of selfless action or dedicated work. We perform actions all the time. The work done to benefit others, without any desire for the fruits thereof for oneself, is the path of this yoga. We can break the bondage of the chain of cause and effect by performing *nishkaam karma* (selfless action). It is essential to disengage the ego and the individual self and offer the results of all our actions either to a higher being, to self-realization, or in the service of the divine in other beings.

Kosha, sheath or covering. According to Vedanta philosophy, the explanation given for the human framework is that it is encased in five sheaths that cover the real inner Self, the Atman. It is through these five layers that one goes through all experiences. The first is the *Annamaya kosha*, comprising the physical body, which is entirely matter and is nourished by food. *Pranamaya kosha* is the vital energy or life force, but it is not the real identity of the human being. *Manomaya kosha* is the level of the mind and all its activities. *Vijnanamaya kosha* is the fourth layer of reality, cognition, intellect, and wisdom. That again is the activities of the human being, not the true Self. The fifth, the *Anandamaya kosha*, is the innermost layer, or the level of bliss. The person is in a state of harmony and health without the vacillations of thoughts and emotions.

Kundalini, said to be the cosmic power in individual human beings. It is a vital life force, an energy, a spiritual potential of a dynamic energy, that is concentrated as a dormant force, 'located' at the base of the spine. It is visualized as a sleeping serpent lying coiled up there. It is awakened and raised through yogic methods of meditation as specified in Raja Yoga. This energy rises through the spine in a channel called the *Sushumna*, in the subtle body and it activates the seven centres of energy in the body, the *chakra*s.

Mahavakya, Great Statement, a declaration found in the Upanishads. The great sayings or aphorisms of Vedanta convey the same thought of the oneness of all, and that the Atman and Brahman are the same. The self-realized individual being, in its pure essence, is the same as the universal reality. The four primary *mahavakya*s are:

Tat Tvam Asi—That Thou Art, from the Chandogya Upanishad of the Sama Veda.

Aham Brahmasmi—I am Brahman, from the Brihadaranyaka Upanishad of the Yajur Veda.

Prajnanam Brahma—Consciousness is Brahman, from the Aitareya Upanishad of the Rig Veda.

Ayam Atma Brahma—This Self, the Atman, is Brahman, from the Mandukya Upanishad of the Atharva Veda.

Maya, in the Advaita Vedanta school of Indian philosophy, is a fundamental concept that means 'appearance' or 'ignorance'. It's a powerful 'veiling' force of Nature, that creates the illusion that the phenomenal world is real. At the individual level, Maya's veiling power is the *jiva*'s lack of knowledge of the real Self, Atman-Brahman, and its identification with the body-mind complex. Maya is a simple statement of facts — what we are and what we see around us. Maya's revealing power gives absolute reality to what we perceive, which is at best the seeming reality. The delusion that clouds our understanding of reality is Maya.

Moksha, freedom or liberation from *samsara*, the cycle of birth and death. Psychologically, *moksha* means being free from ignorance of our identity, knowing who we are, being clear about the purpose of life, and moving towards the state of self-realization. *Moksha* is the final goal as we progress in the four paths of life; it is the goal to move towards, even as we work our way through the world and all our commitments.

Nadi, channel or pathway of *prana*, vital energy, in the subtle body. There are three main *nadi*s within the spinal cord, through which the vital force of energy, the *Kundalini* rises: *Ida* (right) and *Pingala* (left) and they run parallel to the *Sushumna nadi.* The *Kundalini*, the supreme energy, rises from the *mooladhara* through the *Sushumna nadi.* When it reaches the *sahasrara*, then the Yogi gets detached from the body and mind. There are said to be 72,000 *nadi*s in the human system that work as channels or pathways of energy in the system.

Nastika, atheist, the one who does not believe in God; non-acceptance of the "Self". There are three *nastika* schools of philosophy: Charvaka, Buddhism, and Jainism.

Nirvana is the supreme goal, the same as *moksha* and *kaivalya.* The Sanskrit term *nirvana* is associated with Buddhism. Gautama Buddha sat under the Bodhi tree in deep meditation till he could understand and realize the purpose of life. He attained *nirvana*, which means the complete wiping out of desires, of all emotions, and of all suffering. It is the ultimate state of self-realization and ends the cycle of birth and death. In his sermons, the Buddha talked of the cessation of suffering and that leads to *nirvana*. The concept of cessation of suffering is explained by the Buddha in his tale of the two arrows. We get pierced by two arrows in life. One that the world or nature throws at us, of physical suffering, ageing, what people cause, and ultimately death. The results of that arrow cannot be avoided. But the second arrow is our response or reaction to the impact of the first arrow. That is in our control. With the cessation of emotional reactions and all negative emotions we overcome everything and thus are free.

Pancha Mahabhoota, the five basic, fundamental elements, of the universe: earth, water, fire, air, and ether. Knowing of the five elements, the practitioners of Yoga try to understand the laws of nature and focus on assuring greater health, power, knowledge, wisdom, and happiness.

Para Vidya, the higher or spiritual knowledge essential for understanding the universe, our lives on Earth, and ourselves.

Pitta is a *dosha* associated with the qualities of fire. *Pitta*-dominant individuals are ambitious, focused, and passionate. They tend to have a medium build, fair or ruddy complexion, and strong appetite. *Pitta* governs the digestion, metabolism, and regulation of the body temperature.

Prajnana, meta-science; the term stands for higher order, inner profound knowledge, wisdom, ultimate reality, or Brahman.

Prajnanam Brahma, a *mahavakya*, Great Statement, meaning "consciousness is supreme".

Prakriti, nature, the basic cosmic material, the root of all beings. The general meaning of Prakriti is nature, constitution, and general characteristics. In Sankhya philosophy, Prakriti is the root of all beings and combines with Purusha, to create the universe. Prakriti is also seen as the First Cause, all aspects of the reality of the manifested material world. It is the principle of matter.

Prana, in Sanskrit, literally means "life force" or "life energy", and that which connects all beings, suggesting an underlying unity in the diversity of life. *Prana* is the subtle energy that permeates all forms of life. It is the principle of vitality that flows through living beings. It is often associated with breathing but is considered more than just the physical aspect of breathing.

Pranayama, the technique of controlling the vital flow of energy, through breathing.

Prayopavesha, the resolve to die through fasting. Those persons who have completed their responsibilities and have no desires left, opt to leave the body. This finds approval in cases of terminally ill people.

Preyas, that which brings immediate gratification. It is that which gratifies the senses and is instantly pleasurable. This is taken in contrast to *Shreyas*.

Purusha, the cosmic entity or Being, exists beyond the realms of time and space, the pure consciousness or conscious energy that governs life and reality. The Self, the universal principle, is synonymous with Brahman. It is the eternal, indestructible formless and all-pervasive Universal Principle. In Sankhya philosophy, Purusha is the principle of consciousness, the conscious witness, the cosmic principle that manifests and unites with Prakriti and projects nature and the material world with which all human beings relate.

Raja Yoga, a scientific approach of controlling thought waves, with a focus to transforming mental and physical energy into spiritual energy. The main practice of Raja Yoga is meditation. According to Raja yoga, *kaivalya*, *moksha*, or *nirvana* is the final stage of enlightenment that a yogi can reach. In this state, the yogi becomes completely fearless and free. Although sometimes wrongly perceived as negation or annihilation, these are the ideal states of total awareness.

Samadhi, a state of body-mind stillness, attained in deep meditation, when the individual is completely absorbed in contemplation or in an undisturbed state of 'thoughtless-ness'. With attention focused on the Absolute, the mind is undisturbed and remains single-pointed. It is also a state of joyfulness, calmness, but remaining in full mental alertness.

Samsara, (also *sansara*), is central to the concept of reincarnation, metempsychosis; it represents the cycle of birth, death, and rebirth, which is dependent on the individual's past and present deeds (*karma*). The goal of every *jiva* is to get out of the bondage of *samsara* and attain *moksha*.

Samskara, (also *sanskara*), impression or imprint left on the mind-body by education, upbringing, nurturing, and experiences. The imprints on the mind can be positive or negative and they shape predispositions, behaviour, and even the future life paths. *Samskara* also involves a more metaphysical transmission through *karma* and rebirth.

Santhara, a tradition in Jainism of fasting until death.

Shiva-Shakti, the concept is that Shiva represents the masculine energy and is associated with consciousness, while Shakti represents the feminine energy and is associated with energy and power. Together, they create the dynamic interplay of opposites that is essential for spiritual growth and self-realization. Shiva represents the basic elements of the universe, and Shakti makes the elements come to life and act

Shreyas, that which brings long-term joy, happiness, and well-being. This is taken in contrast to *Preyas*.

Shunya, or *shunyata*, the concept of empty, nothingness, or void. The Buddha pointed to the emptiness of the world, that it was momentary, transitory and that it was full of *dukkha*, sorrow.

Siddhi, extraordinary and extrasensory power or ability that can arise from deep spiritual and yogic practice. The Sanskrit word *siddhi* means "perfection", or to attain. These powers are said to be the natural consequence of very austere and sincere yogic practices and are acquired in the advanced stages of concentrated yoga practices. These powers can range from telepathy and clairvoyance to more "magical" profound abilities like changing the physical form, or levitation, reading other minds, and divine vision. However, within the yogic tradition, *siddhi*s are addressed with strong words of caution, as they can become distracting powers that take the yogi away from the path of self-realization and *moksha*, if they become focal points in boosting the individual ego.

Sushumna, the central *nadi*, channel, connects the seven *chakras* starting from the *mooladhara* (at the base of the spine) to the *sahasrara chakra* (the top of the head) in the subtle human body. With the concentrated practice of Ashtanga Yoga, meditation, and psychic control, the practitioners can raise the vital force of the *Kundalini* through the *Sushumna* and open up the *chakras*, to be able to go beyond the limitations of the psychophysical system and attain *moksha*.

Swasthya, good health and well-being, or a state of "balanced within the self". To stay in good health, all the *doshas*, the energies that *define* the individual's physical makeup, must be balanced. The digestive system, the body tissues and waste must function normally, the sensory and motor organs must be in harmony and the mind must be in a tranquil state.

Upanishad, the philosophical teachings of the Vedas. These are a collection of Hindu scriptures, the knowledge sections of the Vedas, the core of the Vedanta Philosophy, and the texts found at the end of the Vedas (Ved-anta). Literally, it means, "being seated next to the teacher", indicating the *guru-shishya parampara*, the method of teacher-student tradition, following the mentoring, which was originally an oral teaching tradition, with the student staying near the teacher. There are said to be 108 or even more Upanishads but among the primary ten are: Brihadaranyaka, Chandogya, Katha, Kena, Aitareya, Taittirya, Prashna, Isha, Mundaka, and Mandukya.

Vasudhaiva Kutumbakam, an Upanishadic phrase, which means the world is one family. It emphasizes the global perspective of universal well-being.

Vata, a *dosha* characterized by qualities of air and ether (space). Those with a dominant *Vata* constitution tend to be lively, creative, and enthusiastic. Typically, they have a light build, dry skin, and irregular digestion. *Vata* governs movement in the body, including circulation, breathing, and elimination.

Veda, ancient and sacred Hindu scripture, said to be the oldest in the world, written in Sanskrit. Even today, no definite date is ascribed to these scriptures. The classic dating is of about 1500–1200 BCE. While several scholars today are increasingly dating these scriptures as 8,000 years old. There are four Vedas: Rig Veda, Sama Veda, Yajur Veda, and Atharva Veda, each Veda consists of four parts: Samhitas (hymns), Brahmanas (rituals), Aranyakas (meditations), and Upanishads (philosophical teachings). The Rig Veda is said to be the knowledge of the verses; Yajur Veda is the knowledge of the sacrifice; Sama Veda is the knowledge of the chants. These three Vedas are collectively known as the *trai-vidya*, or the threefold knowledge system. The fourth Veda, Atharva Veda is a collection of hymns, incantations, and even magic spells.

Vidya stands for information and knowledge.

Vijnana, (also *Vigyan*), the specialized worldly knowledge, empirical science.

Yoga, the theory and practice based on the concept of the union of the body-mind, a series of physical, mental and spiritual practices. The focus is to still the mind and ensure inner balance and detachment.

Yoga Sutras, a collection of 195 aphorisms focused on the theory and practice of Yoga, compiled by Sage Patanjali, in the early centuries of the Common Era.

Bibliography

Alexander, Eben. *Proof of Heaven: A Neurosurgeon's Journey into the Afterlife*. UK: Piatkus, 2012.

Aristotle. *De Anima*. Kindle edition. Penguin, 2004.

Bakshi, G.D. *The Sarasvati Civilisation.* New Delhi, India: Garuda Prakashan Pvt. Ltd, 2019.

Bodeker, Gerard and Chi-Keong Ong. *WHO Global Atlas of Traditional, Complementary and Alternative Medicine*. Vol. 1. World Health Organization, 2005.

Bostrom, Nick. *Superintelligence: Paths, Dangers, Strategies*. Oxford, UK: Oxford University Press, 2016.

Bronowski, Jacob. *The Ascent of Man*. Paperback edition. London: BBC Books, 201. https://www.bbc.co.uk/programmes/b00wms4m

Capra, Fritjof. *The Tao of Physics*. India: HarperCollins, 2007.

Chalmers, David J. *The Conscious Mind: In Search of a Fundamental Theory.* New York, USA: Oxford University Press, 1996.

Chalmers, David J. *The Character of Consciousness (Philosophy of Mind)*. Illustrated Edition. USA: Oxford University Press, 2010.

Clynes, Manfred and Nathan S. Kline. "Cyborgs and Space," 1960. Cyborgs-manfredclynes-nathan s-kline-in-astraunotics.pdf

Cronin, Lee. *Future Life*. University of Glasgow. https://www.gla.ac.uk/research/beacons/futurelife/professorleecronin/

Danino, Michael. *Lost River: On the Trail of the Sarasvati*. India: Penguin, 2010.

Darwin, Charles. *On the Origin of Species*. 1859.

Dawkins, Richard. *The Selfish Gene*. 4th edition. UK: Oxford University Press, 2016

Descartes, Rene. *The Complete Works of Rene Descartes*. Kindle edition.

Divyanandaprana, Pravrajika. *The Science of Happiness*. New Delhi, India: Ramakrishna Sarada Mission, 2020.

Eagleman, David. *Incognito: The Secret Lives of the Brain*. Reprint edition. New York, USA: Vintage Books, 2012.

Einstein, Albert. *The World as I See It*. New York, USA: Citadel, 2006.

Erikson, Erik, H. *Identity, Youth and Crisis*. New edition. New York, USA: W.W. Norton & Company,1995.

Freud, Sigmund. *The Unconscious*. UK edition. UK: Penguin Classics, 2005.

Freud, Sigmund. *Beyond the Pleasure Principle*. The standard edition. New York, USA: W.W. Norton & Company, 1990.

Ganesan, A., Gauthaman, J., and Kumar, G. "The Impact of Mindfulness Meditation on the Psychosomatic Spectrum of Oral Diseases: Mapping the Evidence." *Journal of Lifestyle Medicine* 12, 1 (2021), 1–8. https://doi.org/10.15280/jlm.2022.12.1.1

Hancock, Graham. *Fingerprints of the Gods: The Evidence of Earth's Lost Civilization*. Canada: Anchor, 1998.

Hawking, Stephen and Leonard Mlodinow. *The Grand Design*. UK: Bantham Books, 2011.

Holt, Jim. *Why Does the World Exist?* London: Profile Books Ltd, 2012.

Hegde, B.M. *What Doctors Don't Get to Study in Medical School*. 4th edition. New Delhi: Paras Medical Publisher, 2019.

Infurna. F.J., et al. "Loneliness in midlife: Historical increases and elevated levels in the United States compared with Europe." American Psychological Association, 2024. https://psycnet.apa.org/record/2024-59761-001?doi=1

Jung, Carl, G. *Man and His Symbols*. 12th edition. Kathmandu, Nepal: Rhus Publishers, 1968.

Jung, Carl. G. *Memories, Dreams, Reflections*. Re-issue edition. Kathmandu, Nepal: Rhus Publishers, 1989.

Koch, Christof. *Biophysics of Computation: Information Process in Single Neurons*. Computational Neuroscience Series. Oxford University Press, 2004.

Koch, Christof. *The Feeling of Life Itself: Why Consciousness Is Widespread but Can't Be Computed.* Audiobook. 2020.

Kurzweil, Ray. *The Singularity Is Near: When Humans Transcend Biology.* London: Penguin Books, 2006.

Maharishi, Sri Ramana. *Who Am I?* CreateSpace Independent Publishing Platform, 2016.

Mashelkar, R.A. and R. Pandit. *From Leapfrogging to Pole-vaulting.* Gurugram, India: Penguin Viking, 2019.

Miki, Keira. *Ikigai: Japanese Art of Staying Young, While Growing Old.* New Delhi, India: Diamond Books, 2022.

Nagel, Thomas. "What Is It Like to Be a Bat?" *The Philosophical Review* 83, 4 (Oct. 1974), pp. 435–450.

Nagel, Thomas. *Mind and Cosmos: Why the Materialist Neo-Darwinian Conception of Nature Is Almost Certainly False.* New York, USA: Oxford University Press, 2012.

Oparin, A.I. *Origin of Life.* New York: Dover Publications, Inc., 1953.

Patwardhan, Bhushan, Gururaj Mutalik, and Girish Tillu. *Integrative Approaches for Health: Biomedical Research, Ayurveda and Yoga.* Netherlands: Academic Press, Elsevier Science, 2015. https://www.sciencedirect.com/book/9780128012826/integrative-approaches-for-health#book-info

Penrose, Roger. *Cycles of Time: An Extraordinary New View of the Universe.* London, UK: Vintage, 2011.

Pert, Candace B. *Molecules of Emotions: Why You Feel the Way You Do.* New York: Scribner, 1997.

Prabhakar, V.H. *Vedic Astronomy* (*Vedaanga Jotisha*). Nagpur: Shri Babasaheb Apte Samarak Samiti, 1989. https://archive.org/details/VedangaJyotisha/page/n5/mode/2up

Prigogine, Ilya. *Being to Becoming: Time and Complexity in the Physical Sciences.* New York: W.H. Freeman & Co, 1980.

Prigogine, Ilya. *The End of Certainty. Time, Chaos and the New Laws of Nature.* New York: Free Press, 1996.

Rajaram, N.S. *Sarasvati River and the Vedic Civilization.* Aditya Prakashan, 2007.

Reich, David. *Who We Are and How We Got Here: Ancient DNA and the New Science of the Human Past?* Oxford: Oxford University Press, 2018.

Ridley, Matt. *Genome: The Autobiography of a Species in 23 Chapters*. HarperCollins, 2006.

Riehl, John. "The Roots of Human Awareness." Iowa Now, 22 August 2012. https://now.uiowa.edu/news/2012/08/roots-human-self-awareness

Sagan, Carl. *Cosmos*. London: Abacus, 1995.

Sagan, Carl. *Pale Blue Dot: A Vision of the Human Future in Space*. Balantyne Books, Inc., 1997.

Samples, Bob. *The Metaphoric Mind*. 2nd edition. Fawskin, USA: Jalmar Press, 1993.

Saradananda, Swami, ed. *Sri Ramakrishna the Great Master*. Trans. Swami Jagadananda. Chennai: Sri Ramakrishna Math, 2004, p. 385.

Sarkar, Dr Jaladikumar. *Dr Mahendrlal Sarkar His Life and Thoughts*. Trans. M. Sivaramkrishna. Kolkata: Advaita Ashrama, 2007, p. 56.

Schrodinger, Erwin. *What Is Life? With Mind and Matter and Autobiographical Sketches*. UK: Cambridge University Press, 1992.

Séralini, Gilles-Éric and Jérôme Douzelet. *The Monsanto Papers: Corruption of Science and Grievous Harm to Public Health*. Simon and Schuster, 2021.

Sharma, Kavita A. and Indu Ramchandani. *Life Is as Is: Teachings from the Mahabharata*. New Delhi, India: Wisdom Tree, 2018.

Skinner, B.F. *Verbal Behavior*. Kindle edition. Barakaldo Books.

Skloot, Rebecca. *The Immortal Life of Henrietta Lacks*. London, UK: Picador, 2011.

Tandan, Madhu. *The Logic of Dreams*. New Delhi: Speaking Tiger, 2022.

Vachani, S. and J. Usmani, eds. *Adaptation to Climate Change in Asia*. Cheltenham, UK: Edward Elgar Publishing, 2014.

Valiathan, M.S. *The Legacy of Caraka*. Hyderabad, India: Orient Blackswan, 2003.

Vivekananda, Swami. *The Complete Works*. 25th impression. Volumes 1–9. Kolkata, India: Advaita Ashrama, 2005.

Warmington, Eric H. and W.H.D. Rouse, trs & eds. *Great Dialogues of Plato*. New York, USA: Signet Classics, Reprint 2015.

Watson. John B. *Behaviorism*. Kindle edition. New York: Routledge, 2017.

Scientific Papers, Reports, and Relevant Websites

American Museum of Natural History. "Georges Lemaître, Father of the Big Bang." https://www.amnh.org/learn-teach/curriculum-collections/cosmic-horizons-book/georges-lemaitre-big-bang

American Museum of Natural History. "Cosmic Microwave Background Discovered 50 Years Ago Today." 2014. https://www.amnh.org/explore/news-blogs/news-posts/cosmic-microwave-background-discovered-50-years-ago-today

American Museum of Natural History. "How Has the Earth Evolved?" https://www.amnh.org/exhibitions/permanent/planet-earth/how-has-the-earth-evolved

Archaeological Survey of India (ASI). "Dwarka." Ministry of Culture, Government of India. https://asi.nic.in/pages/Underwater-archaeology/HQ

Carney Institute for Brain Science. "BrainGate Research Program." Brown University. https://www.brown.edu/carney/research-project/braingate

CBC Radio. "Stephen Hawking's Final Theory Could Prove the Existence of the Multiverse." *As It Happens* with Carol Off. 20 March 2018.

CERN. "A Short History of the Web." https://home.cern/science/computing/birth-web/short-history-web

CERN. "Untangling the Origin of String Theory." https://home.cern/news/news/physics/untangling-origin-string-theory

Chakravarti, Ankita. "Robot Commits Suicide in South Korea Because It Was Made to Do a Lot of Work." *India Today*, 5 July 2024. https://www.indiatoday.in/technology/news/story/robot-commits-suicide-in-south-korea-because-he-was-made-to-do-a-lot-of-work-2562775-2024-07-05

Clynes, Manfred E. and Nathan S. Kline. "Cyborgs and Space." *Astronautics*, September 1960. https://web.mit.edu/digitalapollo/Documents/Chapter1/cyborgs.pdf

Danino, Michael. “The Aryan Invasion: Myth or Fact.” Archaeological Sciences Centre, IIT Gandhinagar. https://asc.iitgn.ac.in/assets/publications/popular_articles/The_Aryan_Invasion_Myth_or_Fact-Michel-Danino.pdf

Elhacham, E., et al. “Global Human-made Mass Exceeds All Living Biomass.” *Nature* 588, 442–444 (2020). https://doi.org/10.1038/s41586-020-3010-5

European Environment Agency. *Late Lessons from Early Warnings: Science, Precaution, Innovation*. Summary EEA Report No 1/2013. https://www.eea.europa.eu/publications/late-lessons-2

European Space Agency. “Cosmic Microwave Background (CMB) Radiation.” Science & Exploration. https://www.esa.int/Science_Exploration/Space_Science/Herschel/Cosmic_Microwave_Background_CMB_radiation

European Space Agency. 2024. https://www.esa.int/ESA_Multimedia/Images/2024/03/Gaia

Goodreads. “Carl Sagan.” Quotable Quotes. https://www.goodreads.com/quotes/601581-the-hindu-religion-is-the-only-one-of-the-world-s

Harvard Kenneth C. Griffin Graduate School of Arts and Science. “Human-made Materials Outweigh All Living Beings on Earth.” Science in the News, Blog, 23 December 2020. https://sitn.hms.harvard.edu/flash/2020/human-made-materials-outweigh-all-living-beings-on-earth/#

Indian University Bloomington. “Artificial Cells Demonstrate that “Life Finds a Way.” Department of Biology, 5 July 2023. https://biology.indiana.edu/news-events/news/2023/lennon-minimal-cells.html.

Institute of Human Origins. “Lucy’s Story.” Arizona State University. https://iho.asu.edu/about/lucys-story

Internet Encyclopedia of Philosophy. “Charles Darwin.” https://iep.utm.edu/darwin/

Josephson, Brian. “The Challenge of Consciousness Research.” 1992. https://www.academia.edu/67202522/The_challenge_of_consciousness_research?uc-sb-sw=10424024

J. Craig Venter Institute. "First Self-Replicating Synthetic Bacterial Cell." https://www.jcvi.org/research/first-self-replicating-synthetic-bacterial-cell

Khan Academy. "Hypotheses of the Origin of Life." https://www.khanacademy.org/science/ap-biology/natural-selection/origins-of-life-on-earth/a/hypotheses-about-the-origins-of-life

Kim, M.J., et al. "Birth of Clones of the World's First Cloned Dog." *Scientific Reports* 7, 15235 (2017). https://doi.org/10.1038/s41598-017-15328-2

Lane, N. "The Unseen World: Reflections on Leeuwenhoek (1677) 'Concerning Little Animals'." *Philosophical Transactions of the Royal Society B: Biological Sciences* 370, 1666 (2015). https://doi.org/10.1098/rstb.2014.0344

Lozovska, A., et al. "Tgfbr1 Controls Developmental Plasticity between the Hindlimb and External Genitalia by Re-modeling Their Regulatory Landscape." Nature Communications 15, 2509 (2024). https://doi.org/10.1038/s41467-024-46870-z

Michaelsen, M. M., et al. "Mindfulness-based and Mindfulness-informed Interventions at the Workplace: A Systematic Review and Meta-Regression Analysis of RCTs." *Mindfulness*, 1–34. https://doi.org/10.1007/s12671-023-02130-7

Moger-Reischer, R.Z., et al. "Evolution of a Minimal Cell." *Nature* 620, 122–127 (2023). https://doi.org/10.1038/s41586-023-06288-x

Nagel, Thomas. "What Is It Like to Be a Bat?" *The Philosophical Review* 83, 4 (October 1974), pp. 435–450.

NANOGrav. "An International Collaboration Dedicated to Exploring the Low-frequency Gravitational Wave Universe through Radio Pulsar Timing." 21 June 2024. https://nanograv.org/

NASA. "Big Bang Cosmology." Universe 101. https://map.gsfc.nasa.gov/universe/bb_theory.html

NASA. "Edwin Hubble." https://science.nasa.gov/people/edwin-hubble/

NASA. "Large Scale Structures." https://science.nasa.gov/universe/galaxies/large-scale-structures/

NASA. "Origin of Everything: Hot Bang or Ageless Universe?" https://imagine.gsfc.nasa.gov/educators/programs/cosmictimes/educators/guide/1955/origin.html

NASA. "Superstrings." https://imagine.gsfc.nasa.gov/science/questions/superstring.html

NASA. "Webb, Hubble Combine to Create Most Colorful View of Universe." https://www.nasa.gov/missions/webb/nasas-webb-hubble-combine-to-create-most-colorful-view-of-universe/#

National Geographic. "HMS *Beagle*: Darwin's Trip around the World." https://education.nationalgeographic.org/resource/hms-beagle-darwins-trip-around-world/

National Human Genome Research Institute. "The Human Genome Project." https://www.genome.gov/human-genome-project

National Institute of Health. "The Human Microbiome Project." https://commonfund.nih.gov/hmp.

National Institute of Oceanography. "Dwarka." https://www.nio.res.in/galleries/show/dwarka

Neuralink Corporation. "Brain–Computer Interface (BCI): Prime Study Progress Update." https://neuralink.com/blog/prime-study-progress-update/

North Arizona University. "Islam Creation Story." https://www2.nau.edu/~gaud/bio301/content/iscrst.htm

PBS. "Jean Baptiste Lamarck." Evolution. https://www.pbs.org/wgbh/evolution/library/02/3/l_023_01.html

Perrigo, Billy. "Why Timnit Gebru Isn't Waiting for Big Tech to Fix AI's Problems." *Time*, 18 January 2022. https://time.com/6132399/timnit-gebru-ai-google/

Physics World. "String-theory Calculations Describe 'Birth of the Universe'." 6 December 2011. https://physicsworld.com/a/string-theory-calculations-describe-birth-of-the-universe/

Plato. "Allegory of the Cave." Trans. by Shawn Eyer. https://scholar.harvard.edu/files/seyer/files/plato_republic_514b-518d_allegory-of-the-cave.pdf

Rincon, Paul. "Stephen Hawking's Warnings: What He Predicted for the Future." BBC, 15 March 2018. https://bbc.com/news/science-environment-43408961

Roslin Institute of Scotland. "Dolly the Sheep." The University of Edinburgh. https://www.ed.ac.uk/roslin/about/dolly

Science Daily. "Einstein's Conversion from a Belief in a Static to an Expanding Universe." Science News, 17 February 2014. https://www.sciencedaily.com/releases/2014/02/140217102545.htm

Stanford Institute for Theoretical Physics. "Cosmology." Stanford University, School of Humanities and Sciences. https://sitp.stanford.edu/research/cosmology.

Stanford Encyclopedia of Philosophy. "Identity." https://plato.stanford.edu/entries/identity/

Stanford Encyclopedia of Philosophy. "Kant's Moral Philosophy." https://plato.stanford.edu/entries/kant-moral/

The International Genome Sample Resource. "The 1000 Genome Project." https://www.internationalgenome.org/

The Nobel Prize. "The Nobel Prize in Physics 1903: Henry Becquerel" https://www.nobelprize.org/prizes/physics/1903/becquerel/facts/

The University of Arizona. "Orch OR." Center for Consciousness Studies. https://consciousness.arizona.edu/orch-or

United Nations. "The 17 Goals." Department of Economic and Social Affairs. https://sdgs.un.org/goals.

Yong, Ed. "Out-of-body Experience: Master of Illusion." *Nature* 480, 168–170 (7 December 2011). https://doi.org/10.1038/480168a

Acknowledgements

The idea of this book emerged almost two decades ago. Since then, this was a topic of conversations with many. Over the years new concepts emerged, genome research advanced, and understanding of omics and epigenetics became more evident. Several colleagues, students, teachers, and friends offered valuable critique and suggestions for further improvements in the approach and content. It is very difficult to name everyone who helped us in this process. However, we wish to mention a few who made substantial contributions for this book.

In the initial phases, Alex Hankey and I (Bhushan) closely worked on this concept. We discussed this idea with Professor Brian Josephson at University of Cambridge. During this period, Girish Tillu, Kalpana Joshi, and colleagues from the CSIR NMITLI team and AyuGenomics project helped enrich the Om and genome interlinking. Later, the idea and structure of the book started taking shape during engaging discussions with Avinash Patwardhan, Gururaj Mutalik, Ashok and Rama Vaidya, Darshan Shankar, Sharad and Medha Deshpande.

By now Indu Ramchandani had come in and both of us took the book forward. After we drafted a brief about the book and prospective table of contents, we bounced this concept off a wider and diverse group of experts to receive comments and critique. The Brainstorming Meeting organized at the Interdisciplinary School of Health Sciences was very useful in revising the initial structure. This was attended by over 15 curious individuals representing different generations and disciplines. We requested participants to share their comments. Akash Saggam, Sanchita Sangle, Kalindi Kale, Rahul Shidhaye, Sarika Chaturvedi, Anagha Lavalekar, Eha Kulkarni, Chandan Chatterjee, Sham Diwanay, and Vishnu Joglekar offered valuable comments and support.

We then shared a revised concept note, table of content, and a sample chapter with a few experts to receive broad comments. In this phase, Julia Arnold, Santhosshi Narayanan, Renee Mehrra, Indu Kering, Sujata Kelkar, Ajit Kulkarni, Sanjay Nagarkar, Mukund Chorghade, Anil Deshpande, and K.P. Mohanan offered valuable comments. Based on this exercise, we created the first draft of the full book. Again, we invited diverse experts to review the draft manuscript. In these semi-final and final rounds, the following experts offered valuable comments and constructive critique: G. Jagadeesh, Mahadevan Seetharaman, Ravi Ghooi, Anuradha Shah, Raju Kelkar, R.L. Deopurkar, Supriya Kulkarni, Sumeet Goyal, Rathnam Chaguturu, Sachin Chaturvedi, Madhav Deo, and Pravrajika Divyanandaprana. This final book in its present form was possible only because of valuable inputs received formally and informally from countless friends and well-wishers. In addition to names mentioned here, several others have helped us in this endeavour. We remain grateful to all of them.

We especially thank Gururaj Mutalik and R.A. Mashelkar for continuous mentorship and for writing inspiring forewords. We are grateful to K. Kasturirangan, Gerard Bodeker, Vikas Sukhatme, Julia Arnold, B.M. Hegde, Georg Seifert, H.R. Nagendra, Rajesh Kotecha, Madhuri Kanitkar, Suresh Garimella, Santishree Pandit, Carani.B. Sanjeevi, Ikhlas Khan, Ravindra Ghooi, Roy Upton, Anil Sahasrabuddhe, Vishvanath Karad, Anil Kakodkar, David Frawley, Amarjeet Bhamra, Madan Thangavelu, Rajiv Kumar, Shekhar Mande, Suresh Prabhu, and Vijay Bhatkar, who have graciously written endorsements for the book.

We thank all scientists who have extended kind permissions to cite their work and reproduce some interesting images. We acknowledge key resources such as Wikipedia and Encyclopaedia Britannica and search engines such as Google and Bing. We have sparingly used large language models mainly for consolidating data.

We thank Shalaka and the team of Urmi Designs, Gautam and the team of MyAmCorp, and Advisory Pvt. Ltd for all the graphics and artwork. Last but not the least, we would like to thank our respective families who very silently supported our efforts and endured the months, days, and all odd hours when we remained lost behind the computer screens with total disregard to their share of our time!

We thank the publisher, BluOne Ink, for bringing out this work in a quick time frame keeping in mind the evolving dynamics of the subjects covered.

Index

About the Authors

Dr Bhushan Patwardhan, National Research Professor—Ayush, is an accomplished and highly cited biomedical researcher ranked among the top scientists globally with over a three-decade stint in health research cutting across trans-disciplinary fields. He is an elected fellow of the National Academies of Sciences and Medicine in India. Dr Patwardhan is a distinguished professor at the Savitribai Phule Pune University, India, and an adjunct professor at Western Sydney University, Australia. He is a member of the Lancet Citizen's Commission on reimagining India's health system and an advisor to the WHO's Global Traditional Medicine Centre.

Indu Ramchandani has spent most of her working years as an editor, transcriber, compiler, and book publisher. After five years as the South Asia editor-in-chief at Encyclopaedia Britannica, she moved on to being a freelance editorial consultant. The passion for writing developed with the studies of the ancient scriptures and the epics. Indu Ramchandani has transcribed and published 14 handbooks in the Understanding Vedanta Lecture Series, written several articles on Vedic philosophy, and co-authored *Life Is as Is: Teachings from the Mahabharata.*